U.S. ZINC INDUSTRY

A History,
Statistics, and
Glossary

James H. Jolly

American Literary Press, Inc.
Baltimore, Maryland

U.S. ZINC INDUSTRY
A History, Statistics, and Glossary

Copyright © 1997 James H. Jolly

All rights reserved under International and Pan-American copyright conventions. No part of this book may be reproduced, stored in a retrieval system, or transmitted in any form, electronic, mechanical, or other means, now known or hereafter invented, without written permission of the author. Address all inquiries to the author.

Library of Congress
Cataloging in Publication Data
ISBN 1-56167-356-0

Library of Congress Card Catalog Number
96-080084

Published by

American Literary Press, Inc.
8019 Belair Road, Suite 10
Baltimore, Maryland 21236

Manufactured in the United States of America

To Mom and Dad

"In Science, as in all other departments of inquiry, no thorough grasp of a subject can be gained unless the history of its development is clearly appreciated."

The Founders of Geology
Sir Andrew Giekie (1897)

PREFACE

This book is an attempt to give the reader an overall understanding and summary of the U.S. zinc industry from its beginnings to the present. It is not an attempt to predict the future direction of the industry or provide a review of the latest technology. The primary aim is to provide the reader—whether a novice (with regard to zinc), researcher, investor, banker, or zinc industry professional—with a well-rounded introduction to the development of the domestic industry, its history, statistics, terminology, and processing technology.

The book is divided into seven major sections: (1) Chronology; (2) Ancient History and World Trends; (3) History of the U.S. Zinc Industry, including the histories of the individual States; (4) History of Zinc in the National Defense Stockpile; (5) Zinc Basics, a summary of zinc sources, processing, production, products, and uses; (6) Zinc marketing, including tariffs, cartels, and prices; (7) Statistical tables; and (8) Glossary of Terms. The Glossary is an attempt to preserve and list in one source many of the words and expressions historically used by the domestic zinc industry. Because many words have multiple meanings, only those meanings that apply to or are associated with zinc or the industry are listed. I alone shoulder the burden of responsibility for any errors of fact or interpretation and/or omission of pertinent data or definitions.

This undertaking marks the culmination of a long and friendly association with the U.S. zinc industry, primarily in my former capacity as the zinc commodity specialist with the U.S. Bureau of Mines. However, this book probably would not have been attempted nor completed had my wife, Janice, not accepted a position in 1993 with the International Copper Study Group in Portugal. The move to Portugal freed up the necessary time to complete the book.

I should like to thank those whose contributions made this book possible: Janice, who encouraged and advised throughout the effort but also who put up with my incompetence in learning and operating the computer (my former colleagues at the Bureau would not be surprised at this); my daughter July Ann for editing, formatting, and laying out the manuscript; my other daughter, Jennifer, for advice on preparing the manuscript for printing; Steve Jasinski for preparing the graphs; and V. Anthony (Tony) Cammarota, Jr., former Assistant Director, U.S. Bureau of Mines, for critically reading and advising on the material presented.

One would be remiss if all the authors of zinc articles, dictionaries, and zinc industry technical papers were not also acknowledged for a book of this sort. It is difficult to choose among the many fine and informative authors and works used in this effort, but the following general sources seemed to me to be exceptional: The Mineral Resources of the United States/ Minerals Yearbook series; A Glossary of the Mining and Mineral Industry (1920) by Fay; A Dictionary of Mining, Mineral and Related Terms (1968), USBM staff; Lead and Zinc in the United States (1908) by W. R. Ingalls; Metallurgy of Lead and Zinc (1936) by AIME; Materials Survey: Zinc (1951), USBM; and Zinc (1959) edited by C.H. Mattewson.

James H. Jolly
February 27, 1995

Table of Contents

PREFACE v

ZINC HISTORY CHRONOLOGY 1

HISTORICAL BACKGROUND 21
    ANCIENT HISTORY OF ZINC 21
    MODERN HISTORY OF ZINC 15
    TRENDS IN WORLD PRODUCTION OF ZINC 17

HISTORY OF THE U.S. ZINC INDUSTRY 21
    INTRODUCTION 21
    FROM 1860 TO 1900 22
    FROM 1900 TO 1915 32
    FROM 1915 TO 1919 35
    FROM 1920 TO 1939 38
    FROM 1940 TO 1946 40
    FROM 1946 TO 1960 44
    FROM 1961 TO 1975 46
    FROM 1976 TO 1995 48

U.S. STATE HISTORIES 51
    ARKANSAS 51
    ALASKA 53
    ARIZONA 53
    CALIFORNIA 54
    COLORADO 54
    IDAHO 56
    ILLINOIS 57
    INDIANA 58
    IOWA 59
    KANSAS 59
    KENTUCKY 60
    MAINE 60
    MARYLAND 60
    MISSOURI 61
    MONTANA 63
    NEVADA 64
    NEW HAMPSHIRE 64
    NEW JERSEY 65
    NEW MEXICO 67

NEW YORK	68
NORTH CAROLINA	69
OHIO	69
OKLAHOMA	69
OREGON	71
PENNSYLVANIA	71
TENNESSEE	72
TEXAS	74
UTAH	74
VIRGINIA	76
WASHINGTON	77
WISCONSIN	78

ZINC AND THE NATIONAL DEFENSE STOCKPILE 81
- WORLD WAR I 81
- POST WORLD WAR I AND THE SECOND WORLD WAR 82
- KOREAN WAR DECADE 83
- POST 1960'S ERA 83
- SUMMARY 86

ZINC BASICS 89
- SPECIFICATIONS AND STANDARDS 89
- GEOLOGY AND RESOURCES 94
- MINE DISCOVERY 97
- FLOW OF ZINC IN THE UNITED STATES 98
- MINING 98
- MILLING AND CONCENTRATING 100
- SMELTING AND REFINING 103
- RECYCLING 105
- CONSUMPTION AND USES 107

ZINC MARKETING 111
- U.S. TARIFF HISTORY 111
- CARTELS AND SYNDICATES 115
- ZINC METAL PRICING 118
- MARKETING OF ORES, CONCENTRATES, AND ZINC PRODUCTS 122

BIBLIOGRAPHY 129

STATISTICS 137

GLOSSARY 179

FIGURES

Figure 1: Zinc Mine Production, United States and World, 1895-1994 19
Figure 2: Slab Zinc Production, United States and World, 1895-1994 19
Figure 3: Recoverable Zinc Mine Production, by State, 1850-1994 52
Figure 4: Zinc Smelter Production by State, 1859-1994 52
Figure 5: Zinc Flow Diagram 99
Figure 6: U.S. Secondary Zinc Recovery and Form, 1907-1994 110
Figure 7: Slab Zinc Consumption, 1900-1994 110

TABLES

Table 1: World Slab Zinc Production by 25-Year Periods, 1801-2000 20
Table 2: Apparent World Per Capita Consumption of Zinc Metal, Selected Years 20
Table 3: Principal Zinc Producing Districts in the United States 80
Table 4: Zinc National Defense Stockpile, Inventories, Receipts, Releases, and Goals, 1946-1994 88
Table 5: Commercial Zinc Metal Grades in the United States 91
Table 6: Average Annual Zinc Prices in the United States, 1850-1994 126
Table 7: Average Zinc Prices by Decade, 1900-1994 128

STATISTICS TABLES

Table A1: U.S. and World Zinc Mine and Metal Production, 1800-1994 137
Table A2: U.S. Historical Salient Zinc Statistics 140
Table A3: Mine Production of Recoverable Zinc in the United States, by State, 1906-1994 145

Table A4: Estimated Oxidized Zinc Ore
Production in the United States, by State, 1850-1994 151
Table A5: Lead Slag Fuming Plants
in the United States, 1927-1995 152
Table A6: Recoverable Zinc Extracted at
Slag Fuming Plants in the United States, 1927-1994............................153
Table A7: Primary Zinc Smelter Production,
by State, 1859-1994...154
Table A8: Estimated Primary Zinc Metal Production by
State in Various Time Periods, 1860-1994 158
Table A9: Spelter Companies Operating
in the United States, 1860-1906..159
Table A10: Spelter Companies Operating
in the United States, 1907-1920..162
Table A11: Spelter Companies Operating
in the United States, 1921-1940..166
Table A12: Spelter Companies Operating
in the United States, 1941-1960..168
Table A13: Spelter Companies Operating
in the United States, 1961-1994..169
Table A14: Vertical Retort Companies Operating
in the United States, 1927-1995..170
Table A15: Electrolytic Companies Operating
in the United States, 1915-1994..171
Table A16: Secondary Zinc Recovery and
Form of Recovery, 1907-1994 172
Table A17: U.S. Apparent Consumption of
Slab Zinc, 1900-1994 175

ZINC HISTORY CHRONOLOGY

1541 Death of Paracelsus, who is considered the first European to publish information on zinc metal and may have been the first to call the metal zinc (zinck or zinken (German) or zinzi (Latin)). *The latter point is questionable in that Agricola refers to zinc in published works soon after Paracelsus' death as "what the Noricians and Rhetians, who lived in the Alps south of Salzburg, call zinc".*

1721 Henckel publishes his discovery of the fact that zinc could be obtained from calamine.

1739 John Champion, an Englishman, obtains a patent for a process to produce zinc by distillation downward or "per descensum".

1740 Champion erected a works at Bristol and began the production of spelter. In this process, zinc was distilled from a mixture of ore and carbon in a sealed pot, in the bottom of which was a hole to which an iron pipe abutted. The zinc vapor condensed in the pipe and dripped into a pan. This basically was the first "English" furnace.

1741 The galvanizing of iron is recorded in France; however the first patents on the galvanizing process were not issued until 1836, to Sorel in France, and 1837, to Crawford in England.

1743 Marggraf publishes a method of smelting calamine to produce zinc metal.

1758 The Champion brothers obtain patent for the extraction of zinc from blende, including the roasting of blende.

1772 Lord Stirling ships several tons of red ore (zincite) from Sterling Hill, NJ. to Swansea, Wales for testing but the tests were unsuccessful in recovering zinc. *It is not certain that the ore was sent to test for zinc; it may have been for copper. The chemical nature of zincite was unknown at that time, but that of the somewhat similar red copper mineral, cuprite, was commonly known. Supporting this possibility is the fact that the tract of land where the deposit occurred was at that time known as the "Copper Tract" in land records.*

1779 Courtois, a French chemist, discovers that white zinc (zinc oxide) pigment does not blacken under the conditions that white lead blackens. *A short time after this, the French chemist, De Morveau, proposes that white lead pigment be replaced by white zinc for health reasons. Painters tended to get lead poisoning from grinding pigment and poor hygiene, whereas artists commonly were poisoned by the practice of licking or mouthing their brushes to shape them.*

1780 Discoveries by Luigi Galvani of Bologna (Italy), about current electricity lead to the origin of the word, galvanism, the science dealing with the action of electrical currents. This led to similar word derivatives later utilized by the zinc industry.

1781 James Emerson, an Englishman, patents a method for making brass by directly alloying copper with zinc metal.

1786 John Atkinson is granted a British patent for the manufacture of white paint, in which a method of producing zinc oxide is described whereby zinc blende or calcined calamine is mixed with charcoal, heated in a muffle furnace to liberate zinc fume which is then burned to yield zinc oxide.

1798 Johann Ruberg of Silesia produces

zinc in large muffles, giving rise to the Silesian zinc production method. Unlike the method employed by Champion where the distilled zinc vapor descended downward to liquefy, the vapor in Ruberg's retorts was discharged horizontally into a condenser extending through the furnace wall.

1800 Volta invents the voltaic pile, the first true battery in which chemical energy is converted into electrical energy. *Volta's battery consisted of piled-up pairs (like poker chips) of coin-sized zinc and silver discs separated by pasteboard or leather that had been soaked in salt water or an alkaline solution. Strips of metal alternately connected the "piles"; the strips on the two end piles were dipped into cups of mercury, which when connected completed the circuit.*

World production of zinc is less than 1000 metric tons.

1805 Hobson and Sylvester in England discover that zinc could be rolled if it was heated to 100-150 degrees C.

1806 Zinc smelting begins at Liege, Belgium near the Vieille Montagne Mine, the ores of which had been used since the early 1400's for making brass.

1808 Abbe Dony of Belgium produces zinc by a horizontal retort method that is in some ways superior to the Ruberg method. Dony uses a large number of smaller retorts supported only at their ends and set in rows, thereby gaining more heating area per unit of charge and reducing fuel requirements. *Dony is credited with being the inventor of the Belgian Method, although this is questioned by some.*

1817 Cadmium was isolated and named by Strohmeyer.

1818 First description of a Belgian furnace, based on a furnace at Stolberg is published; the designer of the smelter is not known.

1824 Sir Humphry Davy concisely outlines the principles of cathodic protection and employs zinc to protect copper-sheeted hulls of Royal Navy ships. Davy described his test results thusly:

> "A piece of zinc as large as a pea, or the point of a small iron nail, were found fully adequate to preserve forty or fifty square inches of copper; and this, wherever it was placed, whether at the top, bottom, or in the middle of the sheet of copper, and whether the copper was straight or bent, or made into coils. And where the connections between different pieces of copper was completed by wires ... the effect was the same; every side, every surface, every particle of copper remained bright, whilst the iron or the zinc was slowly corroded."

1832 George Muntz obtains a patent for the production and rolling of Muntz metal, 60% copper and 40% zinc. Because the alloy is resistant to seawater corrosion, it finds widespread use for sheathing the hulls of wooden ships.

1836 First spelter is produced in the United States. One lot is used in making brass for Government weights and measures standards.

Sorel receives a patent in France for a hot-dip galvanizing process.

The wet cell is invented by Daniell.

1837 A patent for the zincing of iron by dipping the iron article in a bath of molten zinc is granted in England to H.W. Crawford.

1840 German production is about 12,000 tons, 65% of world total.

1842 Karsten of Germany discovers that lead bullion gives up its silver content to zinc.

1845 Zinc ore is discovered at Friedensville, PA.

1847 Le Clair, a Frenchman, develops a process to manufacture zinc oxide from zinc metal, by burning the metal vapors issuing from a horizontal retort and collecting the fume in settling chambers. *This later became known as the indirect or French process. A few years later, the direct (from ore) or American process was developed in the United States.*

1848 A small plant is established at Newark, NJ, for recovering metallic zinc, but the attempt is not successful.

1849 Diethyl zinc is the first organometallic compound to be prepared.

The compressed-air rock drill is invented.

Tyler and Helm obtain a U. S. patent for the use of zinc oxide as a compounding agent in rubber, ten years after Goodyear discovers rubber vulcanization.

1850 Alexander Parkes of England obtains his first patent for desilverizing lead using zinc (the Parkes process).

The first workable ores are discovered at Joplin. Lead and silver were recovered but not zinc until the early 1870's.

The Sussex Zinc and Copper Mining and Manufacturing Co., which in 1852 became The New Jersey Zinc Co., attempts to commercially manufacture zinc oxide pigment at its plant in Newark, NJ. Considerable amounts of the oxide are made but their reverberatory-furnace process proves uneconomic. *Subsequent experiments at this plant result in the commercial development of the Wetherill process for producing zinc oxide.*

1851 Samuel Wetherill introduces the grate of cast iron plates that forms the furnacing basis for the American-process for zinc oxide production.

Samuel T. Jones invents the muslin-bag apparatus for the collection of furnace fume. Coupled with the Wetherill grate, the two inventions gave rise to commercial zinc oxide production directly from ore in 1854.

1852 The Newark NJ plant begins to make zinc oxide directly from ore.

1854 The first zinc mining begins in Tennessee at Mossy Creek (Jefferson City). Four years later zinc is discovered at Mascot. The deposits and mines about and between the two discovery areas later form the East Tennessee zinc mining district, one of the major zinc-producing areas of the United States.

1855 The first commercial recovery of sulfuric acid from roasting sphalerite takes place in Germany.

Samuel Wetherill is issued a patent for a zinc oxide production process.

The "Horse Head" trade mark of The New Jersey Zinc Co. is first used on a label of "Number 1 White Zinc, Ground in Oil". *The origin of the "horse head" was in 1786 where it was first minted on the 1-cent coin of the Colony of New Jersey and later was placed on the State's Great Seal; it was from the Great Seal that the trademark was taken.*

The first spiegeleisen alloys are made from the iron-manganese residues remaining after zinc oxide is extracted from New Jersey franklinite.

1858 Experimental spelter is produced at South Bethlehem using Pennsylvania ore.

Jaw crusher invented in the United States by Blake.

Mattheissen and Hegeler begin to erect a zinc works at LaSalle, IL.

The first ready mix paints are prepared, owing in large part to the good suspension properties of American process zinc oxide.

1859 Lehigh Zinc Co. constructs a Belgian furnace at Bethlehem, PA., to process zinc ore mined at Friedenville.

The first commercial operation of the Parkes process to desilverize lead begins in Wales.

1860 F. W. Mattheissen and E. C. Hegeler begin production of spelter at LaSalle, IL., processing Wisconsin ore.

1864 The beginning of the galvanizing industry in the United States. *By 1900 U.S. Steel is the principal galvanizer of steel wire and sheet products and nearly all iron products.*

1865 First commercial production of spelter is made from New Jersey ore at Newark, NJ. It is the first spelter to have the "Horse Head" brand.

1866 The first zinc rolling plant in the United States is established by Mattheissen and Hegeler in Illinois.

The Leclanche wet cell is patented.

Zinc ore is discovered at Bertha, VA., but not worked until 1878.

Zinc oxide is found to be a satisfactory base for face powders.

1867 D.R. Averill of Newburg, OH patents the first ready-mix paint, the success of which depended mainly on using American process zinc oxide. *He began its manufacture and sale which gave rise to the term for such paint as "patent paint".*

Production of zinc begins in Missouri at Potosi.

The first barbed wire is made in Ohio.

1868 Australian mines begin production of zinc.

1869 First transcontinental railroad completed, creating the possibilities of zinc production in the western States and Territories.

The Glendale works is established at St. Louis.

First recognition of zinc as necessary for the growth of fungus.

1871 Ingersoll Rand invents an improved rock drill.

1872 First zinc ore from the Joplin district is shipped.

Hegeler's first large furnace (40 feet long containing 408 retorts in 5 rows) is erected at LaSalle, IL, by Mattheissen and Hegeler Zinc Co. *Long furnaces led to the introduction of labor-saving devices for charging and discharging retorts, which subsequently became a distinctive feature in American zinc-smelting practice.*

The Mining Law of 1872, which established the first comprehensive set of laws and procedures required for mine discovery and development on Federal public lands, is enacted into law.

1873 Zinc smelting begins at Weir, KS.

1874 Lithopone is invented by John Orr of Scotland.

1875 Association of U.S. producers attempts to control zinc prices. The effort fails in 1876 as the higher prices stimulate higher production elsewhere.

1876 D.W. Brunton, in Colorado, discovers that silver recovery is improved in chloridizing processes when zinc is present in the treated concentrate because zinc chloride forms in processing and acts as an effective solvent of silver. *At this time, however, the zinc from the process is considered to have no value and is run to waste.*

The first ball mill patented by Bruckner in Germany.

Bell invents the telephone, which led to widespread use of the Leclanche wet cell. *The telegraph (invented in 1837) also became a large user of the Leclanche cell at about this time.*

1879 Bertha Mineral Co. begins smelting calamine ore at Pulaski, VA.

German producers agree to limit production mainly as a measure to improve zinc prices.

The U.S. Geological Survey is established.

1880 First zinc smelter built in Tennessee at Clinton. The smelter was a Belgian type consisting of 76 retorts. *In 1883 Eades, Mixter, and Heald bought the smelter, expanded it in 1888 to a daily capacity of 3,220 pounds, then permanently closed it in 1894).*

Development of the zinc-carbon dry cells. The zinc anode becomes the container for the cylindrical-shaped cell, rather than an immersed anode as in the wet cell.

1881 First experiments in electrolytic separation of zinc from ores. Leon Letrange, a Frenchman, obtains a patent for the invention of a zinc sulfate electrolytic process. Letrange's process, although not successful, incorporates the use of purified zinc solutions, lead anodes, and spent solution to digest more ore.

Invention of the Hegeler roasting furnace, results in the first sulfuric acid production in the United States as a by-product of blende roasting.

Rail line connects Salt Lake City, UT and Butte, MT, opening the possibility of zinc mining in the north-central Rocky Mountains.

1882 European smelters organize into a syndicate (cartel) to limit production.

The Mattheissen and Hegeler smelter at LaSalle, IL, with a capacity of 28 to 30 tons per day, is the largest in the United States.

The first commercial meters to measure the amount of electric current used by customers are installed in New York City by the Edison Electric Lighting Company. *The meter consisted of a glass case with two small plates of zinc placed in a solution of zinc sulfate. The meter, on being connected with the house current, recorded the usage of current by the electrodeposition of zinc from the anode plate to the cathode plate. The plates were weighed monthly; their difference in weight represented power consumption. A few years later the zinc meters were replaced by mechanical meters.*

1884 First guided balloon (dirigible) flight occurs in France in August using a large propeller powered by a zinc-chlorine battery.

1892 The first use of natural gas in zinc smelting in the U. S. takes place at Marion, IN.

1893 First zinc ore produced in Oklahoma (Indian Territory).

The Dor hydraulic press for manufacturing molded retorts of high density are introduced in New Jersey and Pennsylvania. *Dor presses were used in France as early as 1877 and were common at European smelters when first used in the United States*

The first Brown horseshoe furnace, a mechanical roasting furnace that had been successfully used for roasting copper ore, is erected in Illinois by the Collinsville Zinc Co. *It is considered a great advance in mechanical roasting in that labor cost per ton roasted was one-third that of other zinc roasters in use.*

The New Jersey Zinc Co. begins the first manufacture of French-process zinc oxide in the United States.

1894 Miners at Joplin, MO propose and institute the selling of ore based on assay rather than merely by weight as was done up to this time. The ore buyers and smelters oppose the change but to no avail.

1895 Sherard Cowper-Coles of England uses aluminum cathodes in an unsuccessful attempt to produce electrolytic zinc.

Lanyon builds the first Iola-type furnace in Kansas using natural gas.

1896 Hoepfner of Germany builds a commercial electrolytic zinc plant in England using zinc chloride solutions. *The plant had an annual capacity of 5 tons, later doubled, and was said to have remained in operation as part of a chlorine chemical operation until 1924. Since then, no commercial electrolytic zinc plant using zinc chloride solutions has been put into production.*

Invention of the Wetherill separator for separation of weakly magnetic minerals by magnets. *The Wetherill separator is one of the premier inventions of the mining industry and is instrumental in the successful utilization of the ore at Franklin, NJ as well as recovery of zinc and other metals from complex sulfide orebodies, old tailings and residues in Wisconsin and in the West.*

Invention of the Wifley table. *Considered one of the great advances in ore dressing, in part, because it could process fine-grained ore. It was first used for zinc recovery at Leadville, CO in 1899.*

Forrest and McAuthur in Australia develop the process to use zinc shavings to extract gold from cyanide solutions.

1898 First smelter production of zinc sulfate in the United States. *The Consolidated Kansas City Smelting and Refining Co. recovers 135 short tons from zincy lead-silver ore at its works at Argentine, KS.*

Gustave De Laval develops an open-arc or electric furnace for zinc production at Trollhattan, Sweden. *The process which could continuously smelt roasted blende, using coke as a reducing agent and limestone as a flux, was the first commercial electrothermic zinc process.*

The Lanyon shield to protect workers servicing retort furnaces is developed.

1899 First lithopone manufacture in the United States by Cawley, Clark & Co. in Newark, NJ. and N. J. Graves in Camden, NJ.

1900 Sir William Crookes demonstrates the phosphorescent nature of zinc blende by exposing a zinc blende screen to a scrap of radium. Close observation reveals that the glow produced is due to a multitude of luminous specks, darting and flashing, with each representing the impact of an alpha particle.

The flashlight is invented. *It was so named because the zinc-carbon dry cell of that time had a very short life such that the flashlight was only flashed intermittently to conserve battery life.*

World zinc metal production totals about 479,000 metric tons.

1901 Blake and Morscher patent electrostatic separation machine. *First successful zinc application was in 1907 when sphalerite-pyrite middlings in Colorado are separated.*

The United States replaces Belgium as the world's second largest zinc producer, with Germany in first place.

1902 The Lanyon Zinc Co. is the second largest zinc producer in the world, after Vieille Montagne Co. of Belgium. Other U.S. companies, Edgar Zinc Co., Prime Western Spelter Co., and Mattheissen & Hegeler Zinc Co., rank 4th, 5th, and 7th, respectively.

1907 The United States becomes the world's largest producer and consumer of zinc.

The U.S. Zinc Industry

1909 The Macquisten Tube flotation process successfully separates zinc in the ores at the Morning Mine in Idaho. This ranks among the first zinc flotation processes used in the United States.

A new tariff places specific duties on zinc, zinc oxide and zinc white pigment, as follows: slab zinc, 1.375 cents per pound; zinc contained in ores, less than 10% admitted free; from 10% to 20%, 0.25 cent per pound; from 20% to 25%, 0.5 cent per pound; and over 25%, 1 cent per pound; zinc oxide and white pigment containing zinc, 1 cent per pound; zinc pigment ground in oil, 1.75 cents per pound; zinc sulfide and white sulfide of zinc, 1.25 cents per pound; and chloride and sulfide of zinc, 1 cent per pound.

The "Zinc Convention" is formed to control zinc production and prices. English producers join in 1910.

1910 First use of crude oil in zinc roasting and smelting, in Kansas. The crude was from Oklahoma.

Solid rubber tires utilizing large quantities of zinc oxide (as much as 150 pounds of zinc oxide to 100 pounds of rubber) are developed. *This was important especially to the development of the early heavy trucking industry in that the load was limited by the strength of the pneumatic tires and/or solid tires then in use. Tires could support heavy weight but, in flexing, they generated heat which deteriorated the rubber, causing rapid failure. Zinc oxide, which generated less heat and dissipated heat faster than any other reinforcing pigment, especially in solid tires, kept the tire sufficiently cool preventing failure owing to heat.*

The Joplin Ore Scale, a system for setting the sale prices of ore or concentrate in the Tri-State, is set up at Joplin, MO.

The recovery of zinc and lead by volatilization in a rotary kiln has its origin in a patent by Edward Dedolph, a Canadian. *However, the actual Waelz process is independently developed in Germany in 1923.*

The U.S. Bureau of Mines is established.

1911 First significant zinc mining in New York, in St. Lawrence County at the Edwards Mine. *Although zinc was known to exist in the County as early as 1835, zinc mining in the region was delayed until the very early 1900's, largely because the ores were complex and companies had difficulties in obtaining the mineral rights. This latter problem stemmed from the long established practice of separate ownership of mineral rights established by the English Crown in Colonial times and continued by New York State. This practice resulted in numerous complications in ownership, which, after a lapse of several generations, made it almost impossible to obtain legal titles to mineral rights.*

Hyde flotation process first used successfully at Butte, MT, for separation of zinc in complex ore and from old tailings. *By 1915 most zinc at Butte is produced by flotation methods.*

Herbert C. Hoover, later President of the United States, is Joint Managing Director of Zinc Corporation in Australia.

1912 First zinc mine production in North Carolina.

1913 The first legislation providing for a mineral depletion allowance, 5%, for mining companies is approved. *This same legislation also provides for the beginning of the federal income tax.*

Tariffs changed to ad valorem rates, 10% for zinc content in ores and 15% on slab zinc.

1914 The first electrolytic zinc plant is established at Great Falls, MT, with a daily capacity of 10 tons.

Saeger retort charging machines are installed at National Zinc's Bartlesville, OK, smelter reduc-ing charging time by two-thirds.

The LME introduces its first zinc contract, which traded in "virgin zinc".

1915 Jack Morgan, the son of J. Pierpont Morgan, agrees to become the U.S. metal purchasing agent for Great Britain and France. U.S. high-grade zinc price reaches 40 cents per pound or about $5 per pound in 1987 dollars by June 1915 owing to the "European" War.

1916 First Canadian electrolytic zinc production at Trail, BC.

First large-scale electrolytic refining of zinc at the Great Falls, MT plant by Anaconda.

Radium is used to activate zinc sulfide paints in the military to fluoresce watch dials, compass points and aiming points for night operations.

1917 G.H. Clevenger, an American chemist, discovers cobalt in zinc electrolysis is a serious problem and patents a process to remove it from the solution.

The first recorded use of the sintering process for zinc concentrates at Port Pirie, Australia.

The American Zinc Company begins production of American process lead-free zinc oxide at Hillsboro, IL, using Mascot, TN zinc concentrate. *Up to this time, the New Jersey Zinc Co. had a virtual monopoly on the American process lead-free oxide market by virtue of its ownership in the only lead-free zinc ore in the United States.*

1918 The American Zinc Institute (AZI), a trade association composed of mining, smelting, and manufacturing interest, is organized. *AZI was disbanded in the mid 1980's.*

1919 Traveling-grate furnace, similar to that used in coal stoking, is adapted by The New Jersey Zinc Co. for the manufacture of zinc oxide.

1920 The first radio stations begin broadcasting in the United States, resulting in explosive growth in radio receiver sets. *The early receivers were all battery powered because they required direct current to operate, whereas most house current was alternating current. As a result, the demand for zinc-carbon batteries mushroomed; however, in the late 1920's, sets using alternating current were developed and batteries for radio purposes declined rapidly.*

The development of radio also creates a demand for zinc oxide crystals for use as detectors in "cat's whisker" radio sets by do-it-yourself radio makers. *When rigged into a radio circuit and touched in sensitive spots with a fine copper "cat's whisker" attached to the antenna, the zinc oxide crystal's semiconductor properties converted the incoming radio waves from alternating to direct current which was necessary for audio. Because crystals of sufficient size were not produced, virtually all of the crystals used in the above application were recovered from old cinder piles where, under the right conditions, zinc oxide crystals up to 4 inches in length grew.*

The Mineral Leasing Act of 1920 is passed by Congress.

Large zinc-air batteries are used to power buoys and railroad signals. *In the mid-1970's, small zinc-air batteries are developed for hearing aids.*

1922 German zinc-producing territory granted to Poland.

Specific tariffs replace ad valorem rates imposed in 1913. Rates are higher on slab zinc and ores compared with those of 1909.

1925 The widespread application of selective flotation of zinc minerals begins, leading to increased zinc production and recovery. This also leads to zinc production from complex ores from which the zinc could not be recovered previously.

The Waelz process first used in Upper Silesia.

1926 The New Jersey Zinc Co. obtains a patent for improved zinc die casting alloy and a trademark for the name, Zamak.

Zinc is recognized as essential for the growth of higher plants. *The same is found to be true for higher animals (rats) in 1934.*

The St. Joseph Lead Co. purchases the Edwards Mine and the Balmat deposit in New York State. *Mining begins at Balmat in 1928 and it soon becomes one of the largest zinc producers in the United States.*

1927 The first commercial application of coal-fired, lead slag fuming begins at East Helena, MT by The Anaconda Co.

1928 European zinc cartel formed.

The first commercial plant to use the Tainton process for the electrolytic zinc production is opened by the Sullivan Mining Co. at Kellogg, ID. *Of interest, the first zinc bar was poured on November 6, election day, and was inscribed, after the date, "the day Herbert Hoover was first elected". The use of "first" in the inscription was said to have caused comments later on.*

1929 Introduction of vertical retorting at Palmerton, PA.

1930 Electrothermic zinc smelter built by St. Joseph Lead Co. at Josephtown, PA. (near Monaca*). The smelter is used for zinc oxide production initially but becomes a metal producer in 1936.*

Increased tariff on zinc ores containing less than 25% zinc, high-strength lithopone, and zinc sulfide. Duties are 1.75 cents per pound on slab zinc; 1.5 cents per pound on zinc content of ores; 2 cents per pound on sheet zinc.

1931 T. Sendzimir in Poland develops a continuous method of galvanized sheet production in which cold-reduced steel strip is fed continuously through annealing and cleaning facilities and then through the zinc bath.

1932 First widespread use of high concentrations of zinc dust in paint. *The subsequent availability of polystyrene soon gave rise to zinc-rich primers.*

1934 The New Jersey Zinc Co. introduces its continuous, vertical redistillation process for production of high-purity zinc metal.

1935 Bright plating cyanide baths introduced for electrogalvanizing.

1936 St. Joe Lead Co. begins metal production at its electrothermic smelter in Pennsylvania. *Excessive formation of blue dust, which heretofore had delayed metal production, is solved by the development and use of the Weaton-Najarian vacuum condenser.*

The first commercially successful continuous sheet galvanizing line is installed at Armco Steel Corp's plant at Butler, PA. The Sendzimir Coating Process is used.

1937 Funds to begin a strategic stockpile are appropriated.

1938 Formation of an "International Zinc-Sheet Cartel".

1939 Canadian Trade Agreement implemented January 1, resulting in a 20% decrease in tariffs on Canadian zinc ore and metal.

The United States becomes a major importer of zinc.

1940 Special High Grade zinc production is reported as a separate U.S. statistic for the first time.

1942 Asarco begins production at its new Corpus Christi, TX. electrolytic zinc plant. *This is the only new zinc plant approved for construction during WWII.*

The Premium Price plan becomes effective in February.

The Ruben-Mallory or mercury cell is developed, finding widespread use in World War II as the power source for Walkie-Talkies and other equipment for communications. *This was the first successful alkaline dry cell; it employs zinc powder rather than sheet metal as the anode. In 1960 the zinc/mercuric oxide cell became the first successful implantable battery for cardiac pacemakers; it was used for 10 years before superior cells were developed.*

1943 A trade agreement with Mexico reduces tariffs on imported Mexican zinc ores and metal by 50%.

Zinc recovery from lead slags begins at the Bunker Hill smelter in Idaho.

1945 The antibiotic, zinc bactracin, is developed. *The first substantial use of penicillin, discovered in 1929, begins in the same year.*

1946 Price controls end in November.

The Strategic and Critical Stock Piling Act is enacted. *By the end of 1959 the U.S. Government stockpiling of zinc metal totaled almost 1.6 million short tons. Sales in the 1964-1974 period, however, had reduced this to slightly less than 0.4 million tons by 1975.*

It is discovered in the Netherlands that additions of zinc oxide to nickel-iron ferrite increase the resulting ferrite's magnetic properties, which are useful for electronic applications such as transformers, recording heads, tapes, etc.

1947 Import duties established under the Mexican Trade Agreement in 1943 are made permanent at Geneva Trade Conference.

1949 The New Jersey Zinc Co. places in operation the first commercial splash condenser for use with vertical retorts in the continuous distillation of zinc. *A motor driven graphite impeller within the condenser is used to generate showers of molten zinc for scrubbing and cleaning zinc vapor from the gas-vapor stream emitted from the retorts.*

1950 First plant-wide use of retort charging machines in the United States. *Mechanical retort charging was tried a number of times on an experimental or short-term basis, but economic factors did not justify permanent adoption at any plant.*

1951 The price of zinc is controlled by the Office of Price Stabilization in response to Korean War shortages. *Price controls were terminated in February 1953.*

1952 The first zinc production in an Imperial Smelting Process (ISP) smelter begins at Avonmouth in the U.K.

1953 The first mill application of the Cook-Norteman process for hot-dip galvanizing of sheet in continuous lines.

The LME resumes dealings in zinc for the first time since it closed on August 31, 1939 because of WWII.

1954 RCA introduces the "Electrofax" process, a means of photocopying that is highly successful. *Although still used, it gradually lost out to plain paper copying, xerography, in the 1970's. The "Electrofax" process utilizes the photoconductive and electrostatic properties of zinc oxide which when coated on printing paper is made light sensitive by giving it a negative electrostatic charge. The item to be reproduced is placed on top of the* *sensitized paper and exposed to light. The negative charge is dissipated where the light passes through, with the least discharge where images are. The image is "developed" by applying a positively charged resin which adheres to the negatively charged areas; the resin is then heated to fuse it to the paper yielding a permanent reproduction.*

1957 President Eisenhower imposes import quotas on zinc ores and metal after a finding that the domestic industry was seriously injured by imports. The quotas limited imports to 80% of the average imports of 1953 through 1957.

Union Carbide (Eveready) introduces the first commercial, cylindrical alkaline dry cell battery, also known as the alkaline manganese oxide-zinc cell.

1958 Canada mines more zinc than the United States. *This is the first time since 1906 that the United States was not the world's leading miner of zinc. The United States regained the top spot for the next five years, but since 1964 has not been the leading producer.*

The International Lead and Zinc Study Group (ILZSG) is formed.

1959 The first full-scale, zinc-lead blast (ISP) furnace comes into operation at the Swansea Vale Works in Wales. *Although a number of these plants have been built around the world, none have been built in North America.*

1960 The first commercial twin-belt casting and rolling line for the production of zinc strip is commissioned.

1962 Government program to provide stabilization payments to small producers of lead and zinc ores and concentrates is authorized in June. *The program is allowed to expire at the end of 1969.*

Zinc, lead and copper are removed from the minerals eligible for exploration assistance through the Office of Minerals Exploration.

1964 First jarosite process patents are issued.

1965 Peak year for U.S. consumption of zinc for zinc die-casting use, mainly owing to strong consumption by the automobile industry. *In the 1960-1973 period die-castings use was the principal consumer of zinc metal in the United States. Downsizing and weight reduction programs undertaken by the automobile industry and competition from plastics and aluminum from the mid-1970's sharply reduced consumption in this area.*

1971 The East St. Louis, IL basing point for pricing Prime Western zinc is eliminated in January when a major producer announces its price includes delivery.

The Goethite process is developed.

The Jarosite process is commercially used.

1973 Economic Stabilization plan is abolished at end of the year.

1976 The last horizontal retort smelter closes in the United States.

The Resources Conservation and Recovery Act (RCRA) is passed by the Congress.

1978 The Commodities Exchange of New York (COMEX) begins a contract on zinc metal. *Because of inactivity, COMEX delisted the zinc contract in 1982.*

1980 In December the President signs the Comprehensive Environmental Response, Compensation, and Liability Act, better known as "Superfund".

The last operating vertical retort (electrothermic) smelter in the United States closes, to reopen one year later at only one-fourth its former capacity.

U.S. tariff rates on most zinc ores, metals, scrap and compounds are scheduled to decline over the next eight years as a result of approval of the Tokyo round of multilateral trade negotiations.

1981 Superfund taxes placed on zinc sulfate and zinc chloride production.

1982 Production of the new "zinc" penny begins. *The new penny is 19% lighter than the copper penny and is composed of 97.6% zinc and 2.4% copper, some of which is a surface plating. The copper penny was composed of an alloy containing 5% zinc and 95% copper. About 180 zinc pennies are made from one pound of zinc. By the end of 1987 the Mint had produced the equivalent of 300 zinc pennies per each U.S. citizen.*

The last two operating lead-slag fuming plants in the United States close.

1984 U.S. steel companies begin rapid expansion of electrogalvanized sheet production mainly to meet the demands of automobile manufacturers for increased corrosion protection. *By 1988, shipments of electrogalvanized sheet and strip by U.S. steel companies totaled almost 2 million tons.*

1986 The last remaining Wetherill-type zinc oxide plant in the United States closes.

1987 Only remaining U.S. ore-based American process zinc oxide plant closes at Palmerton, PA. *The closure took place because the source of the ore, the Sterling Mine in New Jersey, was permanently closed.*

1989 The first ore is processed at the Red Dog Mine in Alaska in November.

The Free Trade Agreement to reduce tariffs between Canada and the United States is approved by both countries.

The Harmonized Trade Schedule (HTS) replaces the Trade Schedule of the United States (TSUS).

1990 The American Zinc Association (AZA) is organized.

1991 All U.S. zinc producers switch over to an LME pricing basis.

In Baltimore, MD, the LME establishes the first of six LME warehouses in the United States where zinc and other metals can be delivered or withdrawn.

1992 The President authorizes the sale of all the remaining 378,000 short tons of zinc in the National Defense Stockpile.

1994 The North American Free Trade Agreement (NAFTA) is approved by the United States, Canada, and Mexico. NAFTA is set up to eliminate and/or reduce tariffs among the three countries.

The World Trade Organization (WTO), which will implement the trade terms agreed upon in the Uruguay round of multilateral trade negotiations, is approved. WTO replaces GATT.

1995 The U.S. Bureau of Mines is closed. Commodity and statistical functions are transferred to the U.S. Geological Survey.

HISTORICAL BACKGROUND

"(Alchemists)…believed that the base metals could be changed in nature by becoming deeply colored…Brass was regarded as metal on the way to becoming gold."

The Lure and Romance of Alchemy (1932)
C.J.S. Thompson

ANCIENT HISTORY OF ZINC

Although zinc historically has been one of the five most used metals (the others are iron, copper, aluminum and lead) and has been used by man in alloys for more than 5,000 years, it was not recognized as a distinct metal or element through most of that history.

The technology required for the production of zinc is considerably more difficult compared with the relatively simple furnacing required in producing some of the earlier known metals, such as copper, lead, tin and iron. This difficulty stems from the volatility of zinc in the furnace and the near-instant oxidation of zinc vapor on contact with air. Reduction of zinc oxide by carbon takes place at about 1,100 degrees C., whereas the boiling point of zinc is only 907 degrees C. Because the boiling point is exceeded by the reduction temperature, elemental zinc, when formed, instantly vaporizes and if not kept in a reducing atmosphere, reacts with air forming zinc oxide rather than coalescing into metal. The ancient metallurgists and alchemists therefore, rarely, if ever, saw zinc metal and for a long period in history were not able to associate calamine ore and furnace crusts with zinc metal.

The first deliberate use of zinc is not known. Zinc has been found in copper and bronze artifacts made more than 5,000 years ago; this zinc is believed to have been present in the ores smelted and is thought to be an accidental or incidental component in the artifacts. The early brasses were also likely to have been made accidentally. However, over time, calamine and zinc oxide furnace crusts were recognized and deliberately added to copper ores and metal to produce brasses. Thompson (1936) accepted as certain that the Babylonians of the 3rd millennium B.C. were capable of making a compound of copper, tin and zinc by reduction with charcoal and further that the Assyrian word "elmesu" meant brass. Paliwal, Gurjar, and Craddock (1986), based on descriptions of ores and smelting in India during the reign of "Rama", about 1,600 B.C., suggest that brass may have been in use at that time. The same authors report that later records show that brass was used to manufacture surgical instruments in India as early as 1,000 B.C. Partington (1935) reported that bronzes found at Gezer in Palestine and dated between 1,400 and 1,000 B.C., contain up to 23% zinc. Bracelets made of brass were found in the ruins of Kameiros, a city on the island of Rhodes that was destroyed in the 6th century B.C. (Rickard, 1932). The Romans were well acquainted with brass as early as 200 B.C., according to Carus (1959), but Healy (1978) states that there is no conclusive written account or surviving objects to indicate that either the Romans or Greeks knew of brass until the first century B.C. However, from 20 B.C. through the next two centuries, brass was an important metal for Roman coinage. All brass made in pre-Christian times is thought to have been made by the "cementation process" or by the smelting of zinc- and copper- containing natural ores.

The oldest known piece of zinc metal (87.5% zinc) was an idol found in a prehistoric Dacian settlement at Dordosch, Transylvania (Hoffman, 1922). Two bracelets filled with zinc reportedly were found in the ruins of Kameriros (Hoffman, 1922); however there may be possible confusion with the brass bracelets indicated above by Rickard. Jagnaux (1891) reported that the front upper part of a fountain in the ruins of Pompeii, destroyed in 79 A.D., was covered with zinc; this finding has not been verified and remains in dispute (Dawkins, 1950). According to Healy (1978), there is no indisputable evidence to support the existence or use of metallic zinc in the ancient world or indeed that the Greeks and Romans refined zinc ores. It is possible that in rare instances zinc metal formed in some ancient smelting operations. However, because of the rarity of zinc relics, any such metal was thought to be accidentally made or, if purposely made, the technology was short-lived and soon forgotten.

The Chinese are said to have produced articles of zinc metal in the 7th century and are generally credited with being the first to develop the technology for metal production. However, Li (1948) suggests that as late as the 14th century, the Chinese did not know how to isolate zinc metal, even though, from about the 7th century, the Chinese made "brass" ornaments in which zinc was the chief component. Zinc may have been smelted in the Zawar area of India as early as 1,000 A.D. It was produced there on a relatively large scale beginning in the 14th century (Paliwal, Gurjar, and Craddock, 1986).

It is not definitely known whether the Chinese or the Indians were the first to have learned the technology of smelting zinc ores to metal. Interestingly, there were differences in early processing technology of the Chinese and Indians. While both employed small retorts for production, the Chinese method involved reduction and accumulation of the metal in the retort itself, whereas the Indian method was considerably more sophisticated in that the distilled zinc in the retort was directed, cooled, and condensed in a condenser neck that protruded below the retort and subsequently was collected in a vessel located below the condenser.

Europeans, apparently, were not familiar with zinc metal until the 16th century. Despite minor zinc metal occurrences at some European smelters and imports of the metal from China by the Portuguese, European alchemists were slow to learn the properties of zinc metal, its relationship to brass and its ores, and the secret of its production from ore. Zinc metal was not recognized in Europe until small beads of the metal were separated from zinc slag or furnace calamine at a silver smelting works on the Rammelsberg in 1509 (Rickard, 1932). Some of the zinc collected reportedly was later used to make brass. The imported metal went by the name spiauter, from which the commercial name for the metal, spelter is thought to have been derived.

Paracelsus thought zinc was a "semi-metal" and in 1520 described a few of its properties. Paracelsus is credited by some as first naming the metal "zinken", but Agricola, soon after Paracelsus' death in 1541, attributed the first naming of the metal, "zinck", to Norician and Rhetian miners, who lived in the area of present-day Austria. Clear classification of zinc as an actual metal did not occur in Europe until the early 17th century.

"Although zinc has served us for nearly 2000 years as the agent which converts copper into brass, no metal has ever been rewarded by so many bad names."

Zinc and Spelter (1950)
J.M. Dawkins

MODERN HISTORY OF ZINC

Virtually no effort was made to produce the metal in Europe until the second quarter of the 18th century when it was widely recognized that zinc could be produced from the calamine used to make brass, and that it was the same metal as that imported from India and China (Dawkins, 1950). In the 1740-43 period the first commercial plant to produce metallic zinc in Europe was started by William Champion at Warmley, near Bristol, England. The Champion process was based on the same downward distillation process as used at Zawar, but was a much advanced and larger-scale operation (Morgan, 1985). It is not known if Champion developed his process independently or if he obtained the necessary information from seamen, traders, or sources in Asia.

Toward the end of the 18th century, the technology to produce zinc metal became more widespread and newer types of smelter furnaces began to appear. The modern era of zinc production began with the development of the zinc smelting process (English method) by Champion. Johann Richberg of Upper Silesia made the next great advance in zinc smelting with the development the Silesian-type retort furnace in 1798. About ten years later in Belgium, the first Belgian-type retort furnace made its appearance. It was also about this time that wars in India appear to have led to the abandonment of the Zawar mines and smelting operations. The Indian mines remained closed and did not reopen again until the early 1940's.

By 1820, the zinc industry was well established in Europe. Germany and Belgium were the leading producers but Austria, France, Great Britain, Holland, and Poland also were important centers for zinc production. In virtually all cases, producers used local calamine ores, clays, charcoal, coke and coal. Significant amounts of sulfide ores were not used until the middle of the 19th century, despite the fact that Champion and his brother had patented a process for making zinc from roasted blende in 1758.

In the United States, the first zinc metal was produced about 1836 at the Arsenal in Washington, D.C. The Government brought in Belgian workers to build and operate a small spelter furnace to produce zinc from zincite ore from New Jersey. The primary purpose for the zinc was to provide alloying metal to make brass for the manufacture of National standard weights and measures, which at the time were different in some States. The smelting process proved to be too costly and was not considered a means for commercial production.

The U.S. zinc industry began about 1850 with the production of oxide from New Jersey and Pennsylvania ores. Experiments to produce metal from these same ores were initially unsuccessful; however in 1859, Pennsylvania calamine ore was successfully smelted in a 45-retort Belgian furnace operated by the Lehigh Zinc and Iron Co., and as such, the company became the first to produce commercial metal in the United States. The Lehigh Valley in Pennsylvania where the experiment took place, has been called the "cradle of the American zinc metal industry". The following year, zinc metal was produced at smelters in South Bethlehem, PA., and LaSalle, IL.

By the turn of the century, U.S. mining and smelting activity had grown rapidly, developing mainly in Illinois, Kansas, Missouri, and Pennsylvania because of closeness to zinc ores, suitable retort clay, and coal or natural gas for smelting fuel. The finding of natural gas in Indiana in 1892 led to a short-lived smelting industry near Marion, which died 12 years later. In Kansas, however, the finding of natural gas in 1895 led to the rapid development of a large zinc smelting industry, such that Kansas became the leading U.S. zinc-producing State four years later.

> "...in respect to aggregate (zinc) yield Upper Silesia and Broken Hill (Australia) are the only ones to rival it (the Tri-State) in any way."
>
> W.R. Ingalls (1931)

Although world zinc metal production was centered in Europe in the 18th and 19th centuries, toward the end of the latter century the United States had arisen as a second major world production center.

After World War I, a number of other countries developed substantial zinc industries. Despite increased production elsewhere, Europe, except during the two World Wars, and the United States, until the early 1970's, continued to be the principal smelting centers. Beginning in the late 1960's, zinc production in the United States began to decline owing to the closure of numerous smelters for environmental, economic, and/or technical reasons.

Zinc smelting, however, expanded rapidly after WWII in Japan, the USSR, Canada, Australia, and elsewhere, widely distributing the zinc production around the world and greatly diminishing the percentages of world output of Europe and the United States.

TRENDS IN WORLD PRODUCTION OF ZINC

World production of zinc metal rose from a few metric tons per year in the early 1740's to 7 million tons in 1987. From 1740, 173 years (1913) lapsed before world zinc metal production reached 1 million tons and 203 years (1943) to reach 2 million tons. Three million tons was attained in 1956; 4 million in 1964; 5 million in 1968; and 6 million in 1979. A production of eight million tons may be attained by 2000 if the past trend of 1-million-ton increases continues.

Trends in U.S. and world zinc mine and smelter production over the past 100 years (1895-1994) are shown in Figures 1 and 2, data for which is given in Table 1. For the most part, the trends are similar. U.S. mine and smelter output of zinc was dominant in the world until after World War II, but proportionally declined thereafter, owing largely to new production in many countries. In 1952, for example, 37 countries mined zinc ores and 21 countries operated primary smelters; in 1994 the comparable numbers were 51 mining ore and 41 smelting ore.

The magnitude of the increasing use of zinc in the world economy can be illustrated by the cumulative metal production of 25-year periods since 1800. Every 25-year period, except the 1926-1950 period, has at least doubled the cumulative metal output of the previous period. This means, as shown in Table 1, that the cumulative metal production in any 25-year period exceeds the entire world cumulative zinc metal production of all years prior to the start of the basis period. This fact prompted Pehrson in Bu-

Mines Economic Paper 2 in 1929 to question its continuance in the future by stating the following:

> "If the trend of increases ...continue indefinitely, the annual (zinc) production would be over 2.4 million (short) tons by 1950, and the production for the quarter century (1926-1950) would be in the neighborhood of 49 million tons, an increase of about 120% from the production of 1901 to 1925. Since the continuance of this rate involves such enormous tonnages, it can be hardly doubted that the rate must diminish in the near future, although the amount of yearly output may continue to increase indefinitely."

Interestingly enough, the annual world production speculated above by Pehrson for 1950 was attained in 1952, and his speculative forecast of cumulative production, though not quite attained for the 1926-1950 period, was more than doubled in the following quarter century. Cumulative world output for the period 1976-2000 may not double that of the previous period (93 million tons) but is on line, through 1994, to exceed the world cumulative metal production (about 162 million tons) of all years prior to 1976.

From 1800, the world per capita consumption of zinc, as shown in Table 2, trended sharply upward from an estimated 0.002 kilograms per person to about 0.34 kilograms in 1900 and to 1.37 kilograms in 1985. In the past decade per capita consumption has trended slightly downward owing mainly to slower world economic growth and rapidly rising world population. U.S. per capita consumption for 1995, based on apparent industrial demand, is about 10 pounds, or about 4.5 kilograms.

Per capita consumption in developing economies ranges between 0.1 and 1 pound, so there is great potential for ever increasing zinc demand. Resources do not appear to be a problem. Surprising (to some) is the fact that the reserves and reserve base of zinc continue to increase, and also surprising is the fact that many of the newly discovered deposits are large and high grade. The major challenge for the mining sector is not available material to mine but whether or not they will be permitted to mine it.

Commercial ore reserves are determined as much by political, environmental, and economic factors as by geological measures.

Figure 1. Zinc Mine Production[1], United States and World, 1895-1994

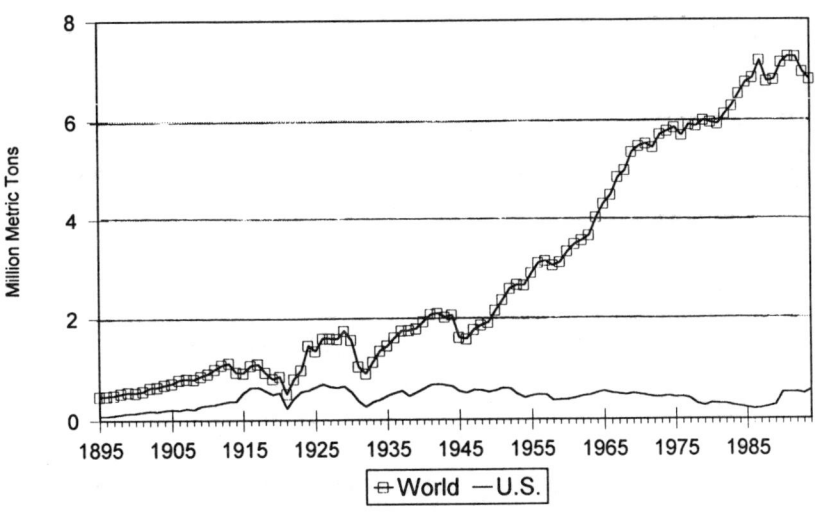

[1] Recoverable content.

Figure 2. Slab Zinc Production, United States and World, 1895-1994

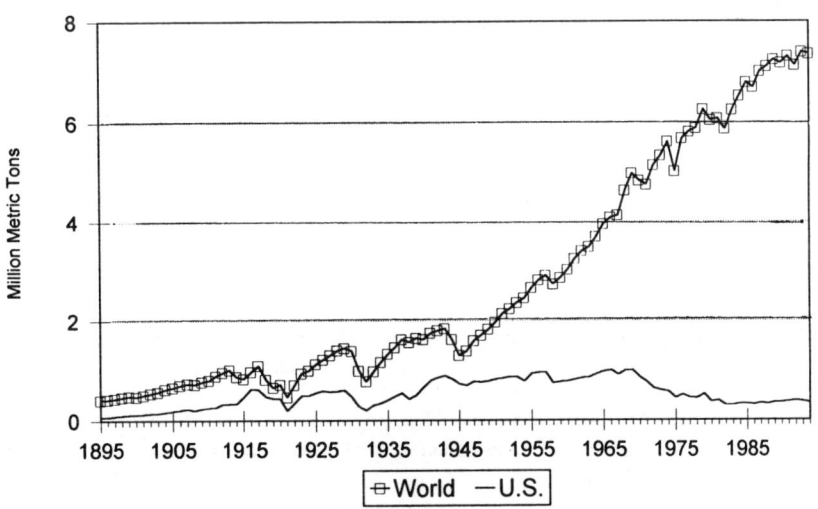

Table 1: World Slab Zinc Production by 25-Year Periods, 1801-2000

25-year Period	World Production (Thousand metric tons)	Percentage Increase over the Previous Period	Cumulative World Production (Thousand metric tons)
1801-1825	58	NA	1/ 70
1826-1850	518	890	588
1851-1875	2,659	510	3,247
1876-1900	8,322	310	11,569
1901-1925	19,670	230	31,239
1926-1950	36,958	190	68,197
1951-1975	92,415	250	160,612
1976-2000e/	2/ 170,000	180	330,612

e/ Estimate
1/ Assumed 12,000 tons production prior to 1801.
2/ Production in the 1976-1994 period is about 126,000 thousand tons. Assumes an average annual production of 7.3 million tons in the 1995-2000 period

Table 2: Apparent World Per Capita Consumption of Zinc Metal, Selected Years

Year	World Population	Zinc Production 1/ (kgs)	Per Capita (kgs)
1800	2/ 910,000,000	e/ 2,000,000	0.002
1850	2/ 1,130,000,000	e/ 49,000,000	0.043
1900	2/ 1,600,000,000	494,000,000	0.310
1920	2/ 1,800 000,000	618,000,000	0.340
1940	2/ 2,200,000,000	1,673,000,000	0.760
1950	3/ 2,564,000,000	1,980,000,000	0.770
1960	3/ 3,050,000,000	3,044,000,000	1.000
1970	3/ 3,721,000,000	4,848,000,000	1.300
1980	3/ 4,476,000,000	6,130,000,000	1.370
1985	3/ 4,882,000,000	6,673,000,000	1.370
1990	3/ 5,329,000,000	7,227,000,000	1.356
2000	3/ 6,285,000,000	e/ 8,000,000,000	1.275

e/ Estimated
1/ Three-year production average, except for 2000, centering on the year of population estimate. For example, for 1980, the production for 1979, 1980 and 1981was averaged.
2/ Source: World Almanac and Book of Facts
3/ Source: U.S. Bureau of Census, Center for International Research

HISTORY OF THE U.S. ZINC INDUSTRY

"It has been said that American-process zinc oxide was first accidentally revealed by the unintentional use of franklinite, by mistake, for sand in sealing a smelter furnace." (The oxide formed near where the furnace gases leaked through the sand.)

Paint Industry Reminisces (1931)
George B. Heckel

INTRODUCTION

Zinc was not mined or smelted in the United States at the time of its birth in 1776. A small brass industry existed, but depended mainly on imported brass and recycled brass scrap (although some brass was probably manufactured by the "cementation" process). The early brass workers were largely artisans who repaired broken brass items or made practical and decorative items, such as andirons, buckles, hardware, scientific instruments, and weather-vanes. During the Revolutionary War period, the domestic brass industry expanded, mainly to produce brass cannons, which were made largely from scrap, damaged cannons and, sometimes, captured British cannons. Despite growth in size and expertise, the brass industry after the War continued to depend on imported brass and calamine. Calamine deposits, associated with copper, silver, gold, iron and lead ores mined previously or being mined at that time in the original States, were not recognized until much later. In 1810, the chemical nature of zincite (naturally-occurring zinc oxide) in the northern New Jersey zinc-iron-manganese deposits was determined and some of the mineral was produced and ground for paint pigment for a number of years thereafter. Some of the zincite from these deposits also is said to have been used to make brass by the cementation process during the War of 1812. In 1836, New Jersey zincite was shipped to Washington, D.C. and converted to metal in a small Belgian retort furnace. The metal was used mainly to make standard weights and measures, which had been ordered by Congress. This smelting process proved to be very expensive, precluding any idea of commercialization.

The U.S. zinc industry is considered to have begun in the early 1850's with the first production of zinc oxide as a commercial product. The Sussex Zinc and Copper Mining and Manufacturing Co., which was organized in 1848 to exploit the

Franklin and Sterling Hill zinc deposits in New Jersey, built a small plant at Newark in that year to produce metallic zinc, but the process was unsuccessful. They then decided to try to manufacture zinc oxide pigment from the ore. The metal-based (French) zinc oxide processes of Sorel and LeClair, which had been developed only a few years earlier in France, were high cost, but yielded a desirable white pigment that was suitable as a replacement for hazardous white lead pigment in paints. Believing a sizable market existed if the price of zinc oxide was lower, Sussex Zinc developed a reverberatory furnace process to produce zinc oxide, and although a considerable amount of oxide was made in 1850, the process remained uneconomic. The following year the Wetherill grate and the Jones muslin-bag apparatus were combined, resulting the first successful commercial process to produce zinc oxide directly from ore; this became known as the Wetherill process and later, as the American process. The Newark plant began production of commercial quantities of zinc oxide in 1852 and in 1854 the process was put into use at the Passaic Zinc Works in Jersey City, NJ.

While the New Jersey mines were the first to be worked for zinc in the United States, the mines at Freidensville, PA provided the first zinc ore for the production of commercial spelter. These ore deposits were discovered in 1845 and were worked for ore beginning in 1853. A smelter and oxide plant were erected near the mines at Bethlehem; but the early smelting processes performed poorly being hindered by the high iron and lime contents in the ore, which reacted badly with the retorts. By 1859 the smelting problems were partially solved and the Lehigh Zinc Co. erected a smelter in South Bethlehem to process the Friedensville ore.

In 1860, Lehigh Zinc became the first to produce commercial spelter in the United States. In the same year, F.W. Mattheissen and E. C. Hegeler completed construction on a zinc smelter at LaSalle, IL and began spelter production using calamine produced in Wisconsin.

FROM 1860 TO 1900

The industry began rapid expansion, especially after the Civil War both in the east and in the Midwest. In the mid 1860's, zinc plants to manufacture both metal and oxide were built at Newark, Jersey City, and Bergenpoint, NJ to process NJ and PA ores. In the Midwest, the Glendale smelter was built at St. Louis, MO in 1869 to exploit the calamine ores found with the lead deposits in southeastern Missouri; and in 1870, the Illinois Zinc Co. built a smelter at Peru, IL to process zinc ore from Wisconsin. In 1871, the first zinc ores from the Joplin, MO district, herein referred to as the Tri-State district (Kansas, Missouri, and Oklahoma), came into the market, providing vast new supplies of zinc for smelting. Although the Tri-State ores had been mined since 1850 for lead, zinc minerals had no market value owing mainly to little, if any, demand and high transportation costs. As a result, the zinc was discarded in the waste rock and tailings heaps around the mines. In 1870 the St. Louis & San Francisco rail line was extended to Pierce City, MO; this, plus the nearness of new zinc smelters in St. Louis and in Illinois, created the economic conditions for the exploitation of the zinc ores in the district. Interestingly the first zinc ore to be shipped from the Tri-State district was said to have been calamine from a waste dump at Granby, MO In 1873 a zinc smelter was erected at Weir, KS and in 1878 another was built at Pittsburg, KS following the development of mines at Galena, KS. South of the Tri-State district in Arkansas, some local ores were mined as early as 1857 but attempts to smelt these ores locally in the 1860's and 1870's failed.

In the eastern United States, calamine ores at Bertha, near Austinville, VA, and at Mossy Creek, TN. were mined beginning in 1879 and 1883, respectively. The Virginia deposits were first mined for lead beginning in 1756 and were the major suppliers of lead to the Continental Army during the Revolutionary War. The major eastern zinc districts of this period were comprised of calamine ores that were essentially free of lead and other impurities and, as such, yielded higher-grade spelters that commanded premium prices. The most famous of these high-grade spelters were the "Bergenpoint" and "Bertha" brands. The zinc mines in both Tennessee and Pennsylvania ceased operations in the early 1890's. In Virginia, most ceased by 1898. Mines in each State eventually reopened but the emphasis was on the production of sulfide ore rather than on the near-surface oxide ores mined in earlier periods.

Zinc minerals were known to exist in many deposits in the western States and Territories from the early days of exploration and mining for gold, silver, lead, and copper. But, owing to low value and lack of rail transportation, zinc minerals were considered to be both waste rock and a nuisance if found in the ore. Transcontinental railroads opened the path to exploitation of zinc resources in the west in the early 1880's but it was not until the end of the century that large shipments of zinc ore came from western mines, mainly in Colorado and New Mexico. In 1899, substantial amounts of zinc ore from Colorado began to be smelted in Kansas and exported to Europe.

Up to the late 1870's, most zinc mined and processed came from surface and near-surface calamine deposits. Thereafter, increasing amounts of sphalerite were mined; by the end of the century, most zinc was produced from sulfide ores. The New Jersey ores were the exception.

Mining processes were generally simple and crude. The rich pockets of calamine ore were mined as surface pits, which followed the contours of the enclosing wall rock and around any boulders of limestone (generally) within the ore body itself. In Pennsylvania, when the surface ores were exhausted, the ore was followed downward in crevices between boulders aligned with the near-vertical bedding of the limestone. As the mineralized crevices were followed, the ore changed from calamine to sphalerite. The deepest crevices were mined to a depth of 225 feet and could have gone much farther but massive inflows of water prevented deeper mining and caused these mines to shut down. In Virginia, the calamine deposits rested against the sides of and between limestone pinnacles, cones, and columns (locally called chimneys, some of which are over 100 feet in height) in deeply weathered limestone. The limestone, the chimneys, and the ore are buried beneath varying thickness (up to 150 feet) of clay. Surface pits were used to extract the very near-surface ore and ores accessible on hillsides, but the deeper deposits were mined by sinking shallow shafts between chimneys into the ore and working out from them until the nearby ore was removed. When chimneys were encountered, the chimney was encircled with a timbered barrow drift to win the ore. A second drift and others were run one on top of the other contouring around the chimney until all the side ore was extracted. In 1890 a portable "pipe-shaft system" was introduced which attained greater efficiency and increased ore extraction. The system was so named because the shaft was constructed of iron-sheathed sections with an inside working diameter of only 3.5 feet (the size necessary for a bent-over miner to work) and when assembled, resembled a pipe. In this system numerous shafts could be sunk into ore as deep as 125 feet for as little as 22 cents per foot. This resulted in shorter ore-tramming distances and in reduced timbering and ventilation costs.

In New Jersey, mining of the Mine Hill ores (Franklin orebody) and a few miles away, the Sterling Hill ores (Sterling orebody) was initially carried out in surface pits. Surface hemimorphite (calamine) ores were extensive at Sterling Hill and were the main ores mined there up to 1870. At Mine Hill, considerably less calamine was present, and from the start primary ores constituted the main ores mined. Early underground mining involved entry through vertical shafts and adits on the hillsides. Mining was carried out by shrinkage stoping and square-set timbering where necessary. The primary ores at Sterling Hill were only sporadically worked to 1896 and then ceased until mining was resumed just before World War I. The Franklin ores were also not worked to any great extent until 1896 when the Wetherill magnetic separator was invented. Thereafter, with the easy separation of franklinite from the zincite and willemite and production of high-grade zinc concentrates, the Franklin Mine became one of the world's premier zinc producers.

The U.S. Zinc Industry

The Tri-State district developed its own unique system of leasing, mining, and marketing. The system of leasing was partially adapted from the procedure used in the southeast Missouri lead belt. The owner of a tract of land would lease it to a party or company for exploitation for a percentage of the mineral obtained as a royalty. The first lessee then would sublet smaller parcels to miners who would generally pay an advance royalty. In the first years from the beginning of zinc mining in 1871, the miner would sink a shallow, rudely timbered shaft into the ore, hand drill the ore, blasting only where necessary, and hoist ore up the shaft in small buckets, barrels, or kegs by whim or windlass. Only the second class or rich ore was mined; the first class or disseminated ore that later formed the great bulk of the district's production, was treated as waste or country rock. The ore was crushed with crude breakers and concentrated with hand jigs. Until about 1891, virtually all ore was concentrated by hand-jigging. In the 1880's some steam engines were introduced for hoisting, but even in the 1890's, horse-and manpower hoisting was common. Although a number of mining companies increasingly assumed large holdings and carried out mining for their own account as time progressed, most mining through the 1890's was carried out in "small" mine operations owing to the distribution and character of the deposits. In order to gain an appreciation of the methods and conditions that existed in the Tri-State district in this period the following has been excerpted from a report by F. L. Clerc in Mineral Resources of the United States for 1882, pp.369-373:

"The most important zinc mines of this region are those at Carterville and Webb City, which are really parts of the same deposit. They produce more than half of the zinc ore raised, and are worked principally for zinc ore. As these deposits are very regular in their formation, and in a measure rule the ore market, and as the method of mining them is essentially the same as prevails throughout their entire district, a general description of them and the method of mining them will here be given. These mines lie in the open prairie, which was once cultivated in farms, about five miles northeast of Joplin, near the head of a small branch of Center Creek. They were discovered about the year 1877. Here, at a depth of from 40 to 100 ft., and often under a cap of limestone and flint 60 ft. in thickness, has been found an immense deposit of zinc blende, which has been worked continuously for over half a mile. The deposit is in the form of a bed of flint, traversed in various directions by solid bars of barren flint, but in general resembling a breccia of sharp, angular pieces of barren flint, closely cemented by crystallized blende, with occasional masses of bright crystallized galena. With the exception of these bars and occasional pillars to support the roof, the whole body is blasted out. Draining is difficult, and rock drifts are sometimes necessary to unwater the ore. ... Where the cementing material is pure blende, the blende breaks freely from the chert and can be almost entirely cleaned by crushing and jigging; where the cementing material is quartzite, or black sand, as the miners call it, crushing is difficult, and a satisfactory separation is impossible. The mines spread over about a section of land, 640 acres. Their weekly output is about 700 tons. The method of working them is as follows: When a good prospect is discovered in new ground the land around it is leased from the original owners for royalties ranging from 10 to 25%, by a number of individuals, who organize various mining or, as they would more properly be called, land companies. The companies have the land divided up into lots 200 ft. square and a plat of it is made; select certain lots for themselves, and throw the others open to miners. They usually start a shaft on one of their own lots, and put in a pump. If the indications continue good, many of the lots, particularly those near the pump shaft, are quickly taken up by parties of miners, who sink shafts upon them, timber the ground, put up hoisting contrivances, furnish all supplies, and bear all expenses.

When ore is struck, it is drifted on and followed in all directions up to the boundaries of the lot in question. The ore is raised to the surface and crushed and washed by the miners, and is sold to one of the zinc or mineral buyers. It is weighed over the company's scales, and paid for to the company, which deducts a royalty of 25% on zinc blende, and 50% on 'mineral' (galena); and if it has pumps running, a pump rent of $1 a ton on zinc ore and $2 on 1000 lb. of galena; and pays over the balance to the miners. The royalties, of course, vary with circumstances, but the above are general. The holders of lots hire other labor to do the mining, at from $1 to $1.50 a day; and usually put up crushing and washing machinery on their lots. Very often the same parties control two or three adjoining lots or fractions of lots, and sometimes neighbors go into partnership. Most companies do not allow ores to be taken from the lots on which they are mined until they have been cleaned and have paid royalty. The machinery is usually of the simplest description—a farm or small stationary engine, covered by a shed of rough boards, a small-sized Blake's breaker, set over a pair of rolls, and a horse whim or a whip. The jigs are ordinary hand-jigs, with an overhead breakstaff, working a sieve 2X3.5 ft. up and down in a box of water. The jigging is usually done by contract, and is paid for by the ton of cleaned ore. It is common to see from 10 to 20 of these jigs grouped together under a shed of poles, covered with branches of trees or rough boards. The ore as crushed yields from 10% to 50% of cleaned ore, and the No. 1 grade assays about 60% to 62% of metallic zinc. The tailings must in most cases be piled up on the lot from which they have come; they are drawn up into a mound with two-horse scrapers or belt elevators, and it is not an unusual sight to see jigs and crushing machinery perched on top of these mounds 15 to 20 ft. above the surface level, the shafts being timbered up to a corresponding height. Several land companies have put in fairly effective pumping machinery. Plunger pumps working in pairs, with wooden walking beams or bob cranks, and driven by gearing and a crank shaft, are the most common; but direct-acting steam-pumps, like the Worthington or Blake, have been largely introduced of late, notwithstanding the disadvantages they labor under from the gritty water of the mines. From a distance these mines, with their swarms of busy men and heaps of tailings piled around the shafts, remind one strongly of gigantic ant-hills, and present a sight not soon to be forgotten. No one can fail to be struck with the glaring defects of such a method of mining, the absence of system, the useless duplication of machinery, the cheap yet expensive expedients, and the crowding together of conflicting operations. Below ground the effects are if possible worse. Each lot is affected by the policy of its neighbors; pillars are left only where they are thought to be absolutely necessary; each miner tries to get as much as possible out of his own lot, is only interested in it as long as he expects to work it, and is not disposed to improve the value of adjoining lots by unwatering them or proving their ore. The roof and pillars are badly trimmed, and in many cases dangerous, fatal accidents being distressingly common. The officers of the land companies are generally individually interested in one or more lots, and all sorts of questions are continuously arising from the conflicting interests of the company and the miners.

For the discovery and working of shallow deposits the present system seems the best that can be devised, but it is clearly not adapted to solve the problems of discovering deeper deposits or working them to advantage. No system can be defended which involves extravagant expense in mining and preparing the ore for market, forces the sale of it without regard to its value, and renders worthless large bodies of ore that might be profitably worked by a better system; and no basis for a great industry, like the zinc industry, which makes it depend on a hundred chances independent of the price of metal, the cost of smelting, or the known deposits of ore, can be considered very safe to build upon. The caving in of a single mine, the breaking down of a pump, less activity in lead mining, or the scattering of the miners to richer camps, may cause a falling off in the output of ore from which it would be very difficult to recover. That mining has on the whole been very profitable in this region is established from the fact that the country around has steadily and rapidly increased in wealth and population. With the last few years, three railroads, the St. Louis & San Francisco, the Missouri Pacific, and the Kansas City, Fort Scott & Gulf, have built branches through the ore field to each of the three towns, Joplin, Webb City, and Galena."

Mining of calamine (largely smithsonite) in the upper Mississippi Valley (Iowa, Illinois, and Wisconsin) began about 1860 to supply the Matthiesson and Hegeler smelter in Illinois. The calamine occurred with lead in shallow surface deposits. The same types of mining procedures as used in the Tri-State ores, were used to mine the ores in the Wisconsin lead-zinc fields. Many of these deposits had been worked out down to water level by the late 1870's and mining was not deemed profitable if continued pumping was necessary. The smithsonite occurrences merged at depth near the water table with sphalerite and galena ores, but zinc production tended to languish in the 1880's and 1890's owing largely to the inability of separating marcasite, an iron sulfide, and sphalerite in the ore, which resulted in a low-valued, low-grade zinc concentrate product. It was not until about 1905 that a slight-roast magnetic separation process was successful in producing good zinc concentrates from these marcasite-rich concentrates.

The smelting practices used by the U.S. zinc industry, except those in New Jersey, were initially adopted from those used in Europe.

Thereafter, the technologies developed on the two continents were mutually available. But locally, the U.S. industry generally tended to keep the old technology, foregoing improvements to achieve short-term profit. Even so, U.S. smelting practices, tended to remain diverse because of the variance in ore types and their smelting characteristics, and differing labor and energy costs.

The U.S. zinc industry rapidly became a large world producer of zinc despite being slow in some cases to widely adopt or utilize new technologies for zinc production. Although U.S. smelting technology gradually improved, especially in Illinois, some areas showed little progress. For example, the smelting furnaces used in Pennsylvania, New Jersey, Kansas, and Missouri in the 1890's were not significantly different from the relatively crude Belgian furnaces originally built in those regions in the 1870's. The eastern furnaces were direct-fired Belgians and generally of better construction than those in the Tri-State area, but they had no cellars into which ashes and residues could easily be dropped and removed. Instead the furnaces were built from the floor, and ashes were raked out from a shallow pit under the firing grate, with retort residues dropped to the floor in front of the furnace, both to be removed by shoveling. The retort charge was mixed on the furnace floor, although in later years, the charge was prepared in pits and brought to the retorts ready for charging. Retorts were hand-made, although later auger machines were used. The Dor press, which molded superior retorts by great hydraulic pressure, was already widely used in Europe by the time it was first tried in the United States in 1893. The operations at Tri-State smelters were generally similar to those of the eastern smelters. The character of the Tri-State smelters during this time were described by F. L. Clerc (same reference as above). This is thought to fairly describe smelting in that region for the 1870-1890 period, portions of which are given below:

> "The furnaces are built with ash-pits above the ground, with a sloping bank of earth or cinder leading up to the furnace floor. The buildings are scarcely more than sheds, and are huddled together with little regard to their mutual relation. The first cost of the building is inconsiderable. In the smelting process the cheapness of the fuel renders economy in this direction unimportant, and the cheapness of living makes labor obtainable at wages as low as anywhere in the country.
>
> ...The furnaces, roughly constructed of inferior material will not long sustain the heat required to exhaust the zinc from the cinder, and it is excepted opinion that there is no economy in "butchering" the furnace for the sake of a small additional percentage of metal; it is preferred to increase the production of the furnace and to reduce the cost of labor and fuel by increasing the charge of ore—in other words to butcher the ore and save the furnace. At the same time the personal supervision of the proprietors and their intimate knowledge of the business makes possible results that could not be expected by a company operating on a larger scale. ... Occasional good sales, or cheap lots of ore, give uncertain profits; and a slight rise in the price of metal is sufficient to "fire in" new furnaces. Under such circumstances no real advances in the metallurgy of zinc are to be expected."

The smelting companies in the Tri-State district did have an additional advantage to make profits in that the zinc concentrate (called "ore" in this district) was sold by weight and not by metal content. The miners were often ignorant of its value, and a smelter could sometimes buy "ore" containing 5% or more zinc for the same price as the standard concentrate. The local ore buyers, especially, had an advantage over the miners, as well as over buyers from distant smelters, because of their intimate knowledge of the mine operations and the behavior and yield of that particular "ore" in the furnace. It was not until about 1894 that the selling of ore by assay became general practice.

The general quality of smelting in the Tri-State area did not initially improve with the discovery and use of natural gas in Kansas in the mid-1890's or the building of "Iola" furnaces there in 1896. The Iola furnace was essentially a

Hegeler furnace fired by natural gas rather than producer gas. However, because the gas distribution and reaction with air were not understood, severe sooting occurred, reducing both spelter recovery and production. The natural gas used was virtually free of cost, but despite this, the cost of spelter production for both the coal-fired and natural gas fired smelters was about the same because of the poorer recovery in the natural gas fired plants. By 1899, despite extensive research to solve the sooting problem, zinc recovery from Tri-State ore at the best smelters was still only slightly better than 80%, whereas in the east, the recovery was better than 85%. Part of the higher recovery in the east was due to the higher purity of the ores used, as well as to the use of "prolongs" to increase recovery. The mid-western smelters tried prolongs but abandoned them on the ground that the extra zinc recovered was less than the cost of labor required to use them.

Despite slow progress in some smelting regions, several important advances in U.S. zinc smelting technology were made that raised American standards significantly. Efforts to increase the capacity of the retort furnaces were attained by adding rows of retorts, as many as nine, in the furnace; this, however, created problems in the charging and discharging of the highest retorts in that a man could only work on five rows from the floor. In order to surmount the problem, E. C. Hegeler of the Mattheissen & Hegeler Zinc Co. in LaSalle, IL. conceived the idea of making a long furnace fired by producer gas rather than a shorter, tall furnace fired by coal or gas. In 1872, Hegeler built his first long furnace, a unit 40 feet long and containing 408 retorts (204 per side) 5 rows high. Although the Hegeler long furnace found limited use outside LaSalle, it led to the introduction of mechanization, labor-saving devices, and the team concept of charging, tapping and discharging retorts, which subsequently became a distinctive feature in American zinc-smelting practice. In 1881, Hegeler developed the Hegeler roasting furnace, a huge, mechanically rabbled, multiple-hearth muffle furnace that eventually became successful in the United States but not in Europe where low-priced labor was available for traditional hand rabbling. The Hegeler roaster was also the first to manufacture sulfuric acid from blende-roasting gases, and, until 1925 it was virtually the only zinc roaster used in the United States for that purpose.

Despite the successful use of the Hegeler roasting furnace, the search for simple mechanical roasters continued in part because there was little interest in acid production. In addition, the Hegeler roaster was closely covered by patents and was expensive to build. In 1893, the rights to use the Brown horseshoe furnace, which had previously been used to roast copper ores, were obtained by the Collinsville and the

S. C. Edgar zinc companies for the States of Indiana, Illinois, Missouri, Kansas, and Arkansas. These two companies immediately gained a significant cost advantage: the cost of labor accounted for about 50% of the cost incurred in using hand-rabbled roasting methods. The labor costs per roasted ton of concentrate in the Brown horseshoe furnace at that time was 45 cents, one-third that of the old hand-rabbled kilns and reverberatories. Because other smelters were excluded from using the Brown furnaces, they sought out other types of mechanical roasting furnaces to counter the cost advantages of the smelters using the Brown and Hegeler roasters. This hunt gave rise to the Ropp and Cappeau roasting furnaces, resulting in one of the most costly litigation cases involving a metallurgical patent.

The owners and zinc lessees of the Brown patents brought suit against the Lanyon Zinc Co., claiming that the company's Ropp furnace was an infringement on their patent. The Brown furnace contained a slotted recessed chamber on each side of the hearth that protected the mechanical parts of the rabble carriage from heat and gases. In the Ropp furnace, a tunnel was constructed under the hearth with a slot in the center of it through which the rabble carriages were driven. The tunnel in effect protected the mechanical parts of the apparatus, as in the Brown furnace. The case was heard in a number of courts and all of them held that the Ropp tunnel was a recessed chamber and consequently infringed on the Brown patent. This led to the development of the Cappeau furnace, which was merely the Ropp furnace with the side walls below the hearth removed. The hearth was supported on legs. Because the area was open under the hearth, the court ruled that this was not an infringement because there was nothing resembling a chamber.

Despite a slow start technologically, with the exception of the Wetherill process, the United States had become a major world producer by the end of the century and was poised to become the leading producer and a leader in zinc technology in the 20th century. The above result was set up to some extent by the technological progress that the industry made in the 1890's. The processing of mixed and fine ores was vastly improved by the development in 1896 of the Wilfley table which furnished a new and efficient means for the washing of fine material, and in the same year, the Wetherill magnetic separator, which led to vastly increased exploitation of the Franklin Mine in New Jersey and later to the development of zinc sulfide deposits in Wisconsin and the West. The roasting of concentrates by mechanical means, especially by the Brown horseshoe furnace that debuted in 1893, increased efficiency and throughput and lowered roasting costs. In the smelting process, retort manufacture improved and retort life was increased by the use of hydraulic pressure; heat recuperation was introduced leading to economy of fuel, improved extraction, and the use of larger furnaces; and natural gas was utilized as smelting fuel, first in Indiana in 1892 and in Kansas in 1896. The efficiency of natural gas smelting, though relatively poor at the start, gradually improved and by the turn of the century had matched and, in some cases, exceeded the zinc recovery attained by the coal-fired smelters.

FROM 1900 TO 1915

From 1900 through 1914 U.S. mine production rose from 150,000 metric tons (tons) per year to 377,000 tons recoverable zinc in ores and concentrate, and U.S. smelter output of metal from 112,000 tons to 320,000 tons. In 1907 the United States became the world's largest producer and consumer of zinc, accounting for slightly more than 30% of the world total in both categories.

The Lanyon Zinc Co. became the world's second largest zinc-producing company in 1902, and two other U.S. companies were in the top five. The Belgian company, Vieille Montagne, was the leading producer. Mineral separation processes vastly improved, and U.S. smelting technology, considered relatively backward in the previous century attained European levels and in some areas was superior. Zinc recoveries in smelting continued to rise and the use of skims, drosses, and scrap increased to meet zinc chemicals demand.

Women were rarely employed in mining and smelting in the United States. In 1903 the mines in Upper Silesia employed 8,597 men and boys and 2,640 women; the smelters, 6,792 men and boys and 1,275 women.

Data Source: Mineral Resources of the United States (1903)

The uses of zinc expanded. Galvanizing of wire, sheet, and manufactured items such as buckets, fencing, poles, and pipe consumed about 60% of the zinc metal produced; brass, about 20%; and sheet zinc, about 15%. Sheet zinc was mainly used for roofing and gutters, linings in ice boxes, and consumer items, such as wash boards, fruit jar lids, and batteries. Zinc dusts found uses in desilverizing lead, fireworks, match head compounds, making hydrogen for balloons, textile dyeing, and extraction of gold and other noble metals from cyanide solutions. Zinc oxides were increasingly used for paint pigment and zinc chloride was widely used as a wood preservative, mainly for treating railroad ties, telephone and telegraph poles, and mine timbers. In 1907, more than 19,000,000 pounds of zinc chloride was used for wood preservation purposes.

Strong demand and improved extraction technology led

The U.S. Zinc Industry

to initial zinc mining in numerous States or increased production in those already producing. Colorado and New Mexico began zinc production in the 1890's, and shortly after the turn of the century, Montana, Idaho, Utah, Texas, Washington, California, and Arizona began production. Zinc production was possible in many of the western ores only as a result of the better concentration methods and development of post-smelting methods to recover the lead, copper, gold, and/or silver contained in the zinc ore or concentrate. The zinc in many cases was not economic by itself. In the Midwest, The Tri-State district, led largely by production in Missouri, was, by far, the country's leading producing area, in part, due to increased mining of sheet-ground ores by the larger companies. Zinc mining resumed in 1905 on a large scale in Upper Mississippi Valley district owing mainly to the development of electrostatic and roast-magnetic separation methods to separate marcasite and sphalerite from each other in jig/table concentrates of the two minerals. Production also resumed in Tennessee after a hiatus of about 10 years. Kentucky had a small output associated with fluorite mining. In the east the Franklin Mine in New Jersey remained the single largest zinc producer in the United States and by itself, made New Jersey the second leading zinc-producing State. Small amounts of zinc occasionally were produced in North Carolina, New Hampshire, and New York.

Efforts to improve recovery from ores and old tailings and to increase the zinc content of concentrates received a welcome boost with the introduction of the Wifley table and Wetherill magnetic separator at the end of the 19th century. The Wifley table rapidly became widely used in many mills, whereas the Wetherill magnetic separator was initially used to greatly expand the production capability of the Franklin Mine, in that willemite and zincite could easily be separated from franklinite without any pre-preparation other than concentration by jigging. Poor, at best, separations could be made with Wetherill-type magnetic separators when used on complex sulfide ores. However, experiments using a gentle roast to enhance the magnetic properties of iron-bearing minerals combined with magnetic separation proved to be successful; and as a result, in 1902 and 1903, the combined process was used in Colorado and Wisconsin to make available zinc in a class of ore that hitherto was of little value. To avoid the roasting step mentioned above, a number of electrostatic devices were also developed to separate minerals without pre-roasting. Because of sizing and dryness requirements, electrostatic technology was not extensively used for zinc recovery, although Huff electrostatic machines enjoyed some success in the 1910-1920 period at some mines in Wisconsin and in the Rocky Mountains.

Flotation processes made a successful entry into the domestic zinc industry about 1910, but they did not become a large production source of smelter feed until the early 1920's. The Australians were particularly at the forefront of flotation technology in the early 1900's and, to a large extent, brought the technology to the United States. In 1907, the Sanders flotation process made a successful separation of sphalerite and fluorite in an experimental plant in Kentucky; this process also was used at a plant in New Mexico to separate pyrite and sphalerite in 1908. Beginning about 1910, Mineral Separation and Hyde flotation processes were being used in mills at Butte, MT. And in Idaho, the Macquisten Tube process was successfully used to separate blende in sphalerite-barite-siderite middlings from tabling operations. By the end of 1914, a number of improved flotation processes had provided the means to make both the Butte, MT and Coeur d'Alene, ID mining districts into significant zinc producers and had led to significantly increased zinc output in Colorado, Tennessee, and Utah.

The smelting of Rocky Mountain sulfide ores, beginning in 1899, initially caused severe processing problems for the Midwest smelters, who were used to almost pure Joplin blende concentrates with very low lead and iron contents. Within a few years they learned to cope with these lower-grade, more complex concentrates and, as a result, gained access to abundant new ore supplies by eliminating European exports of these ores and by being able to process imported ores from British Columbia and Mexico. The processing of these concentrates also led early on to the Belgian practice of resmelting the retort residues for the recovery of gold, silver, and lead.

The quality of smelters and their locations changed significantly in the 1900-1914 period. By 1901, the gas-fired smelters had raised their zinc recoveries to the same level as the coal-fired smelters. Because of overall lower production costs, a number of coal smelters closed down over the next few years or upgraded their plants in order to stay in business. In Indiana, the gas pool that gave rise to the State's zinc smelting industry was exhausted by 1903, virtually eliminating the industry there. The gas pool on which the Kansas smelting sector was based, also had weakened considerably by 1910; of 16 smelters operating in Kansas in that year, only 10 remained at the end of 1911. A few new plants were built in Kansas at other gas fields and others obtained gas by pipeline; however, the industry exodus to the rich Oklahoma zinc and gas fields was compelling. By 1912, Oklahoma had a larger smelting capacity than Kansas, and although the latter had a larger production than the former in that year, it was the last time that Kansas zinc metal production would exceed that of Oklahoma. Increased demand for zinc and the changing economics of production also resulted in the erection of six new coal-fired plants in Illinois, Indiana, and Pennsylvania, and two gas-fired

The U.S. Zinc Industry

plants in West Virginia.

The gains made in the quality and character of the U.S. smelting sector, relative to European smelting practices, only a few years after the turn of the century were summed up by W. R. Ingalls in his 1908 publication, "Lead and Zinc in the United States", thusly:

> "In summing up the status of American zinc smelting at the end of 1907, it is no longer to be said that as compared with European practice it is characterized by backwardness. It is rather to be said that while the European practice excels in some respects, the American excels in others. In most matters of mechanical handling, and in roasting blende ore, the American is decidedly ahead. In the construction of the distillation furnaces and in distillation practice, including the recovery of by-products, the European leads. The cost of smelting is about the same in the two continents. The cheaper labor of Europe is offset by its dearer fuel. In percentage of metal extraction, the European smelters on the whole have the advantage, but it is not a large one. In conclusion it may be said that the American zinc-smelting practice, capable as it is of further great improvement, is no longer a branch of the metallurgical industry to be ashamed of."

FROM 1915 TO 1919

The war in Europe gave great impetus to the zinc industry of the United States within days of its start. In the 10-month period prior to the start of World War I (WWI) in August 1914, the U.S. price of spelter had declined well below 5 cents per pound and domestic spelter stocks had risen to about 58,000 metric tons or about 20% of that year's production. With the outbreak of war, U.S. spelter prices immediately rose, then fell until stocks were depleted. They soon resumed sharply upward, reaching war-time highs of 26.5 cents per pound for Prime Western and 40 cents for High Grade in June 1915 before falling off by more than half from these lofty levels by the end of the year. The rise and fall in zinc prices in 1915 were, to a large extent, related to the method of purchase established by the British and French. Each government made their zinc, brass, and munitions purchases individually, and thus bid the prices up on one another. To surmount this situation, both countries had appointed the J.P. Morgan & Co. by May 1915 as purchasing agent, resulting in reduced competitive bidding and sharply lower prices. The price rise and the rise of U.S. dominance of the zinc market during the war was in large part due to the location and structure of the European zinc industry. Of the three main smelting centers in Europe, one was in Germany. The others were in Belgium and northern France along the Meuse River in the line of the German advance. At that time, a major source of concentrate for plants in the above countries came from the mines at Broken Hill, Australia. The Germans were immediately cut off from Australian supplies and the Meuse plants were damaged or idled by being in the war zone. The British zinc industry, the third largest in Europe but small in comparison to that of Germany and Belgium, had access to the Broken Hill concentrates but had nowhere to smelt them. This situation led to the strong demand for American zinc, to large U.S. imports of Australian concentrate, and to the rapid rise in American production.

The U.S. zinc industry responded to the increased demand and good prices and, by 1917, the industry accounted for 60% of the world output and had raised spelter production to a level 90% higher that it was in 1914. This increase was accomplished by increasing capacity at existing smelters, reopening abandoned and obsolete plants and constructing new plants. By the time the United States entered the war in April 1917, the smelting capacity exceeded all requirements. As such, the poorest plants began shutting down and some of the planned new plants never got off the drawing boards. At mid-year 1916 in the United States, 57 horizontal retort smelters with 205,000 retorts in operation, were producing zinc; at yearend 1918, only 36 smelters with 123,000 retorts remained. Despite the closures, smelting capacity was still greater than needed, resulting in several additional plant closures in 1919. The zinc export market had also receded owing to the rapid rise in zinc production in a number of countries during and after the war, to overall reduced world demand in 1918 and 1919, and to the rapid rebuilding of the zinc industry in continental Europe after the war.

German interests in the American zinc industry were substantial when the war began in Europe. The "German trio"—L. Vogelstein & Co., Beer, Sondheimer & Co., and the Merton Group (Metallgesellschaft, Metallbank and Metallurgishe Gesellschaft, and the Merton Co.)—had direct control of about 30% of the U.S. smelting capacity, principally through their ownership of the American Metal Co. and the American Zinc, Lead & Smelting Co. and their subsidiaries and through metal marketing smelting contracts

When the United States entered the war, the Alien Property Custodian took over the German interest in the U.S. zinc industry and in some cases sold the companies' stock to eliminate all alien enemy ownership and control.

The war also spurred the testing of new production methods, the principal beneficiary of which was the development of electrolytic zinc production. A number of electrolytic processes had been tried and failed since the work of Letrange in 1881; the difficulties encountered led the reknowned metallurgist, W.R. Ingalls, to state in 1911 that "I shall not say that the hydrometallurgy of zinc has no future but I feel safe in predicting it will not become a branch of metallurgy of general application." However, in 1914 metallurgists in both Canada and the United States pilot tested some of the promising systems, several of which were put into commercial production during the war, but most died soon after the war ended. The two pioneering processes that survived were those developed in 1914 at Great Falls, MT and at Trail, British Columbia. Commercial plants at both locations came on stream in 1915, processing hitherto uneconomic complex lower-grade zinc concentrates recovered at Butte, MT and the Sullivan Mine in B.C., respectively. Both smelters produced High Grade (99.90%) zinc, which brought premium prices during the war. An additional factor which made the plants and the process economically viable was the recovery of by-product metals.

> "I can hardly claim that we have accomplished anything new, especially new in this line except in so far as we have done on a somewhat larger scale what had often been done by others in the laboratory and in a smaller way."
>
> *Frederick Laist (1916), copatentee with F.F. Frick, commenting on the Anconda electrolytic zinc process*

Although electrolytic production provided significant amounts of high grade spelter (99.90%) for the war effort, most of this grade of metal was made by redistillation of ordinary spelter. Some higher-purity metal was made at electrolytic smelters during the war, but it was not until 1934 that a 99.99% metal could be made commercially by a redistillation process.

The war created a large demand for brass, rolled zinc, and galvanized products. Brass had many uses, but as a war material, it was mainly used in shrapnel and heavy artillery shells. High Grade zinc was required for the manufacture of cartridge brass. The zinc requirements for these materials can be illustrated by considering that a British 18-pounder or 3.3 inch shrapnel had almost 6 pounds of brass containing about 2 pounds of zinc. As about 25 million shells of this size and larger were ordered by European Governments from U.S. manufacturers by 1916, the zinc required would exceed 50 million pounds. The copper casing used in Lebel cartridges, used by the French army by the millions, contained 1 pound of zinc per 125 cartridges. Rolled zinc was also in heavy demand from the beginning of the war for lining packing cases made for shipment overseas, especially cases containing munitions or other materials likely to be affected by salt air. Zinc plate also found wide use in ships as boiler plate and for cathodic protection. Galvanized products, mostly steel sheet, were also big export items. U.S. exports of barbed wire, much of which was galvanized, rose from about 95,000 tons in 1914 to more than 420,000 tons in 1916. The value of exported zinc, zinc semimanufactures, brass and zinc-bearing manufactures, excluding steel products and shrapnel, rose from $23 million in 1914 to $115 million in 1915, to a high of $431 million in 1916, falling off to $335 million in 1917 and to only $64 million in 1918.

FROM 1920 TO 1939

Following the war, economic activity slowed culminating in depression in 1921. Many domestic zinc mines closed and zinc production plunged to only 182,000 metric tons, the lowest output since 1904. With economic recovery, zinc metal production rose substantially, and by the mid 1920's, production was again near the peak levels attained during WWI. Production in 1930 was maintained at relatively high rates despite the onset of economic depression, which reached its low point in 1931 and 1932. Again the zinc industry retrenched; metal production plunged in 1932 to about 184,000 tons, virtually the same level reached in the previous depression. In the five year period of 1930 through 1934, zinc metal averaged only 2.9 cents per pound, and in 1932 sold for as low as 2.3 cents per pound.

All zinc use declined in the early 1930's, except for rolled zinc use for jar tops, said to be due to growth in home food canning.

Zinc production gradually increased in the 1930's but never attained the high levels of 1928 and 1929. Despite the improvement in production in the 1930's, many of the more inefficient mines and smelters did not reopen until after World War II had started.

Although the period between the two World Wars was one marked by two severe economic depressions, the zinc industry made great strides in mining, flotation, roasting, and metallurgy technology. In mining, mechanical loading, detachable drill bits, artificial pillars, improved and faster ore hoisting, and safer explosives vastly improved worker conditions, lowering costs and improving worker productivity. The development of heavy media separation of sulfide ores in the late 1930's was a significant advance in milling by reducing in half the quantity of ore requiring fine crushing. However, the development of the selective or differential flotation process after the end of World War I for the separation of zinc from other sulfides and other materials was probably the most important event in zinc ore dressing in the 20th century.

Selective flotation was successful at a few operations before

1918; but the flotation process was not effective on many ores and did not attract the general attention of the zinc industry until the early 1920's when more universally suitable flotation reagents and improved flotation machines became available. Its large-scale use in the mid-1920's had a significant and immediate impact on the zinc industry. Not only did it open up previously unworkable complex ores to exploitation but it also raised the value, zinc grade, and overall quality of the concentrates produced. In 1921, only 34,000 tons of flotation concentrate (excluding any from the Tri-State) averaging 40% zinc and almost 12% lead were produced; whereas, in 1928, comparative production of flotation concentrate totaled 500,000 tons grading 53% zinc and only 3% lead. Despite the success of zinc flotation, a number of mining companies found it more economical to continue the use of jigs and tables to concentrate the coarse ore and to use flotation only for the fine grained ore and slimes. However, by the late 1930's, most zinc concentrates, outside the Tri-State, were produced by flotation methods.

The availability of immense tonnages of flotation concentrates introduced new problems for the smelters because they were much finer grained than the coarse-grained jig material the smelters were used to. As such, these concentrates dusted badly—not only when handled with the mechanical ore-handling devices used, but also during the roasting process, especially at those plants using Hegeler roasting furnaces. This led to substantial revision of the ore-handling and roasting methods in most major U.S. plants. With regard to the former, it was found dust losses in ore handling could be minimized if the concentrate was kept moist (dried to 4%-10% water) and fed directly into the roaster or until just prior to entering the roaster. In the latter case, by the late 1920's all the Hegeler roasters were discarded and mostly replaced by furnaces of the McDougall or Herreshoff type or by various forms of shower or suspension roasters. They also had a large dusting penalty, but it was relatively easy to contain and recirculate. Suspension roasting, which was first commercially accomplished at Trail, B.C. in the early 1920's, was not suitable for the roasting of coarse-grained sulfides because of incomplete roasting, but when the finely ground flotation concentrates became available, it proved to be a highly efficient method for the speedy oxidation of zinc sulfide. The fineness of the flotation calcines was also a problem for the horizontal and vertical retort processes, which required a granulated charge with good pore space for proper distillation to occur. To obtain the proper calcine charge, zinc sintering, first developed in Australia in 1917, became a common process in U.S. smelting practice by the mid 1920's. Sintering also provided a means for improving the metal produced by reducing the lead and cadmium contents in the sinter feed.

American-process zinc oxide up to 1920 was virtually all produced by either the western or eastern Wetherill processes, both of which burned the loosely mixed charge on fixed grates in furnaces that had to be operated under seal. However in that same year, some operations began to briquette the furnace charge; this greatly simplified the handling of the material charged, eased furnace operations, and improved the zinc recovery. The melding of briquettes and the continuous traveling grate soon followed, thus resulting in a continuous, open-furnace zinc oxide process rather than a somewhat imperfect, tightly enclosed batch process. The employment of the rotary kiln using a loose mixed charge also began to be utilized at this time. Two developments, the Waelz kiln in 1925 and the slag fuming process in 1927, had significant impact on the extraction of zinc from low-grade ores, scrap, wastes and residues, and slags. The former was versatile in that the kiln could process almost any kind of zinc-bearing feed. The latter process was used almost exclusively to extract zinc, cadmium, and lead from slags derived in lead smelting. Prior to the slag-fuming process, the relatively large quantities (10%-25%) of zinc contained in lead slags were just discarded. In the 1927-1982 period, an estimated 2.7 million tons of zinc was recovered from lead slags in the United States.

U.S. electrolytic capacity rose substantially in this period; new plants were built at Anaconda, MT and Kellogg, ID in 1928 and at East St. Louis, IL in 1929. At the same time a great advance was taking place in continuous retort smelting, which up to this time was highly desired by smelting companies but was largely unsuccessful, owing to excessive "blue" dust formation. The New Jersey Zinc Co. developed its vertical retort process and began zinc smelting using the process in 1929. The blue dust problem was overcome by closely controlling furnace conditions which provided a continuous supply of gases, carrying zinc vapor and other gases of constant composition and temperature. This made it possible to operate large condensers under substantially uniform conditions to collect the vapor and minimize blue dust formation. In 1930, a somewhat different continuous distillation process, the electrothermic process, was placed in operation at Josephtown, PA by the St. Joe Lead Co. This smelter did experience difficulty in condensing the vapor, and was operated as an oxide plant until the condensing problem was solved in 1936. Although a number of processes to refine zinc were available, the American zinc industry continued to produce most of its zinc by the old horizontal retort method.

FROM 1940 to 1946

The war in Europe again thrust the U.S. zinc industry into

prominence as the only major source of zinc available to the British and their allies. In 1939 and 1940, the Germans had taken control of the Belgian, Dutch, French, Norwegian, and Polish zinc industries, and with other Axis powers, controlled 75% of the European zinc smelting capacity and about 50% of the total world capacity.

The British Empire at that time controlled about 18%. Of that, 80% was in Canada and Australia. The United States accounted for most of the remaining world zinc smelting capacity, but unlike most of the other world producers at that time was not embroiled in war. British blockades prevented the Axis-controlled smelters from importing concentrates and initially reduced their production capabilities significantly; however, the Germans increased their zinc mine output in Europe, mainly in Poland, to such an extent that the European Axis powers experienced little, if any, zinc shortages during the war. When the war erupted in Europe in September 1939, the U.S. zinc industry was experiencing a relatively good year in climbing out of the 1938 recession. Prices were rising, and British orders for U.S. zinc metal were increasing. The Allied blockade and German submarine activity greatly reduced the flow of zinc concentrates to Europe; consequently, producers sought other outlets, principally the United States, because of its rising demand and enormous smelting capability, and Canada, in part due to lower tariff rates. In 1940 and 1941 the unprecedented demand for metal for the domestic defense efforts and British orders resulted in U.S. production increases of 33% and 62%, respectively, over 1939 production. When the United States entered the war in December 1941, the domestic smelting industry was in reasonably good shape (producing the necessary metal for the war effort), but the mining sector was not (large amounts of imported concentrates were essential to meet mobilization goals). During the war, the zinc industry was rigidly controlled with ceiling prices, allocation of metal, and export-import controls. In 1944 some controls were lifted and in 1945 all controls were dropped.

In 1941, domestic mine production was up 13% from the previous year but was still 3% less than that of 1926. To encourage higher mine output, the Government authorized extra payments for concentrate production under a Premium Price Plan in 1942. Despite the Plan, mine production increased only slightly in 1942. Thereafter, through the end of the war, U.S. mine production of zinc never exceeded that of 1941. This was attributed, in large part, to the depletion of ore reserves at some mines and to severe manpower shortages. The Government attempted to address the latter situation in July 1943 by authorizing the release of 4,500 men from the Army to work in copper, lead, and zinc mines; about 1,100 of these men were assigned to zinc operations.

U.S. zinc production was centered in five localities—the Tri-State district; Tennessee-Virginia; Sussex County, NJ; St. Lawrence County, NY; and the Rocky Mountain region. The western States, primarily Montana, Idaho, New Mexico, Colorado and Utah produced about 44% of the zinc output during the war; the central States, mainly Oklahoma and Kansas, 30%; and the eastern States, mainly New Jersey, Tennessee, and New York, 26%. Mine production was augmented during the war by the reworking of old tailings, by extraction of zinc from low-grade ores and smelter residues by Waelz kilns, and by slag fuming. Slag fuming operations were started up in Utah and Montana in 1942 and a third one began in Idaho in 1943.

The necessity for ore imports became readily apparent by the end of 1939 and in 1940 ore imports jumped five-fold. The highest import year was 1943, when concentrates containing about 465,000 tons of zinc were imported. In the 1941-1945 period, zinc concentrate imports, mainly from Mexico, Canada, Australia and Peru, averaged about 365,000 tons of contained zinc annually and accounted for an estimated 35% of U.S. primary zinc smelter supply in that period.

The domestic smelting sector that existed throughout the war period was pretty much in place in 1940. In that year, there were 17 active zinc smelters and 3 active electrolytic plants. In 1941 the Evans-Wallower electrolytic plant at East St. Louis, IL, which had been idle since 1931, was reconditioned and put back into production, and in December 1942, a new electrolytic plant was commissioned at Corpus Christi, TX. The above plants, with some modification and expansion, were the only ones that operated in the war period. By 1943, virtually all planned changes in zinc smelting and refining were completed, and the principal problem was one concerned with the available supply of zinc concentrate. In the 1940-1945 period, zinc metal production averaged about 800,000 tons annually, up from a production of 510,000 tons in 1939. Electrolytic plants accounted for about 35% of the war-years' metal production, with the remainder produced by retort and secondary plants.

The metal was mainly used for brass and for galvanizing purposes. Whereas galvanizing dominated pre-war use, brass dominated the war use. The large requirements for brass also was a major factor in the Government permitting the building of the new electrolytic plant as high-grade metal largely for brass production was in short supply and in great demand. The production of High-Grade and Special High Grade metal at electrolytic plants was augmented by redistillation of less pure metals in retorts and distillation columns.

During the war about 45% of the slab zinc produced was used to produce brass, 80% of which went to make cartridges and shell casings. In 1943 more than 50% of the zinc metal consumed was used for brass production. Galvanizing of all types declined to about 30%, but more zinc was used in coating pipes, wire, and shapes, mainly for the military, than for sheet galvanizing, which only accounted for about 30% of this sectors consumption. The use of zinc-based alloys also declined owing to shortages of high-grade zinc and to rigid military specifications that prohibited the use of zinc-based die casting in much of the war equipment. These alloys did find substantial use as stamping and forming dies in fabricating certain airplane parts; an advantage they had on other die materials in this application was that they were easy to remelt and reform. About half of the zinc dusts produced were used to manufacture zinc hydrosulfite and sodium hydrosulfite which were used mainly as reducing agents in the bleaching of textiles, paper and clay. Dust was also used in paints and military smoke-generating compositions.

With the winding down of the war in 1945, reconversion back to a peace-time economy became a major concern. All war-time controls were revoked, although the Premium Price Plan was continued until mid-1946. Zinc mine and smelter production declined but was adequate to meet all requirements. In June, the Reciprocal Trade Agreements Act, permitting as much as a 50% reduction in the duty rates prevailing on January 1, 1945, was extended for three years, much to the dismay of domestic zinc producers.

At the end of the year, legislation to establish a stockpile of strategic and critical metals came under consideration.

The post-war outlook was very positive for the zinc industry, except for the availability of adequate domestic ore reserves. The situation as seen by A.L Ransome and B.A. Estill in the Zinc chapter of the 1943 Minerals Yearbook, published in late 1944, appears to have been pretty much on the mark and is partially quoted below.

"The position of the United States has been preeminent in the zinc industry of the world over a period of many years, owing to vast resources of relatively low-grade ore. This wealth of raw material more than supplied domestic demands in this country until the middle 1930's. Production from our mines was adequate to supply an exportable surplus, and foreign zinc was relatively unimportant to our national economy. The advent of the Second World War late in 1939 precipitated heavy demands for zinc which could not be met by domestic mine production, and imports increased rapidly. The 1942-43 rate of mine production cannot long be maintained, particularly if premium prices should be withdrawn, and it seems certain that the post-war level of domestic mining will be lower, perhaps in the range of 400,000 to 600,000 (short) tons a year. The pent-up consumer demand to be filled in the post-war years in all probability ... (will be much higher than the mine output),

...and be made up by continued imports...The domestic mines conceivably could fill the demand for new zinc for a limited time, given adequate price and tariff protection, but ore reserves have been drawn upon heavily during the war period, and it is doubtful if domestic mines will again completely and continually fulfill the zinc needs of the country, even though zinc will be mined for many years to come."

Interestingly, since then through 1994 the annual domestic mine production has never exceeded that produced during any of the four war years, 1941 through 1944.

From 1946 to 1960

With the war's end, zinc demand remained high, but mine production declined in 1946 by about 100,000 tons owing to strikes and labor shortages caused, in part, by the slow return of miners who had migrated to cities for defense jobs and those in the military. Through 1949 both mine and smelter production tended to remain below WWII levels and overall trended downward. Domestic demand exceeded production in the first five years after the war, resulting in increased imports and liquidation of domestic smelter stocks, which fell more than 200,000 tons at the end of 1945 to less than 20,000 tons by the end of 1948.

In 1950 demand accelerated with the outbreak of war in Korea, and through 1953, mine and smelter output approached WWII levels. Mine production exceeded 600,000 tons in 1951 and 1952, but has never attained these levels since then. High import levels were maintained during and after the war by continued strong domestic demand and Government stockpiling programs. However, world zinc prices, which exceeded the domestic price in 1951, fell sharply in 1952 owing to foreign production in excess of foreign need and saturation of European markets. The average annual domestic zinc price fell from 18 cents per pound to 10.9 cents in 1953. Employment in the lead and zinc mining and milling sector and in zinc smelting fell about 8,000 workers to 30,000 by the end of 1953. The continued dumping of excess production into the U.S. market coupled with lower prices caused some domestic mines to close and others to curtail production in 1953 and 1954. As a result the domestic mining companies petitioned the Tariff Commission in 1954 for relief under the Escape Clause of the GATT and the Trade Agreements Act, claiming damage by reason of the trade concessions granted to foreign governments. The Tariff Commission found the complaint well grounded and recommended that higher tariffs might be implemented. In lieu of withdrawing any U.S. international commitments on trade, the Government opted to increase the stockpiling of domestically produced zinc; this resulted in a modest increase in prices, but still was less than adequate to keep marginal mines in operation. Even the permanent closing of the famous Franklin Mine in New Jersey in 1954 owing to ore exhaustion after more than 100 years of continuous operation did little to improve zinc prices.

In the mid 1950's, U.S. slab zinc consumption exceeded 1 million tons for the first time (1955), and smelter output attained record levels (1955-57). However, mine production continued to languish owing mainly to low prices and world competition, and by 1958 had fallen to 374,000 tons, the lowest since the 1930's depression years. Numerous mines had closed or had reduced production; nearly all the mines in the Tri-State were idle. The smelting sector also was not immune from the changing conditions in the domestic and world markets. In 1957, the U.S. Steel Corp. smelter at Donora, PA and the American Zinc Co. smelter at Fairmont City, IL were permanently closed.

Foreign concentrates, imported at high rates since 1953, constituted more than half of the smelter feed at domestic zinc plants. In 1957, the zinc miners again petitioned the Tariff Commission for relief from imports. In 1958 on the basis of the findings, the President ordered the establishment of annual import quotas on zinc concentrate and metal, to be set at 80% of average annual comparative imports for the 1953-1957 period. This resulted in modest improvement in domestic zinc prices, leading to slightly higher mine production in 1959 and 1960. World production of both zinc ore and metal at the time continued to be in excess of world demand resulting in large stock increases and depressed prices. The United Nations sponsored a conference on lead and zinc in 1958 to address this issue. As a result of this conference various countries agreed to specific voluntary restrictions on zinc production, and to the creation of an intergovernmental body, the International Lead and Zinc Study Group (ILZSG). In 1959, ILZSG was set up to improve the supply-demand balance of the metals through enhanced market transparency. Throughout this period, the U.S. Government continued to absorb much of the excess by adding foreign slab zinc to the supplemental stockpile via various barter agreements until zinc was removed from the exchange list in 1961. From 1956 through 1960 under the barter program, about 290,000 tons of slab zinc had been acquired for the supplemental stockpile.

In the fifteen year period after the war, the domestic zinc mining sector underwent significant change as to location of production (see Figure 1). The percentage contribution of the western States, the west central States, and the States east of the Mississippi River to the zinc mine output of the United States averaged 41%, 29% and 30%, respectively, in the 1941-1945 period; in the 1950-1954 period the respective ratios were 56%, 13%, and

31%; and in the 1956-60 period, 52%, 5%, and 43%. In the last two years of the period, the western and eastern States' mines produced about equal amounts, with less than 2% coming from the central States. The Tri-State mines were mainly reprocessing old tailings during this time. There was also a pronounced shift in the rankings of the leading-producing States. During and immediately after the war, Oklahoma, Idaho, New Jersey, Kansas and New Mexico were the leading mine producers of zinc. By the early 1950's, Montana was the leading producer, followed by Idaho, New Jersey, Colorado and Oklahoma. By the late 1950's Tennessee had assumed the top spot, followed by New York, Idaho, Utah, and Colorado.

FROM 1961 to 1975

This period was one of extremes and transition, and probably, the most difficult that the producing sector of the U.S. zinc industry has had to undergo, although the painful adjustment and downsizing of the domestic zinc industry was to continue for another decade. In the early years of the period, domestic mine and smelter production had increased each year through 1965 and 1966, respectively; import quotas remained in force and legislation to assist small lead and zinc miners was passed. However, zinc was dropped in 1962 from the list of minerals eligible for Office of Minerals Exploration funding, and the liquidation of the Government's stockpile began in 1964. One year later, import quota controls were dropped. Mine production tended to diminish thereafter, and in 1964 Canada became the world's leading mine producer, a position the United States had held since 1906. Despite declining domestic mine output, smelter production rose to all-time highs in the 1965-1969 period, only to decline by the end of the period to a low not seen since the depression years of the 1930's. Domestic zinc consumption reached an all-time high in 1973, but numerous smelter closures had reduced capacity to such an extent that the domestic smelting sector was capable of meeting only about half of that requirement. Imported metal gained a strong foothold in the domestic economy in the early 1970's and has maintained a strong position ever since. Recession, brought on to a large extent by the "energy crisis" in late 1973, caused zinc production and consumption to fall in 1974 and, severely so, in 1975.

The sources of mine production continued the transition from the old zinc fields and mines, many of which were played out or contained only subeconomic resources or which were operating at lower levels owing to poorer economics. The Tri-State mines ceased all production; Kansas and Oklahoma had no production after 1970, but Missouri, owing to new production in a new lead belt, the Viburnum Trend, had again become a major producing State. Montana became a minor producer after 1966 resulting, in part, in the closure of both electrolytic smelters in that State in 1969 and 1972. Other States that experienced diminished mine production within the above period (and were totally out of production within a few years after the end of the period) were California, Illinois, Kentucky, Maine, Nevada, New Mexico, Utah, Washington, and Wisconsin. Pennsylvania, after a 70-year hiatus, again became a significant producer with the opening of a new mine in the Friedensville area, and Tennessee enhanced its leading-producer position with the opening of a mine in the newly discovered zinc district in central Tennessee. This shift in production centers had the effect of concentrating zinc production more in the eastern States than in the west, and in reducing the number of mining operations. By 1973 fewer than 70 mines accounted for the total U.S. production and the top 25 producing mines accounted for 93% of that output, whereas in the early 1950's, more than 400 mines accounted for the U.S. zinc output and the top 25 mines, for less than 50% of production. Employment in mining zinc also fell by three-fourths.

Of the 22 primary smelters producing zinc during WWII, 16 were still in operation at the start of 1960. No new conventional smelters had been built since the war; however, most had undergone some modernization and expansion, such that the primary plant capacity was slightly more than 1 million tons. Many of the smelters were old plants using costly labor intensive pyrometallurgical technology and several were obsolete and inefficient. Others began to lose their feed sources or had labor problems and most required large capital infusions to meet the then newly passed environmental regulations. Compounding the domestic smelters' problems was foreign competition that, for the most part, was producing metal in new plants using the latest technology. As a result of the above, as well as effects from Government actions, the domestic smelting sector began to close down. Two smelters closed permanently in the early 1960's and two others in the late 1960's. The 1971-1973 period was the most traumatic for the U.S. zinc smelting industry. Six smelters closed down mainly for economic and/or environmental reasons. In addition, the electrolytic zinc plant at Sauget, IL. also closed in 1971, but it was rebuilt and resumed production in 1973. Asarco Incorporated's Amarillo, TX. smelter was also expected to close in 1973 for environmental reasons but was allowed to continue operating under a variance until mid-1975. By the end of 1975, despite a period of record zinc consumption, numerous smelters had closed down and domestic primary slab zinc capacity had fallen to slightly less than 600,000 tons.

In 1971, mine and smelter closures, a cost-price squeeze, imposition of a surcharge on imports of concentrates, ceiling price controls, and devaluation of the U.S.

dollar all combined to make it a difficult year for the zinc industry. In 1972 and 1973, consumption of slab zinc rose to all-time highs; however, most of this increased consumption was met by imports and metal released from the National Defense stockpile.

Government actions were significant contributors to the rapid decline in U.S. smelting capabilities in the early 1970's. Principal among these factors were the effects of price controls, stockpile releases, and duties on imported concentrate. In 1971, the Economic Stabilization Plan (ESP) froze the price of domestic zinc at a time of high and rising world prices. Not only did this prevent U.S. smelters from improving profits which may have been used to rebuild and modernize their operations but it also prevented them from competing in world markets for concentrate because foreign companies could pay higher prices owing to higher selling prices for their metal. Pleas to suspend the duty on imported concentrate to help alleviate the latter situation were unheeded. The price controls were lifted at the end of 1973 and duties suspended in 1975 but most of the smelter closures had already occurred. The price controls were in effect at the same time the government was selling vast amounts of metal from the stockpile. These two actions severely limited the economic prospects of many smelting operations and accelerated decisions to close the smelters down. In the period 1964 through 1975, the Government released 1 million tons of slab zinc from the stockpile, 60% of which was released in the 1972-1974 period. In 1974 the "energy crisis" hit and subsequent reductions in automobile production led to reduced zinc demand. Slab zinc imports fell, but sales from the stockpile continued to be large, totaling 240,000 tons despite rapidly falling demand, especially in the latter half of the year. Only in 1975 when the domestic economy had slowed substantially, did the Government cease stockpile sales.

FROM 1976 TO 1995

The domestic zinc industry continued its downward spiral through most of the above period but stabilized and regained some of its lost position in the world economy by the end of the period. Zinc consumption languished throughout most of this period owing to a second "energy crisis" in 1978-79, recessions, the maintenance of a strong dollar, and reduced use of zinc die-cast alloys, especially in automobiles owing to down-sizing and weight reduction programs. The introduction of the "zinc" penny in 1982 and the extensive galvanizing of automobile parts to prevent corrosion beginning in the 1980's were two of only a few bright spots leading to higher zinc consumption in the entire period. The smelter sector was not given any relief from environmental and high cost problems in this period and continued to be ravaged. Two new electrolytic zinc plants were constructed in the second half of the 1970's; however, of the six operating plants at the beginning of the period, only one pyrometallurgical plant and one electrolytic plant survived. Mandatory requirements for the treatment of electric arc furnace dusts from steel-making, however, provided a new source of zinc feed for smelters, resulting in higher slab zinc production beginning in the early 1980's. Several new mines came onstream, including the Red Dog Mine in Alaska, which, by itself, reversed the downward trend in U.S. mine output. However other mines, some with long and large production histories, closed down. U.S. mines and smelters produced 7.2% and 9.0%, respectively of the world's zinc production in 1975, but at the end of 1986, the respective percentages had fallen to 2.9% and 4.7%, but improved thereafter, reaching 7.2% and 5.6% in 1992. Despite the improvements in production, the United States continued to be the world's largest importer of zinc metal and, ironically, because of mining in Alaska, became a major world exporter of zinc concentrate.

The rate of economic activity rose in 1976, domestic mine and smelter production and slab zinc consumption were up substantially. The last of the domestic horizontal retort smelters closed in that year, but in its place, a new electrolytic plant was constructed. In 1977-1979, mine production fell substantially, owing in large part to recession and labor strikes, one of which lasted one year at Balmat, NY. Mine production in 1979 fell to its lowest level since 1932. A new "greenfields" electrolytic zinc plant opened at Clarksville, TN in 1978, improving domestic metal-producing capacity, but in 1979, the electrothermic smelter at Monaca, PA closed; it was the largest zinc plant in the United States and the fourth largest in the world in production capacity. The Monaca smelter reopened one year later but at about one-fourth its former capacity. The slab zinc capacity of the smelter was subsequently raised, in large part, to permit the processing of large amounts of crude zinc oxide extracted from electric arc furnace dusts.

Mine production trended downward through 1986; the output in 1986 was the lowest since 1921. In this period a number of famous, large, zinc-producing mines closed: the Bunker Hill and Star-Morning Mines in Idaho; the Austinville Mine in Virginia; the Sterling Mine in New Jersey; the Friedensville Mine in Pennsylvania; and the mines at Copperhill, TN. Major mines that opened were Gordonsville in Tennessee, Greens Creek and Red Dog in Alaska, Pierrepont in New York, Montana Tunnels in Montana, and West Fork in Missouri. Through 1989 Tennessee had been the leading producing State 29 times in the previous 32 years, but since that time, Alaska has been the leading producer. Generally throughout the period, the top 25 mines accounted for 98% or more of the total U.S.

zinc mine output.

Primary smelter production declined sharply from 1979, as did the number of smelters. In 1980 the vertical retort smelter at Palmerton, PA. closed permanently, although certain portions of the facility remained in operation to process electric arc furnace dusts and ore from the Sterling Mine until it closed in 1986. The large Bunker Hill electrolytic zinc plant, including the mine and lead smelting complex, closed down in November 1981. There were a number of plans to revive the operation, but environmental concerns precluded revival of the smelter, resulting in its dismantling in 1985. The mine was briefly reopened but was reshuttered owing to poor economics. The electrolytic zinc plant at Corpus Christi, TX closed indefinitely in October, 1982, then reopened briefly in 1984. Because of high costs, low zinc prices and lack of feed sources, it was closed permanently in 1985. The two slag fuming plants upon which the Corpus Christi plant depended for about half of its feed both closed in 1982 and were written off by the company in 1984. At the end of 1995, the domestic primary zinc smelter sector consisted of three electrolytic plants, one of which has been temporarily closed since 1993, and one electrothermic smelter, for a total available production capacity of only 380,000 tons.

A number of old-line zinc companies disappeared from the domestic zinc scene in the 1970's and 1980's. The three most historically important companies to depart were Anaconda, St. Joe, and New Jersey Zinc. Other prominent companies to depart were Amax, American Zinc, Bunker Hill, Gulf + Western, Eagle Pitcher, United States Smelting, Refining and Mining, and U.S. Steel. Several new companies, including Big River, Doe Run, Union Zinc, and Zinc Corporation of America, arose out of the bits and pieces of the earlier zinc companies, but two of these, Big River and Union Zinc sold their zinc interests to foreign zinc companies in 1995. Asarco, an old-line zinc company, maintained its strong position in the world zinc industry during this period and expanded its zinc involvement in the United States by taking over most of American Zinc's mines and zinc oxide operations in the early 1970's. Comparatively, though, the U.S. zinc industry of the 1990's was much smaller, and what remained as regards structure and operating companies was considerably different than that which existed before the 1970's.

U.S. STATE HISTORIES

Brief summaries of historical zinc mining and processing in the various States are given below. Table 3 is a listing of the important zinc mining districts in the United States and Figures 3 and 4 graphically show the States with the largest cumulative zinc mine and smelter production through 1994. Additional State data are given in the Appendix.

ARKANSAS

Deposits of zinc ore are common in the northern counties of Arkansas, especially in Marion, Boone, Sharp, and Stearcy Counties. The first attempt to exploit these deposits was in 1857, when a zinc smelter was erected at Calamine in Sharp County. This smelter was unsuccessful and closed in 1861. In 1872 another smelter was built to process local ores but this too failed after a few months of operation. Because of high transportation costs interest in local deposits languished until about 1886 when active prospecting again resumed. By 1899, several large companies were established, and, by 1920, 71 companies were in operation in the northern counties. Despite this activity, zinc mine production in Arkansas was never large and, cumulatively, has only been a minor source of U.S. zinc production. After WWI, zinc mining in the State fell below a thousand tons per year and, for the most part, was below 100 tons until all production ceased in 1962. Calamine accounted for most of the State's early zinc mine output, but sphalerite has been the principal zinc mineral produced.

Despite early failures, a substantial zinc smelting industry developed during WWI owing largely to the presence of natural gas and nearness to the Tri–State zinc fields. Three horizontal–retort smelters were subsequently built, one at Van Buren in 1916, and two at Fort Smith in 1916 and 1917; the Van Buren smelter was idle throughout the 1928-1936 period and closed soon after WWII, and the Fort Smith smelters closed in 1924 and 1963.

The zinc orebodies in Arkansas are Mississippi–Valley–type (MVT), occurring in fissures, fault breccias and solution breccias and disseminations in Ordovician and Mississippian limestones. Most of the immediate host rocks of the zinc mineralization, however, have been altered to hydrothermal dolomite and jasperoid, as in the Tri-State district. The ores were sphalerite and smithsonite with only small amounts of galena. The ore bodies tend to be shallow and, for the most part, were mined by surface methods and small shafts. Beneficiation was accomplished largely by hand sorting and jigging until the introduction of some flotation in the early 1930's.

Figure 3. Recoverable Zinc Mine Production, by State, 1850-1994.

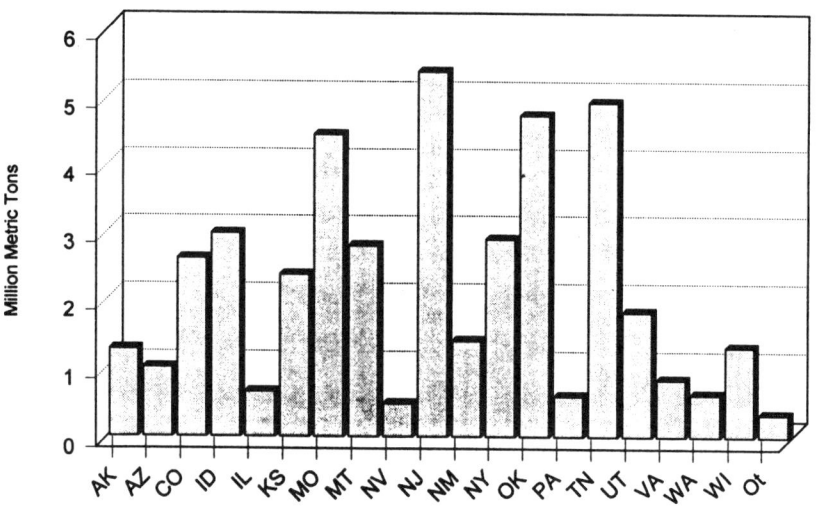

Ot=Other: AR, CA, IA, KY, MD, ME, NC, NH, OR, SC, SD, and TX

Figure 4. Zinc Smelter Production by State, 1859-1994.

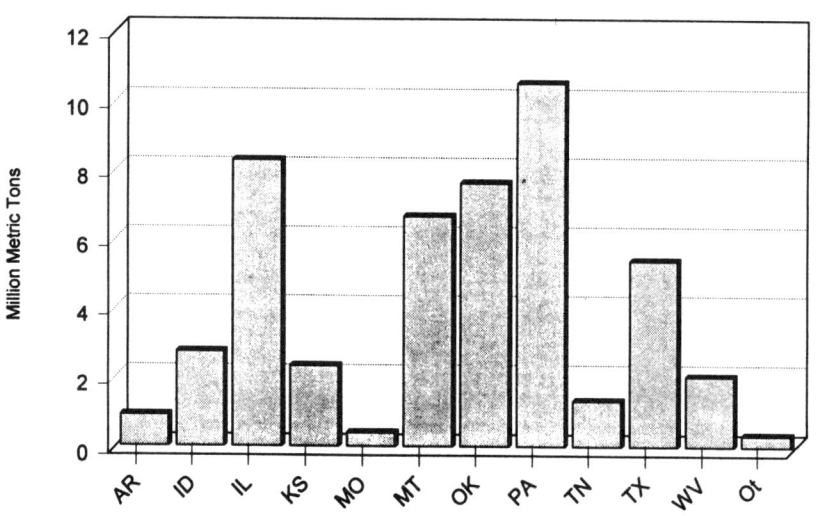

Ot=Other: CO, NJ, UT, VA, and WI

ALASKA

Until the late 1980's, the only recorded zinc production in Alaska was in the 1940's. This production was small and came from various low-grade gold operations and the testing of ores from lead–zinc prospects during WWII. In the early 1970's, sulfide mineralization was reported in the De Long Mountains in the western Brooks Range in northwestern Alaska; exploration in the mid 1970's identified a number of potential ore deposits in the region, of which only one, the Red Dog deposit, has so far been developed. The Red Dog deposit is a volcanogenic, stratiform, massive sulfide body, which ranks among the largest and richest zinc ore bodies so far discovered in the world. Ore, which is exposed at the surface, is mined by open–pit methods. Ore concentration is by flotation using, in part, large column flotation units. Mining began in 1989; the following year the operation was the world's single largest zinc producer in terms of zinc mine output, and single-handedly, almost doubled U.S. mine production of zinc. The concentrates, which are moved 54 miles by truck to a port on the Chukchi Sea, are shipped only during the summer months when the ice pack breaks up.

Another significant zinc producer, the Greens Creek Mine, is located in a National Monument in the northeast part of Admiralty Island. The ore body was discovered in the mid-1970's and, like the Red Dog Mine, was initially mined in 1989. It is a stratibound massive sulfide deposit occurring on the limb of an overturned anticline in complexly folded volcanic rocks. The ore body is polymetallic and is primarily considered to be a silver deposit (averages about 23 ounces per ton), despite having ore that averages about 8% zinc, 3% lead, and 0.2 ounce gold per ton. Zinc and lead concentrates containing the silver and gold are produced by flotation.

ARIZONA

Zinc was present in many of the copper–lead–silver–gold ores mined in the State in the latter half of the 19th century, and, although zinc was recovered in the various concentrates produced, it was considered to be gangue and discarded in the smelting process. The copper concentrates produced from deep ore at the Silver King Mine in the early 1880's, for example, were said to contain about 18% zinc but none was recovered. Zinc production was slow to develop largely because the ores were complex and transportation costly. Many ores contained high zinc portions but unless these zinc–rich ores contained payable silver, gold or copper, they were bypassed or mined as waste. The first zinc ore was produced in the Union Basin district about 1900; one carload was sent in that year and another in 1907 to a zinc oxide plant in Colorado. Neither shipment was profitable. In 1908 the Golconda Mine, which was in the same district, began small but sustained zinc production. Zinc production in Arizona was relatively small up to WWII, and in 1921, 1924 and several of the depression years, none was produced. Production jumped during WWII, and after the war through 1952, Arizona ranked in the top five zinc–producing States. Thereafter, zinc production trended slowly downward with only minor and sporadic output from the late 1970's onward.

Zinc occurs in various types of deposits. Veins and hydrothermal replacement deposits in limestone associated with copper deposits have accounted for the most of the zinc production. These deposits are generally genetically related to granitic and porphyric intrusives. The zinc and zinc–lead rich ores tend to lie toward the outer perimeters of main copper orebodies. Other important zinc–containing orebodies in the State occur as veins and massive tabular, pipelike, and irregular sulfide replacements in schists and rhyolite. Secondary oxidized zinc mineralization is common but has not been extensively exploited.

CALIFORNIA

California has been a minor and sporadic producer of zinc since 1906. Production has come from a number of mines scattered around the State, but mainly from mines in Shasta and Inyo Counties. The deposits mined were generally zinc–lead and copper–zinc sulfide veins and replacements in limestones and schists. Significant tonnages of secondary oxidic ores also were produced at some mines.

In Shasta County, an experimental electrolytic smelter was established in 1916 at Bully Hill in an attempt to extract both zinc and copper because zinc in the local ore was not recovered and the high zinc content in the copper concentrates made it difficult to recover the copper by conventional smelting methods. The process was not economically successful and the plant was closed down in 1917, but continued as an ore testing facility until 1919. It was in this latter period that the plant did gain a small measure of fame, in that, it was here where, in testing zinc ore from Broken Hill, Australia, the presence of cobalt in the leach solution was first recognized as a deleterious element in zinc electrolysis. In 1917, a 25-ton–per–day electrolytic smelter was built at Kennett, but this plant was in operation only a short time and it too closed down soon after the end of WWI.

COLORADO

Colorado zinc production has come largely from the Leadville, Gilman (Red Cliff), Ruby, and Elk Mountain mining districts in the Central Colorado Mineral Belt and the Telluride, Creede, Silverton, Rico, and Eureka mining districts in the San Juan Mineral Belt in the southwestern part of the State. In the former Belt, ore deposits are fissure vein fillings in granite, vein intersection replacement lodes in limestone, and massive sulfide manto and chimney replacement bodies in carbonate rock, often adjacent to or associated with porphyry sheets. Principal ore deposits in the latter Belt occur in and around a highly faulted caldera as fault-controlled fissure veins, chimneys and irregular replacement bodies in limestones, braided–like vein lodes, and veins in intrusive rocks. Primary sulfide ore minerals are sphalerite and galena with high silver values, pyrite, and in some ores, chalcopyrite and tetrahedrite.

Gold, discovered in the 1850's and 1860's, is common in many of the ores and led to the initial mining in the above districts and the subsequent finding of the rich silver deposits in the same districts shortly thereafter. Many of the near–surface ores at Leadville and a number of other places were rich silver–bearing lead carbonates which contained little zinc. As mining proceeded downward, zinc carbonate and silicate ores were often encountered adjacent to and below the lead carbonate deposits. Below the oxidized ores, complex sulfide ores, often rich in sphalerite, were encountered. The oxidized zinc ores, however, were generally ignored before 1900 because there was little or no market for oxidized zinc ores at the time and in some cases, because they were not recognized as being zinc deposits. Zinc minerals were generally avoided in mining if possible, because their presence in lead ores and concentrates brought penalties at the lead smelters and increased shipping and handling costs.

Efforts to make clean lead–silver concentrates from complex sulfide ores to avoid lead smelter penalties, did lead to improvements, such as the development in 1896 of the Wifley table. As early as 1885, galena and sphalerite had been successfully separated at a plant near Leadville, however the sphalerite had no value. But with remarkable foresight, it was stored and, years later, was said to have been sold when the zinc market developed.

After 1900 the zinc carbonate ore bodies became economic and were rediscovered in many of the old workings. The mining of these ores helped to reinvigorated the then slumping fortunes of numerous mines at Leadville and several other mining districts. In the 1910–1915 period alone about 500,000 tons of calamine ores were produced in Colorado. However, after 1900 the mining of complex sulfide ores, some containing significant sphalerite and pyrite, became increasingly important. The first zinc production in the Gillman district began in 1903, but zinc production did not become significant in the district until after the start of WWII. From 1941 through 1977, Gillman was the State's leading zinc-producing district. Interestingly, the ore at the Eagle Mine, the principal mine in the Gillman district, contained considerable pyrite, such that the flotation mill tailings consisted of up to 70% pyrite. Because of the pyrophoric nature of finely ground pyrite, the tailings could not be used for sand filling in the mine workings and were disposed of in a tailings dam.

The first zinc plant in Colorado was built at Canon City in 1890.

A process was devised to recover not only the lead and zinc but also silver, copper, and gold from the sulfide ores. The ore was crushed and fed into a furnace fired with fine coal; the lead, zinc, and sulfur were driven off, whereas the copper, silver and gold were left contained in a matte and subsequently were recovered. The zinc–lead fume was captured in bags, refined of impurities, and sold as paint pigment. In 1902 a plant was set up at Canon City to process by magnetic separation vast piles of iron–zinc middlings that had accumulated from the production of lead and silver at Leadville. With the exhaustion of the middling accumulations and the introduction of selective flotation, the plant was closed in 1932. The roasting and sintering facilities of the plant were later used to process concentrate from the Eagle Mine, followed by shipping the zinc sinter to a smelter at Dupue, IL. Other Colorado zinc plants included zinc smelters at Florence and Pueblo and zinc oxide plants at Canon City and Leadville.

IDAHO

Zinc was known to be present in many of the lead–silver ores discovered in the State in the last quarter of the 19th century, but was only first produced in 1905. Since that time the State has been a large producer of zinc, virtually all of which has come from the Coeur d'Alene region or district in the northern panhandle. The base metal deposits in the Coeur d'Alene district occur mainly as steeply–dipping replacement veins in moderately to highly deformed and faulted, siliceous and argillaceous metasedimentary rocks of Middle Proterozoic age. The major veins are extensive and may continue to depths exceeding one mile. The primary ore minerals are sphalerite, galena, and argentiferous tetrahedrite; the dominant gangue minerals are siderite and quartz. The origin of these ore deposits is not known; however, the earlier lateral and magmatic hydrothermal hypotheses have been replaced by ones suggesting the deposits were the result of metamorphic remobilization in which fluids generated during low grade metamorphism transported the metals to the sites of deposition.

Lana Turner, the movie actress, was born in Wallace, Idaho, in 1920. Her father was a lead-zinc miner; he was murdered there in 1929. My daughter, July Ann, was born in Wallace in 1957.

The first production came from zinc–rich sulfide veins encountered at the Success Mine, up Nine–Mile Creek a few miles north of Wallace. The mine was the State's most important zinc producer up to WWI when the Interstate–Callahan Mine became the leading producer. In 1915 the Interstate–Callahan Mine was the third largest producer of zinc in the United States. The Morning Mine near Mullan also became a significant zinc producer in 1910, when a Macquisten–tube flotation plant was installed to treat sphalerite–barite–siderite middlings from the tables. This was one of the first successful attempts to float zinc in the United States. A few years later hot–acid flotation using Callow flotation machines were installed at the Morning mill and a few other mills in the district. Zinc production soared during WWI but waned thereafter.

In 1924 the huge zinc–lead Star–Morning vein was discovered and all–flotation mills were established to treat the ores from the Morning and Star Mines. The flotation process also made economic the large amounts of zinc–rich ores that had been previously bypassed in mining the Bunker Hill lead–silver deposit owing to mineral separation difficulties. The above three mines have accounted for most of Idaho's zinc production, which at times, was the leading U.S. zinc–producing State.

This new abundant supply and resource of zinc led to proposals to build a zinc smelter in the Coeur d'Alene district. A Tainton–process electrolytic zinc smelter near Kellogg came on–stream in November 1928 and was in production until it was closed in 1982. Prior to 1928, virtually all zinc concentrates produced in the district were shipped either to Portland, Oregon for export to Europe or to the Anaconda zinc plant in Montana.

In 1981 the Bunker Hill Mine and zinc and lead smelters closed in large part, because of labor and environmental problems. This closure also led to the closure of the then combined Star-Morning Mine in 1982. The Star-Morning had been in continuous operation for 92 years and at 8,100 feet, was the deepest mine in the United States. Sporadic but small zinc production in Idaho has since come from the above two mines but, since 1982, most has been derived as a coproduct from the mining of silver and lead at the Lucky Friday Mine near Mullan.

ILLINOIS

The State is the Nation's second largest producer of zinc metal and, in 1860, was the site of the first commercial spelter smelter in the central United States. Illinois has been a consistent, though sometimes small, mine producer of zinc since 1860. Up to that time zinc was known to occur, but no market existed.

Most of the production has come from the northernmost part of the State; in recent decades, however, most has come from the southernmost part. The northern zinc deposits are MVT and are part of the region known as the Upper Mississippi Valley mining district, which includes a small area of Iowa and a portion of southern Wisconsin. Sphalerite, marcasite and galena occur largely as fissure and cavity fillings in Ordovician limestones and dolomites in openings known as crevices, pitches, and flats. The first zinc ore to be mined was smithsonite found in association with lead in the shallow workings above the water table, but many of these deposits were mostly exhausted soon after the Civil War.

Because of the ready availability of Joplin concentrates and the shift toward more production of similar ore in Wisconsin, very little zinc was produced thereafter until the early 1900's, when the underlying sulfide ores began to attract renewed attention. However, significant production of sulfide ore was delayed until about 1905 because sphalerite and marcasite could not be successfully separated by gravity methods, thereby yielding low-grade zinc concentrates of low value. The marcasite problem was solved by electrostatic and/or magnetic separation of the iron sulfide after giving the concentrate a slight roast to make the marcasite slightly magnetic. Since that time sphalerite has been the major source of zinc production in northern Illinois.

The zinc and lead ores in southern Illinois are MVT and occur with fluorite in Upper Mississippian limestones and shales in the form of bedded replacement deposits and fissure veins along joints and faults. Sphalerite (about 1%) and minor galena are recovered as byproducts of fluorite production. Gravity methods, followed by flotation have been the usual methods used for sulfide recovery. Until 1906 none of the sphalerite produced could be sufficiently separated from the fluorite to be used as smelter feed; very little was actually separated until after WWI, despite having the ability to separate the two minerals.

Because of abundant coal, good retort clay, closeness to zinc ores, and good transportation, Illinois has been one of the principal zinc smelting States since Mattheissen & Hegeler brought the first smelter onstream 1860. A second smelter was built at Peru in 1870. Hegeler later developed the gas-fired long retort furnace and the Hegeler roaster in Illinois and established the first rolling mill to manufacture zinc sheet in the United States at LaSalle. Since 1870 a number of horizontal retort smelters, a vertical retort smelter and an electrolytic plant were operated in the State; currently only the electrolytic plant remains in production. Cumulatively, slab zinc production in the State has totaled about 8.5 million tons since 1860.

INDIANA

Indiana has had no zinc mine production, but in 1892 a small zinc smelting industry developed at Marion to utilize natural gas discovered in the area. This was the first use of natural gas in the United States for zinc smelting. The ore was imported from Kentucky, Missouri, Tennessee and Wisconsin and smelted in standard Belgian retort furnaces with the gas burned underneath instead of coal on a grate. The gas pool gradually weakened, and by 1903 zinc smelting had ceased. Some of the smelter facilities were later used to roast and sinter zinc concentrates for smelters in West Virginia.

IOWA

The zinc and lead deposits of Iowa were known to the early French explorers, and the lead deposits were among the first to be mined by Europeans in North America. They are MVT and extend along the Mississippi River for nearly 80 miles in Dubuque, Clayton, and Allamakee counties. The ore occurs in crevices and caves and more rarely in pitches and flats as in the equivalent ores of Illinois and Wisconsin. Production has been largely zinc carbonate ore, although some zinc sulfide ores have been produced. The first recorded production in the State was smithsonite or dry bone mined in 1881; the last recorded zinc production was in 1916. The State has had no recorded zinc metal production, although some was probably produced at an experimental electrolytic plant that operated at Keokuk in the 1914–1918 period.

KANSAS

All of the zinc mined in Kansas has come from the Tri–State zinc district in the southeastern corner of the State. A rail line into Kansas from Missouri was completed in 1871 and was instrumental in the erection of a smelter at Weir in 1873. Zinc–lead ores were discovered at Galena in 1876 and were in production the following year. The development of the Galena district and ready access to Joplin/Granby ore resulted in the construction of another smelter in 1878 at Pittsburg, near the southern edge of the Kansas coal fields. Natural gas was discovered at Iola in the mid 1890's and resulted in a great expansion in smelting activity in southeastern Kansas. In a few years the State became the leading producer of slab zinc in the United States. Several of the Kansas smelting companies soon after the turn of the century attained production levels that placed them among the leading world zinc producers.

However, beginning about 1910, the entire Tri–State district began to undergo substantial change. The zinc ores of northeastern Oklahoma, only discovered a few years earlier, were just beginning to come into production, whereas the gas fields supplying the Kansas smelters began to weaken, causing a number of those smelters to close down over the next few years. After 1912 smelter production in the region had shifted to Oklahoma, with Kansas losing its leading–producer position. The development of the Oklahoma zinc fields and the war in Europe, however, resulted in increased zinc mine production in Kansas. In 1916 production began in the Baxter Springs and Blue Mound areas of the Pitcher district and in the Kansas portion of the Waco (MO) district. However, with the end of the war, zinc mining not only ceased at the Galena fields, but also practically ceased in all the zinc districts in Missouri. The Waco district continued production, but at a lower level. The Pitcher district mines were the large producers and accounted for most of the State's zinc production. In both 1925 and 1926, the above mines in Kansas produced almost 30% of the total U.S. mine output of zinc. All zinc mining in Kansas ceased in 1970. In total Kansas has produced more the 3.1 million tons of recoverable zinc from its mines and about 2.3 million tons of zinc metal at its smelters.

KENTUCKY

Kentucky has been a small and sporadic producer of zinc. Zinc and lead deposits are found in the central and western parts of the State. In the former area, sphalerite and galena with barite, calcite, and fluorite occur in veins in Paleozoic dolomites; however, only small and sporadic zinc production has ever taken place. The latter area is an extension of the southern Illinois fluorite district, and virtually all of the State's zinc output has resulted from the mining of fluorspar–rich veins and bedded deposits, although in the 1890's and early 1900's much of the production came from calamine ores associated with the fluorspar deposits, but often without containing any fluorite.

MAINE

The zinc–bearing deposits are located in the south central part of the State within a belt of metamorphosed volcanic and sedimentary rocks. Ore occurrences are varied in geological relationships, ore grade, and metals present. Virtually all of the State's zinc production has come from two deposits in Hancock County. The Bald Mountain deposit is a massive sulfide orebody containing copper, zinc, gold and silver, occurring in siliceous rocks that have a minor carbonate content. The Blue Hill deposit occurs as long, thin, tabular, stratiform bodies in quartzitic rocks in an area where the Ellsworth schist comes into contact with Paleozoic granitic intrusions. Sphalerite is the principal ore mineral, but bodies of chalcopyrite also occur.

The first mining of zinc may have occurred in 1875 when some copper, lead, and silver were extracted from the Deer Isle Mine in Hancock County before mining ceased shortly thereafter owing to low ore grades. The mine reopened in 1906 and produced about 250 tons of ore grading about 25% zinc before closing permanently in mid 1907. No other recorded zinc production occurred until 1968 when production of copper and zinc began at the Penobscot Mine. Operations ended there in 1972 because ore reserves were exhausted. In the fall of 1972, the Blue Hill Mine was placed in production; however, it was closed in 1977 owing mainly to poor metal prices.

MARYLAND

Minor amounts of calamine and lesser amounts of sphalerite were sporadically produced from lead–copper–zinc ores in Frederick and Carroll Counties in the 1880–1910 period. The State may also have had some zinc metal output owing to the operation of an experimental electrolytic zinc smelter at Baltimore in the 1916–1918 period, but there is no record of that production.

MISSOURI

Missouri's mines have accounted for more than 6 million tons of zinc production, which ranks the State as the third leading zinc-producer after New Jersey and Oklahoma. Zinc has been produced in three extensive mining areas—the Tri-State, Central Missouri, and Southeast Missouri Lead districts. The Tri-State district, which includes a number of sub-districts and covers more than 1,200 square miles in southwestern Missouri, northeast Oklahoma, and southeastern Kansas, has been, by far, the principal producing area.

The three principal zinc mining districts in Missouri are thought to all be MVT, although major differences exist between them. The Tri-State ore consists predominantly of sphalerite and minor galena in a gangue of hydrothermal dolomite and silica that forms jasperoid when replacing the host Ordovician through Pennsylvanian carbonate rocks. The ore deposits occur as relatively shallow (less than 500 feet), nearly horizontal sheets, blanket veins, vertical runs, veins, and lenses that fill and follow solution channels in breccias and along cherty horizons. Sheet ores tend to be extensive and low grade (1–4% zinc), whereas the runs and open-space-filling ores have lower tonnage (100,000 tons) but are high grade (10–30% zinc). Consequently, run ore was almost exclusively mined in the early years.. The typical pattern for an ore run is a central core of dolomite surrounded by a jasperoid zone that grades into unaltered limestone, whereas the sheet ores occur as replacements and fillings disseminated in the altered limestone and jasperoid.

The Central district ores typically occur in small, high-grade deposits in gently dipping, relatively undisturbed rocks that range in age from Cambrian through Pennsylvanian. Most deposits, however, occur as open-space fillings in fractures, faults and solution collapse breccia zones of Ordovician dolomites; replacement sulfides are rare. Sphalerite is the most abundant sulfide. Jasperoid and quartz are absent and hydrothermal dolomite is rare.

The Southeast Missouri Lead district includes the Viburnum Trend and Old Lead Belt districts and is the world's largest lead mining district. The ore deposits occur in Cambrian dolomites and sandstone that unconformably overlie and flank the Precambrian crystalline rocks that form the core of the St. Francois Mountains. Ore controls are varied and complex, and include sedimentary and solution collapse breccias, algal reef structures, abrupt facies changes, faults, and pinchouts at the unconformity boundary. In the Viburnum Trend, solution collapse breccias are the most important ore hosts. Galena and lesser sphalerite are the principal ore minerals, but the lead content in the ore exceeds that of zinc by 5 or more to 1. The ore deposits in this district, unlike the Tri–State, lack extensive silicification and jasperoid formation.

The first zinc production in the State was calamine ore from the southeast Missouri Lead Belt in 1867. The ore was mined to feed a small smelter erected at Potosi in that same year. Two years later, another zinc smelter was established at Carondelet. Zinc metal production from the two smelters to 1870 was about 60 tons. Zinc production also began in the Central Missouri district at this time and was continuous through WWI; however, zinc output in the district was always small. Zinc ore was known to occur in the lead mines around Joplin in the early 1850's, but lack of rail transport prevented economic production until 1872. The completion of the rail line to Pierce City in 1871 and into Kansas in 1872 was a major stimulus that opened up the heretofore lead–only mines in the Tri–State district to economic production of zinc. In the 1875–1917 period, except for 1881 and 1883, Missouri was the Nation's leading mine producer of zinc, most of which was produced in the Tri–State. By the early 1900's, the grade of the ores mined in Missouri had declined substantially owing to depletion of much of the run ore and increased mining of the sheet ground ore. Because sheet ores required larger and more efficient milling to be profitable, this tended to result in more consolidation of mine operations and erection of central and custom mills to process the ores in Missouri. Substantial amounts of calamine continued to be mined, especially in the mines near Granby and Joplin.

Concentration was all by hand sorting, jigs, and tables until flotation was introduced in the 1920's to treat slimes. Whereas mills in most parts of the country eventually adopted all–flotation circuits, the few mills that subsequently operated in the Missouri portion of the Tri–State district after WWI continued to use a combination of gravity–flotation circuits till final closure of mine operations.

The peak production year of the State and the Tri–State district was 1916; however in Missouri, by 1918 zinc production abruptly fell as miners and mining companies abandoned the lower–grade Missouri zinc fields and moved their activities to the much larger and richer zinc fields in Oklahoma. The companies remaining in Missouri were confronted with an additional problem: the high cost of keeping their workings dewatered. The reduced activity in the Missouri fields and interconnected nature of mines increased the amount of water that they had to pump; this extra cost resulted in the subsequent closure of additional mines, especially in the well-developed areas. By 1919 most of the mines in the Missouri portion of the Tri–State were non–producing; the only significant zinc production came from mines in the newly–developed Waco district, which had its initial zinc output in 1916. Thereafter, only small amounts of zinc were produced in southwestern Missouri other than small spurts during WWII and the Korean War. Beginning in 1950 byproduct zinc production from mines in the Southeast Missouri Lead district exceeded that of Missouri's Tri-State mines, except in 1953. With minor exceptions, final zinc production in the Missouri portion of the Tri- State was in 1957.

A new era of zinc production in Missouri began in 1968 when byproduct and coproduct production of zinc began at newly commissioned lead mines in the "New Lead Belt", which is also known as the "Viburnum Trend". One year later through the present, the lead mines in the Trend have been the source for the State's zinc output.

MONTANA

Montana is the sixth largest zinc mine–producing State, having a total recoverable zinc production in ore and concentrate of about 3.2 million tons of which about 2.3 million tons or 71% was produced in the Butte district. Zinc has been produced in many lead–silver–copper–gold mines in six mining districts in the western part of the State. Ore occurrences are diverse, but generally most are fissure fillings and replacement veins in highly–faulted monzonite, aplite, and porphyry phases of the Boulder Batholith and in Paleozoic limestones adjacent to igneous intrusives. The Butte deposits are thought to have been formed by hydrothermal solutions emitted from the magmatic reservoir that supplied the intrusives. The ore minerals in the veins are zonally arranged outward from a central core that contains copper–rich veins. Zinc is dominant in the intermediate and peripheral zones around the central zone. However, in the intermediate zone, silver and lead are common and often with depth, copper may reappear as the dominant metal.

The production of zinc in Montana began in 1905 in the Butte district. Output was slow to increase because many of the zinc ores encountered were too complex to be successfully concentrated. Both production and zinc recoveries sharply improved with the introduction of flotation in the district in 1912. By 1915 virtually all of the State's zinc production was produced by flotation, and Montana, in that same year, became one of the top three zinc producers in the United States.

The development of the electrolytic zinc process and opening of a commercial plant by the Anaconda Company in 1916 also was another factor in increased zinc mining in the State because it permitted profitable processing of heretofore low–valued, low–grade zinc concentrates and recovery of their contained byproduct metals. In 1927 the State's zinc production was supplemented by the opening of the world's first lead slag fuming plant at Great Falls.

Zinc output fell to low levels during the Great Depression but had returned somewhat by the start of WWII. During the war, zinc production continued in most mining districts, but at Butte, zinc mining was largely suspended at Government request so the limited manpower could be devoted to the production of copper.

Production surged after the war and in the mid–1950's, for two years, Montana was the leading zinc-producing State. Because of high cost and low zinc prices, a number of mines closed down in 1966. The sharply lower zinc production in the State was a factor leading to the closure of the State's two electrolytic zinc plants in 1969 at Anaconda and in 1972 at Great Falls, in that the processing of local, company-owned ore was important to the economic viability of the two, rather isolated (in terms of feed sources and markets) zinc plants. Thereafter ,zinc mining sputtered along at low levels, although the opening of the Montana Tunnels Mine in 1987 provided a significant upturn in output.

NEVADA

Zinc has been produced at numerous, mostly small, mines in the State. In 1915, for example, 36 mines produced zinc, but the State's production was less than 6,000 tons. Most zinc deposits are fissure veins, mantos, and replacement lodes in Paleozoic limestones and dolomites and commonly are associated with faulting and some, with intrusive igneous rocks. Ores typically are complex and contain zinc, lead, copper, and/or iron sulfides and appreciable silver and gold. Most primary deposits have been extensively oxidized by ground water, resulting in the formation of extensive calamine and lead carbonate deposits.

Most mines that eventually became zinc producers were originally developed as gold, silver, and lead mines in the latter half of the 19th century. The first zinc production is thought to have been calamine in 1905 in the Yellow Pine district in Clark County. This district accounted for more than three–fourths of the State's zinc output through WWI. Calamine was the principal ore mined in this period, but thereafter sphalerite increasingly became the source of zinc output owing in large part, to flotation. Zinc production in the State fell after WWI but tended to rise thereafter through WWII. In the late 1950's production fell sharply and in the 1970's became sporadic. Some zinc production began at Ward Mountain in the late 1980's but this too has been sporadic.

NEW HAMPSHIRE

The only reported zinc production was during WWI, when small amounts of zinc were recovered from complex sulfide ore in a well defined ore shoot in mica schist in Grafton County. This ore deposit was originally worked in 1906, but was soon abandoned because ore separation was difficult. Interestingly, it was at this mine that the Lungwitz process for smelting zinc in a blast furnace was unsuccessfully tried out in 1906. The smelting process, which was never commercially employed, attempted to recover liquid zinc by keeping the charge under sufficient pressure to prevent zinc vaporization.

NEW JERSEY

New Jersey is the leading zinc–producing State in the United States. All production has come from mining of two ore deposits, one at Mine Hill and the other at Sterling Hill, in the Franklin Furnace–Sterling area of Sussex County. Consolidation under one company of all mining interests on the two hills in 1897 resulted in the forming of two mining operations, the Franklin Mine and the Sterling Mine. These mines became world famous not only for the quantity and quality of the zinc they contained, but also for their unique assemblage of minerals, including the zinc minerals; franklinite, willemite, and zincite. The deposits are hosts for more than 330 different minerals, 70 of which fluoresce under ultraviolet light and 67 for which this is the type locality.

The deposits are believed to have been first noticed in the mid–1600's by Dutch and Huguenot explorers. In 1774, a few tons of red zincite ore from the Sterling deposit were sent to Wales for smelting, believing it to be "copper" ore. The idea that zincite was copper ore may have been based on the fact that the area in or near Sterling Hill was known as the "Copper Tract" in early 18th–century land records. Despite this early designation, only traces of copper ores have ever been found in the Sterling area. The first true zinc mining occurred about 1810 when ground zincite (zinc oxide) was used for paint pigment. In the mid–1830's, zincite was mined for the Federal Government who made spelter in a small Belgian retort furnace in Washington, DC. for manufacture of brass, which was used for standard weights and measures. In 1872 ore was taken from the Franklin orebody for a similar purpose; the place where this ore was mined became known as the "Weights and Measures Opening".

Beginning about 1850, the Franklin ores were used to make zinc oxide, as all efforts to make metal failed. Some primary ore was also extracted at Sterling Hill in 1852. This early mine production in New Jersey through the year 1854 is thought to have been the only commercial zinc ore mined in the United States in that period. The Franklin orebody was worked continuously until ore reserves were exhausted in 1954. The Sterling ores were sporadically mined from 1852 through 1897 when mining was stopped because the "consolidated" company shifted all its efforts to the development of the more easily mined, larger deposit at Franklin. The Sterling Mine was reopened in 1915 and was in continuous operation, except for the 1958–1961 period, until it was permanently closed in 1986.

The mining methods employed were shrinkage stoping, a combination of shrinkage stoping and cut and fill, top-slicing of pillars, and square setting for recovery of crown pillars. Magnetic and gravity mills were employed to concentrate the ore minerals. Franklinite was removed by magnetic separators, and the willemite and zincite were concentrated on jigs and tables. Deposits of hemimorphite, which overlie the primary ores at Sterling, were mined from open pits beginning about 1860.

The origin of the zinc in these orebodies is controversial; most researchers believe that the original zinc in the deposits was the result of either stratibound replacements in limestone or stratiform zinc–iron–manganese deposits of sedimentary origin. The origin of the zinc is obscure in that the deposits were later recrystallized and intensely folded and deformed by more than one episode of metamorphism. At both deposits, ore outcrops in open synclinal folds in marble. Continental glaciation stripped off portions of these orebodies in the Pleistocene, distributing chunks of ore in the downstream moraines. Some of these ore chunks are said to have been recovered and sold as zinc ore and collected as specimens. One large glacial fragment was said to have weighed several thousand tons. The orebodies range from 10 feet to over 100 feet in thickness and reach depths exceeding 1,100 at Franklin and 2,200 feet at Sterling. The zinc ore grades in both deposits typically exceeded 20% zinc.

The unique assemblage and occurrence of the zinc minerals in the Mine Hill ore deposits and wording of certain leases and deeds as to what minerals belonged to what company led to many years of litigation before the participants decided to end the problems through consolidation of their companies into what became The New Jersey Zinc Company in 1897.

Prior to 1848 the whole of Mine Hill belonged to Dr. Samuel Fowler, but in March of that year, he sold to the Sussex Zinc and Copper Mining and Manufacturing Co. "all the zinc, copper, lead, silver and gold ores and also all other minerals containing metals (except the metal or ore called franklinite, and iron ore when it exist separate from the zinc) existing, found or to be found on Mine Hill". That same day, Fowler deeded to the same company the rights to mine franklinite (but not other iron ore) in a tract of land north of Double Rock (DR), which was one of the tracts already controlled by Sussex in the prior deal. Thus Sussex had the rights to mine all ores including franklinite, except iron ore, in the tract north of DR and the rights to mine all ores, except franklinite and iron ore, south of DR. In 1850 Fowler placed the rights to the franklinite and iron ore south of DR in the hands of trustees who then leased portions to various companies. The Franklinite Mining Co. obtained a lease on what was called the "magnetic vein" and attempted to produce iron from the franklinite. This attempt failed, so the company began zinc oxide production in 1854; however to improve results, they sought out ores on Mine Hill that contained the highest percentage of red zincite occurring with franklinite that they could find. Thus the New Jersey Exploring and Mining Co., into which Sussex had merged in 1852, brought suit claiming the ore mined by the Franklinite Co., was zinc ore. This dispute was first tried in 1857, where the court decided that franklinite was the predominate mineral and, therefore, the ore was franklinite and not zinc ore. This judgment was appealed and in 1862, the State Court of Appeals held that the New Jersey company was entitled to all the zinc minerals mixed in with the franklinite and that the vein in dispute was in fact zinc ore, thereby conveying the rights to this ore to the New Jersey company.

The case resumed in the U.S. Circuit Court in 1871, after a new owner, Moses Taylor, took over what formerly was the Franklinite Mining Co. The U.S. Court reversed the 1862 decision in 1877 confirming the rights to the vein as being the franklinite ore excepted in the Fowler deeds of 1848. The then-called New Jersey Zinc Co., not having adequate assets to carry the case to the U.S. Supreme Court, agreed in 1880 to a settlement with Taylor by merging their assets, forming a new company, the New Jersey Zinc and Iron Co. The reorganized company embraced both zinc and franklinite on Sterling Hill and the tract on Mine Hill north of the Rock, and zinc and all other ores, except franklinite and iron ore when it exists separate from zinc, south of DR on Mine Hill.

After the Taylor decision, the surviving Fowler trustee leased a 500 foot section of Mine Hill to C.W. Trotter, who erected a zinc oxide plant at Elizabeth. The oxide plant was soon destroyed by fire, and as a result, Trotter began to sell franklinite ore to the Lehigh Zinc Co. in Pennsylvania. New Jersey Zinc and Iron brought suit to eject Trotter, which was subsequently dismissed. Trotter and Lehigh zinc leased additional parts of the "magnetic iron vein" and the franklinite "vein" in the west side orebody. This brought forth additional litigation from 1889 through 1895. New Jersey Zinc and Iron claimed that zinc ores were removed from the zinc vein belonging to it under the deed of 1848. In 1895, after several hung juries, a verdict was given in favor of the defendants and established the fact that the ore was franklinite not zinc ore as defined in 1848. In 1896, the parties to the case and the Passiac Zinc Co. merged to form the New Jersey Zinc Company, thus ending almost 40 years of litigation.

NEW MEXICO

Zinc has been produced in a number of mining districts in New Mexico but the vast majority has been produced in the Central district in the southwest corner of the State. The Pecos Mine in the Willow Creek district was an important zinc producer in the 1926–1939 period and the Magdalena district, especially in the 1945–1969 period, has also been an important zinc district. The Central district may be the oldest mining district in the United States, since there is evidence that indicates Spanish explorers mined native copper here soon after Cortez conquered Mexico in the 1500's. Many of the other zinc–producing districts were originally worked for gold, silver, lead, and/or copper before any zinc was purposely mined.

The first mining of zinc ore in New Mexico was in 1891 when mines at Hanover produced and shipped calamine assaying about 35% to the zinc oxide plant at Mineral Point, WI. Calamine shipments continued but were very sporadic up to 1903 when regular calamine shipments from the Central and Magdalena districts resumed. In 1906 the Magdalena district was the second largest producer of zinc in the Rocky Mountain region. From 1903 through 1975, zinc production was continuous and nationally significant. Thereafter, it was both small and sporadic.

The ore deposits in New Mexico are varied, and principal zinc mines were developed in oxidized orebodies and in replacement lodes and veins in limestones, skarns, schists, and intrusive igneous rocks. The primary zinc ores in the State are largely related to Laramide intrusives, which appear to have provided the favorable structures for ore deposition and the hydrothermal solutions and gases carrying the ore elements. In the Central district, the zinc orebodies are pyrometasomatic replacements in faulted, late Paleozoic limestones and skarns adjacent to a Laramide granodiorite stock or the dikes associated with it. The ores in the Magdalena district have a similar occurrence and history. The ores in the Willow Creek district are different in that they occur as replacements in schistose bands developed in metamorphosed and sheared pre–Cambrian granite.

NEW YORK

New York ranks as the sixth largest zincproducing State. All of the recorded zinc mine production has occurred in St. Lawrence County, in the northwestern part of the State. The only other area that may have produced zinc is in the Shawangunk Mountains in southeastern New York. Lead was sporadically mined at various lead/zinc deposits in these mountains from about 1830 into the WWI period; however it is not known whether or not any zinc ore was actually marketed.

Zinc was known to occur in ores mined for lead near Balmat in the St. Lawrence mining district as early as 1835; however it was not until the discovery of zinc ore in a road cut at Edwards in 1903 that the district attracted notice. In 1911, production at the Edwards Mine began. The Balmat Mine began mining the first of several orebodies in 1928; the Hyatt Mine briefly opened just after the end of WWI, soon closed, and did not reopen until 1974; and, the Pierrepont Mine, located 28 miles north of Balmat, came on–stream in 1982. The Edwards Mine was permanently closed in 1980 and the Hyatt Mine, in 1984.

The ore deposits in the St.

Lawrence district are mostly stratibound, hydrothermal replacements in contorted, folded and faulted Precambrian Grenville marbles and dolomites. Ores occur in lense-like masses 2 to 50 feet in thickness and 50 to 800 feet along strike, and characteristically, they can be traced down plunge as much as several thousand feet. Ores typically contain 25% or more sulfides, largely sphalerite and pyrite with small amounts of galena. Ore deposits are mined by a variety of methods depending on dip and width of the ore body and character of the host rock. Room and pillar methods are used if the ore body is relatively flat lying; overhand stoping or contour mining, depending on ore thickness, are used on moderately dipping ores; and shrinkage stoping is used on steeply dipping sections. The ores were beneficiated initially by gravity and magnetic means until the late 1920's when an all–flotation mill was built at Balmat.

NORTH CAROLINA

North Carolina has had minor zinc production in only a few years in the history of U.S. mining. Most of the State's total zinc production has come from the sulfide zones of the Silver Hill and Silver Valley Mines in Davidson County. In recent decades, small amounts of zinc have been recovered in mining gold in Davidson County.

OHIO

The State has had no recorded zinc mine or metal production, but in 1921, an American-process zinc oxide plant was established at Columbus, mainly to supply zinc oxide to the rubber tire industry located at Akron. This plant, which closed in 1986, was the last producer of sulfide ore-based zinc oxide in the United States.

OKLAHOMA

Oklahoma ranks as the third largest source of mined zinc and the third largest producer of zinc metal in the United States. Zinc–lead ore was first discovered and mined in Ottawa County in the north-eastern corner of the State near the town of Peoria in 1891. Exploration and mining spread westward in Ottawa County to Quapaw in 1903 and into the Miami area in 1905. During this same period, a number of small lead and zinc deposits were discovered in the Arbuckle Mountains in Murray County; small amounts of zinc ore, mainly smithsonite, were produced sporadically in the area up to the end of WWI. The Pitcher field, the principal zinc field in the Oklahoma section of the Tri–State district, was discovered in 1914.

Mickey Mantle, the New York Yankee baseball star, was born (1931) and raised in Commerce, Oklahoma. His father was a zinc miner.

Zinc mine production in the State, which was slightly less than

13,000 tons in 1915, rapidly increased, and by 1918 had risen 10-fold, easily vaulting Oklahoma into first place as the largest zinc-producing State in the Nation. The rapid development was, in large part, due to the mass exodus of miners and operators, who abandoned lower-grade mines in southwestern Missouri and transferred their activities in 1917 and 1918 to the comparatively richer fields found in Ottawa County. In the Oklahoma zinc fields at this time, 500 to as many as 800 churn drills were placed in operation to find and delineate ore bodies. Oklahoma was the leading zinc-producing State from 1918 through 1944. The depletion of high-grade deposits owing to WWII needs and the general lowering of the ore grade caused a gradual, then marked, decline in the State's zinc output. After 1946, the zinc production in no single year exceeded that of any year between 1917 and 1944. Production fell sharply in 1958 and ceased altogether in 1970.

The Oklahoma zinc–lead ores are similar in origin and character as those found in the Kansas and Missouri Tri–State zinc fields. The deposits occur as narrow, sinuous runs, usually of the boulder–type ore; thin to moderately thick deposits of considerable lateral extent; and tabular deposits in thin bedded cherts, often referred to as sheet-ground deposits. The ore occurs in as many as 16 beds of limestone, chert, and dolomite in the Mississippian Boone formation. The ore minerals are sphalerite and galena in a ratio of about 4 to 1; they occur in masses cementing boulders of chert or jasperoid; crystal masses lining vugs and cavities; fracture fillings in chert; lenses between thin beds of chert; and disseminations in jasperoid, dolomite and shale. Interestingly, the first zinc–lead ores mined near Commerce occurred at a depth of about 100 feet in sandstone breccias and in flint breccias below the sandstone. The sandstone was impregnated with heavy oil, which, on evaporating, left asphalt. The asphalt caused considerable trouble in the concentrating plant because it balled up the ores on the jig beds and tended to float off with imbedded ore particles. It also tended to lower the grade of the concentrates by increasing pyrite and gangue contamination.

All mines were developed from vertical shafts, ranging from 100 to 500 feet in depth. Because of the way these ores are developed, hundreds of shafts have been sunk into these orebodies. Virtually all ore was mined by open stope methods using natural pillars, which accounted for 8% to 20% of the orebody. The ore was only treated in the early years by jigs and tables to make concentrates of both sphalerite and galena. Little effort was made to recover sphalerite from slimes until the widespread introduction of flotation at centralized milling operations in the mid–1920's. Even then jigs and tables were the main methods of concentration; flotation was used almost exclusively to treat slimes.

The first smelting of ores in

Oklahoma took place in 1907 at Bartlesville, where two smelters were established to take advantage of the underlying gas pool. Numerous smelters were subsequently established owing in large part, to the depletion of the natural gas pools in Kansas and the rapid development of the Ottawa County zinc fields. By 1912 Oklahoma had the largest retort capacity in the United States. In 1916, additions to the smelter at Collinsville made it the largest retort smelter in the world. In 1917, the State was the Nation's largest producer of zinc metal, a position held sporadically through 1927. Thereafter, for the most part, the State's spelter production ranked among the top three producers. One of the three remaining active electrolytic zinc plants in the United States is at Bartlesville on the same site as one of the early retort plants.

OREGON

Oregon has had small and sporadic zinc production through 1965 mainly from complex sulfide veins in the Bohemia district in Lane County until recently. In 1991 the Silver Butte copper–zinc–silver mine in southwestern Oregon, produced 751 tons of zinc in concentrate from copper ores, but zinc production has been sporatic ever since. Production at Silver Butte has accounted for virtually all of the State's zinc output.

PENNSYLVANIA

Pennsylvania has been a relatively small producer of zinc ore; but owing to abundant coal, excellent river and rail transport, and closeness to available zinc ores and concentrates and major domestic markets, the State has accounted for slightly more than 21% of the U.S. zinc metal production and ranks as the Nation's leading metal producer. Currently the State is the site of the only remaining operating pyrometallurgical smelter in the United States. This smelter, which is located at Monaca north of Pittsburgh, is the Nation's largest in zinc metal and oxide production and capacity.

Zinc was discovered in the Saucon Valley near Friedensville in 1845. The mining of large near–surface bodies of smithsonite and hemimorphite began about 1853 and continued to 1893, when all mining came to a halt owing to the high cost of pumping ground water. In 1869, the largest mine in the Valley, the Ueberoth, installed the world's largest single–cylinder, walking–beam engine (called locally "The President") to control the inflow of water. In 1876 the mine was making 20,000 gallons per minute and, although "The President" was capable of handing the high flow, the cost were too high and the mine closed. The State had no further zinc mine production until the MVT sulfide ores in Ordovician limestones underlying the calamine ores near Friedensville were developed in 1958. The Friedensville Mine was a substantial zinc producer until 1983 when it was also closed, mainly because of the high cost of pumping water, which accounted for about one–third of the mine's operating costs.

In 1856, the Lehigh Zinc Company of South Bethleham built the State's first spelter furnace near Friedensville to produce spelter from local ore; however, the process failed to yield any zinc metal. In 1858, Samuel Wetherill, the inventor of the American zinc oxide process, was able to produce about 50 tons of spelter made from Pennsylvanian calamine but the process was high cost and soon abandoned. Interestingly, the first rolled zinc produced in the United States was made in 1859 in Philadelphia using some of Wetherill's spelter. In 1859 a Belgian-type smelter was established at South Bethlehem; this proved successful and in 1860, the first commercial zinc metal was produced.

In 1898 a large horizontal retort smelter and zinc oxide plant was constructed at Palmerton mainly for the processing of New Jersey zinc ores. During WWI, large smelters were built at Langeloth (1914) and Donora (1915). In 1929 the world's first commercial continuous zinc smelting process, the vertical retort, was placed in operation at Palmerton, and in 1930, an electrothermic-process smelter opened at Monaca. The electrothermic plant initially was only an oxide producer, but in 1936, it became a slab zinc producer. All of the smelters, except the electrothermic plant, have since closed: Langeloth in 1947; Denora in 1957; and Palmerton in 1980. The electrothermic smelter in 1995 was the largest zinc plant (in capacity) in the United States and was the world's largest processor of calcine derived from electric arc furnace dusts.

TENNESSEE

Zinc production in Tennessee first began in 1854 at Mossy Creek, now Jefferson City, when near-surface oxidized zinc deposits were worked via open pits for calamine for the making of zinc oxide pigment. Similar ores were found and mined at Mascot shortly after mining at Mossy Creek started. Mining in the area ceased in 1858 and the mines were idle throughout the Civil War. In 1867 mining at Mossy Creek resumed to feed a zinc oxide plant that was erected at Jefferson City. However, after only a few months operation, litigation forced the mine and plant to close for the next 13 years. The Eades, Mixter, and Heald Co. bought the Mossy Creek property in 1882, and operated the mine intermittently until 1894. This ore, which was mainly sphalerite, plus zinc ores produced at other mines in the State were generally shipped to Clinton, where a zinc smelter had been erected in 1880. Beginning in the 1880's, zinc ores, mainly calamine, were produced in Claiborne and Union Counties near the Powell River and at Straight Creek in Claiborne County. In 1892 the first zinc ores were mined near New Market; some of this ore was shipped to the Bertha smelter in Pulaski, VA.

Mining was sporadic until about 1911, when many of the mining properties mainly in the Mascot-Jefferson City district were consolidated into what eventually became the American Zinc Co. of Tennessee. In 1913 American Zinc had a 900-ton-per-day gravity mill in operation at Mascot and in 1914, a flotation circuit to treat gravity fines was added. Interestingly, it was at this mill that copper sulfate was first used as a conditioning agent in a commercial zinc flotation process, and it was also here that the first heavy-media, sink-float unit for continuous operation was installed (1936). The sink-float unit replaced jigs that had been used to reject as tailings, coarse, barren and low-zinc gangue particles in unclassified mill feed. In 1927, the first zinc concentrates from the complex, massive sulfide ores at Ducktown were recovered, and in 1930, a U.S. Steel Corp. subsidiary began zinc production at the Davis Mine just south of Jefferson City. A number of mines were reopened during WWII but closed soon after it was over.

The discovery and development of several major deposits in the Mascot-Jefferson City-New Market district in the 1950's resulted in Tennessee becoming the Nation's largest zinc-producing State in 1958. This position was further solidified when the Elmwood Mine in central Tennessee came on stream in 1974. In the 1958-1989 period, Tennessee was the leading zinc producer in 29 of those years; since that time, the State has ranked second after Alaska.

In 1978, the Jersey Miniere Zinc Co. constructed an electrolytic zinc plant at Clarksville primarily to process the high-quality zinc concentrate produced in Tennessee. This was the first smelting carried out in the State since the closure of the smelter at Clinton in 1894. In 1994 the Clarksville zinc plant was sold to the Savage Zinc Co.

All of the zinc deposits in central and east Tennessee are MVT and occur within, and rarely below, the Ordovician Knox Group. Sphalerite is virtually the only ore mineral, although small amounts of galena, chalcopyrite, and pyrite are sometimes present. The sphalerite occurs as either bedded deposits or as collapse breccia breakthrough deposits, resembling expansive asymmetrical networks in which the ore shoots form wandering and interconnecting branches with noncommercial rock between and within. Deposits are generally mined by room and pillar or open stope methods, modified by cut and fill, overhand and/or underhand operations. At some operations where very thick ore sections are encountered, underhand stoping around mill holes leaving tall pillars is used. Many of the MVT deposits have associated oxidized ores which were mined by early and somewhat crude open pit methods.

The ores in the Ducktown (Copperhill) district occur as large massive sulfide lenses replacing certain folded and faulted beds in highly metamorphosed lower Cambrian sediments consisting of gray-wackes, schists, and slates. The shape of the orebodies is controlled by the folding of the country rock. The ore is composed of pyrrhotite, and chalcopyrite as the principal ore minerals, with sphalerite and galena as minor minerals. The deposits, which can be several thousand feet long and have widths of 200 feet, are generally mined by open pitting or underground by open sub–level stoping. The ore and wall rocks stand well such that very little timbering is necessary. The ore minerals are separated by flotation into concentrates of copper, sulfur (pyrite) and zinc, with the first two being produced since the mid–1850's and the zinc since 1927. The Copperhill mines have all closed, the last one, closing in 1987.

TEXAS

Texas has been an insignificant source of mined zinc, but owing to good transport, adequate fuels and nearness to ore supplies, it ranks as the fifth largest U.S. producer of zinc metal, despite the fact that only two smelters have operated in the State. Sporadic zinc mine production has come from zinc and silver–lead mines in Brewster and Presidio Counties in western Texas, Burnet County in central Texas, and in the Quitman Mountains of El Paso County.

Zinc smelting began in Texas in May 1923 at a newly–constructed, gas–fired horizontal retort smelter at Amarillo. The second zinc refinery came about as a result of the start of WWII. The Government was expecting war requirements for high quality zinc to be higher than available from domestic electrolytic plants and approved the building of an electrolytic zinc plant at Corpus Christi in 1940. This was the only new zinc plant approved for construction in the United States during WWII, although other operating or closed, existing plants gained approvals to reopen, expand or modernize. The Corpus Christi plant began operation in 1942, but was closed in 1985 owing mainly to loss in feed sources and high energy costs. The Amarillo plant closed in 1975 owing mainly to environmental problems.

UTAH

Zinc has been produced at numerous mines in the State. Most zinc ores in Utah were mined underground, generally using square–set stoping or cut and fill mining methods and open stoping where possible. The control of water inflow has been a major problem at some mines in Utah, especially in the Park City and Tintic districts; in some cases water problems severely limited the proper development of orebodies and the economic viability of some operations.

Most zinc production has taken place in three mining districts: Bingham, East Tintic and Park City. In all three districts the principal orebodies were discovered by prospectors searching or mining for gold, silver, and/or copper in the 1850–1880 period. The deposits are veins, mantos, and replacements in folded and/or faulted Paleozoic limestones and occasionally veins and replacements in quartzite and intrusive rocks. The deposits appear to be largely the result of hydrothermal solutions, derived from or associated with intrusive rocks, that deposited the ores mainly in the fissures and replacement zones in favorable calcareous beds. Oxidation of these ores has been extensive, and commonly reaches a depth of 500 feet and locally as much as 1800 feet. Large bodies of smithsonite typically occur in the oxidation zone; however, many of these orebodies were not mined and remain as possible future resources. In the Bingham district, zinc replacement deposits tend to be zonally arranged in limestones, quartzites, and sandstones within a cylindrical zone or annulus encircling the main copper porphyry mass; some of the most important zinc–lead ores were mined from eight named cross–cutting fissures at the Lark Mine. At Park City, lode ores tend to follow east– northeast fault zones and likely represent the channelways through which the mineralizing solutions passed that resulted in the replacement or bedded ores found in certain limestones. Bedded deposits, which can extend for a mile or more, tend to lie along continuous master fracture zones rather than in numerous small fractures. At Tintic, there are both fissure and replacement veins, pipes and extensive mantos replacement orebodies, which often form persistent ore runs traceable over thousands of feet.

The first notice of oxidized zinc minerals in Utah was probably at a mine near Alta in 1870; however none there or anywhere else in the State were commercially produced until small production began in the North Tintic district in Utah County in 1905. In that same year small amounts of sphalerite were also recovered at mines near Park City and Frisco. Up to that time zinc in the ore was a problem because it brought penalties at the copper and lead smelters. Even after sphalerite and calamine minerals began to be recovered, many operators bypassed calamine and zinc–rich sulfide orebodies or, if mined, they were put aside, dumped or used as backfill. Sphalerite recoveries were poor until flotation became common in the mid–1920's; prior to that time, large amounts of sphalerite were lost to tailings or contained in lead concentrates and lost in slags when the concentrate was smelted. The scale of these losses can be illustrated by an estimate of initial zinc losses at Park City from the beginning of mining in the latter half of the 19th century through 1914. Mills there were estimated to have discharged in that period more than 2 million tons of tailings averaging about 4% zinc into seven dams across the creek that passes through the town. The largest of these tailings accumulations was 7 miles below Park City and measured 3.5 miles long, 500 feet wide and 2 feet deep. Some of these tailings, however, were later reprocessed to extract additional zinc, lead, and silver. Zinc losses in lead slags have been much greater. In 1923, for example, more that 50,000 tons of zinc was incorporated in lead slags generated at Utah lead smelters in that year alone. Considering the large amounts of Utah lead ores smelted in the State before widespread ore flotation, large amounts of zinc were mined in the early period but not recovered. A slag–fumming plant was built adjacent to a lead smelter at Tooele during WWII and produced a smeltable zinc fume until 1971.

Utah ranks as the Nation's ninth largest producer of zinc ore. The State's zinc production jumped during WWI and in the mid–1920's when flotation became important, and continued at fairly steady rates until it fell sharply in 1978; since that time zinc production has been sporadic and small.

VIRGINIA

Virginia had essentially two periods of zinc production, an early period (1868–1905) of large–scale calamine production and a later period (1927–1981) of zinc sulfide production. Some sporadic production of both ore types occurred in the 1910–1918 period. The great bulk of the State's output of zinc has come from the lead–zinc deposits found near Austinville in Wythe County in southwestern Virginia. These deposits were mined for lead as early as 1756 and were the principal source of lead for the Continental Army during the Revolutionary War and for the Confederacy during the Civil War. During the Revolutionary War, the mines were managed by Colonel Charles Lynch, whose infamous conduct led to the term, *lynching* in American vocabulary. Also of historical note is the fact that from 1780 through 1799 these mines were owned and operated by Moses Austin, after whom Austinville is named and who was the father of Stephen Austin, a national hero and first Governor of Texas. The first zinc mining in the State occurred during the Civil War when several tons of calamine ore were mined in the Austinville area and shipped to a smelting works at Petersburg. Calamine mining resumed in the same area after the war in a small way beginning about 1868; however, large–scale zinc mining only began in Virginia with the opening of the Bertha Mine in Wythe County in 1879. Shortly thereafter, a zinc smelter was built at Pulaski to process Virginia zinc ores. The spelter produced was of exceptional purity, such that the "Bertha" spelter gained a world–wide reputation. By 1905 the calamine deposits near Austinville were largely exhausted and zinc mining ceased. The Austinville deposits were purchased in 1902 by The New Jersey Zinc Co., who instituted an exploration program to test the value of the primary sulfide deposits, which underlay the mostly mined–out secondary calamine ores. Development was slow to occur up to the mid–1920's when some sulfide mining resumed; in 1927 an all–flotation mill was placed in operation at Austinville and production rose rapidly. In 1957, the nearby Ivanhoe Mine was brought on–stream and operations were combined with those at Austinville.

For the most part, the zinc

output from these mines from 1928 remained at high levels until ore exhaustion in 1981. The only other zinc mining in Virginia of significance outside the Austinville area, was near New Market in Brockingham County, where small amounts of zinc sulfides were mined from Ordovician dolomite breccias at the Timberville mine in 1949 and 1950 and at the Bowers–Campbell Mine in 1957 through 1963.

The deposits in the Austinville-Ivanhoe district borders on the East Tennessee district, and the two have often been included in a Tennessee–Virginia district because they have similar characteristics and occur in the same physiographic and geologic province in rocks of the Ordovician Knox Group. The Austinville ores typically occur as fracture fillings and replacements in the form of irregular lenses in fractured and brecciated dolomitic limestone along bedding or paralleling a system of minor strike faults. The sizes and thicknesses of the lenses are affected by the horizon in which they occur. The strike lengths of lenses range from less than 50 feet to 400 feet and dip lengths range from less than 100 feet to over 2,000 feet. The ore consists of sphalerite, pyrite, and galena with dolomite the predominant gangue mineral. Associated with the orebodies are large amounts of recrystallized material, which commonly occurs as a shell around the orebodies except on the footwall side; The recrystallized zone proved to be a valuable guide in prospecting for sulfide ore. Large and extensive deposits of calamine ores formed by alteration of sphalerite occur between highly–weathered and pinnacled, subsurface carbonate rocks and thick overlying residual clay deposits. These secondary ores are composed of hemimorphite, smithsonite and hydrozincite; of these hemimorphite is, by far, the most abundant.

The calamine ores were mined by pits and via small shafts up to about 125 feet in depth. Underground workings followed the ore, timbering where necessary. The sulfide deposits were primarily mined by underhand open stoping and modified underhand stoping with random pillars.

WASHINGTON

Zinc minerals were discovered in Washington in the mid–1880's when lead mining began in Pend Oreille and Stevens Counties. Zinc was not initially recovered owing to the remoteness of the area and lack of transportation. Despite the completion of a railroad to Metaline Falls in 1906, zinc production was slow to develop. The first significant zinc production was in 1915, although about 5 tons and 9 tons of zinc in concentrate, respectively, were produced in the Metaline district in 1906 and 1911. Until 1931, the State's zinc production tended to be small and erratic as to mine source. Zinc output improved in the 1930's, especially after 1937, owing mainly to increasing mill capacities at the Pend Oreille and the Grangview Mines in Pend Oreille

County. Thereafter, through 1968 zinc production continued at relatively high rates before tailing off and becoming intermittent after 1976.

Virtually all of the recoverable zinc mined in Washington has come from more than 200 mines and prospects in four mining districts in the above two counties, with about 70% coming from Pend Oreille County and 30% from Stevens County. The only zinc to be produced outside these counties was a result of copper mining at the Holsten Mine.

Deposits in the two counties vary but are considered to be MVT with most commercial ores occurring as large, irregular, replacements and fillings along faulted, sheared and brecciated zones in Cambrian limestone. Sphalerite and galena are essentially the only ore minerals; gangue consists of jasperoid, dolomite and calcite. The ore is generally associated with dolomitization and, in places, with abundant cavities and caverns. At a few mines, the carbonate rocks were partially converted to silicate minerals during the emplacement of a late–Cretaceous intrusive pluton.

The ore horizons can be very solid and competent, permitting large open stopes to be developed. Some operators employed inverted, rill–bench stoping to produce extremely large open stopes, some of which were several hundred feet high with unsupported stands across widths exceeding 200 feet. Some carbonate ores were mined prior to 1930 in shallow pits; open pitting was also the method used at the Van Stone Mine, a large zinc producer which opened in 1952. Most mines, however, were underground operations.

WISCONSIN

Zinc has been produced largely in Grant, Iowa, and Lafayette Counties in what is called the Upper Mississippi Valley mining district. The ore deposits are MVT and, for the most part, are fissure and cavity fillings (pitches and flats) occurring in Ordovician limestones and dolomites. Sphalerite and galena are the principal ore minerals, although above the water table large deposits of calamine ores, mainly smithsonite, are found with galena and cerussite. The bulk of the mining has occurred underground utilizing room and random pillar and open stope mining methods.

The first zinc ores to be mined were the near–surface calamine ores about 1859; these ores were mined to feed the Mattheissen and Hegeler smelter that opened in 1860 at LaSalle, IL. Most of the large known zinc deposits had been worked down to ground–water level by the late 1870's, and because mining was generally uneconomic below the water table owing to high pumping cost, zinc mining in the State came to a virtual halt. It was not until 1890 that significant ore production resumed in the district. However, once the main sulfide bodies were mined, many operations found that only low-grade zinc concentrates could be produced owing to associated marcasite, which could not be separated from the sphalerite by gravity methods. Wetherill separators and other separators of low magnetic intensity were of limited value until about 1905 when the green concentrate was given a slight roast to drive off a little sulfur thereby rendering the marcasite magnetic.

Wisconsin was a moderately large zinc producer until 1981 when the last mining operation closed down. Overall, the State ranks as the Nation's 12th largest producer of recoverable zinc.

Table 3: Principal Zinc Producing Districts in the United States

State	Mining District	Deposit Discovery Date	Start of Zinc Mining
AK	Red Dog	1968	1989
AZ	Big Bug	1880	1939
	Bisbee	1877	
AR	Northern Arkansas	1818	1857
CO	Gilman	1878	1903
	Leadville	1860	1885
	San Juan Mts.	1848	1917
ID	Coeur d'Alene	1858	1906
IL/IA/WI	Upper Mississippi Valley	1659	1859
KS/MO/OK	Tri-State (Joplin)	1810	1869
MO	Viburnum Trend	1955	1971
MT	Butte	1864	1914
	Barker	1864	1911
NV	Pioche	1864	1914
	Yellowpine	1864	1905
NJ	Franklin-Sterling Hill	1640	1774
NM	Central	1790	1902
NY	St. Lawrence-Edwards/ Balmat	1838	1915
PA	Freidensville	1845	1853
TN	Ducktown	1843	1927
	East Tennessee	1854	1854
	Middle Tennessee	1967	1974
UT	Bingham	1863	1914
	East Tintic	1869	1914
	Park City	1869	1907
VA	Austinville	1756	1863
WA	Metaline	1869	1906

Sources: Koch and Schuemeyer (1979); Ingalls (1908);
U.S. Bureau of Mines.

ZINC AND THE NATIONAL DEFENSE STOCKPILE

"In the future, as in the past, industrial vigor will be the principal factor in deciding contests between people."

T.A. Rickard (1932)

(NOTE: All tons mentioned in this section are short tons to conform with official documentation and legislation)

The National Defense Stockpile (NDS) is a reserve of non-fuel materials that the United States might require in the event of war or national emergency but that might not be available in sufficient quantities from domestic or reliable foreign sources. Zinc was a critical metal in WWI and WWII and has been an important stockpiled metal since the beginning of national stockpiles in 1946.

Since the end of WWII, the government acquired about 1.6 million tons of slab zinc for the stockpile, authorized release of the entire amount and sold or disposed of about 1.2 million tons valued at about $500 million. Virtually all of the zinc in the stockpile was acquired by 1960, whereas virtually all of the disposals (sales) occurred in the 19641974 period. Since 1975 the inventory in the stockpile has changed very little, although Public Law (P. L.) 102484, signed by the President in 1992, authorized the disposal of the remaining inventory of zinc in the NDS over a five year period subject to adjustments based on recommendations of a broad based, market impact committee. Some zinc subsequently was sold from the stockpile, but not at a rate leading to disposal in the desired five years.

WORLD WAR I

The need for a stockpile of critical raw materials for national defense purposes became apparent during WWI when shortages of materials upset production schedules, delayed development programs and slowed exports of vital materials to U.S. Allies. The United States did not experience zinc shortages during WWI, although zinc sheet and plate and some of the higher grades of zinc metal were, on occasion, in short supply. When the war broke out in Europe in 1914, the United States was the world's single largest zinc producer, accounting for about 35% of total production, and was thereby self-sufficient. European countries produced most of the remaining world output, but with the outbreak of war the European zinc production was severely curtailed. Battlefields were in or near major smelting districts and naval action disrupted concentrate imports. As a result, the British and French sought zinc supplies from the United States. Demand and prices soared, leading to rapid expansion of the U.S. zinc industry. When the United States entered the war in 1917, the industry's capacity to produce zinc exceeded U.S. and export requirements to such an extent that the zinc industry began to worry about industry retrenchment after the war ended. At the end of the war, various Government programs had accumulated in excess of 10,000 tons of zinc metal and concentrate at domestic smelters; this material was disposed of in 1919 and 1920. Although zinc was not in short supply during WWI, it was nonetheless a critical material for use as brass in artillery shrapnel and shell casings, metal liners in ammunition and explosive boxes and magazines, coatings on barbed wire entanglements and sheet metal; additionally, it was a component of naval and battlefield smoke, medical tape and rubber tires, and a substitute for other critical metals including aluminum, copper, lead and tin. In the latter part of the war, for example, lead became very scarce, resulting in its substitution by rolled zinc in coffins.

POST WORLD WAR I AND THE SECOND WORLD WAR

Based on the WWI experience, the Army developed a list of 42 essential materials required for military operations in 1921, but there was little interest in creating a strategic materials reserve at that time. Materials stockpiling regained impetus in the 1930's owing to deteriorating conditions in Asia and Europe. The Naval Appropriations Act of 1937 authorized the purchase of $3.5 million in critical materials, and two years later the Critical Materials Stockpile Act of 1939 (P. L. 76117) authorized $100 million for the stockpiling of designated items including basic industrial raw materials for military production and their supporting industries. The outbreak of WWII in 1939, however, prevented establishment of the stockpile because the materials to be stockpiled were needed for immediate industrial expansion and military production. The Government, through various agencies mainly the Reconstruction Finance Corporation and its subsidiary, Metals Reserve Company controlled the production, price, and distribution of zinc during WWII. Again, at the start of the war, the United States was self-sufficient in zinc production. As the war expanded, the demands on the domestic industry increased. The smelting sector expanded rapidly; the mining sector did not. Mine production rose in 1941 and 1942, but declined thereafter, despite efforts to stimulate output with bonuses and the release of almost 1,100 men from the Army, specifically assigned to work in domestic zinc mines. As a results of deficient domestic zinc mine output, dependence on imported foreign ores and concentrates increased as the war progressed and reached 537,000 tons (zinc content) in 1943. That year's imported ores and concentrates accounted for almost one-half of U.S. smelter feed requirements as compared with only 5% in 1939. Thereafter, through the 1960's, imported ores and concentrates continued to account for a substantial portion of U.S. zinc needs.

In 1940 the shortage of zinc became acute, and by the end of the year a voluntary system of supply allocation was initiated by the industry. In March 1941, zinc was declared critical, and in June it came under full allocation control of the War Production Board. Stockpiling during WWII was initially based on a 3yearwar scenario, but in August 1943 the war outlook had improved to the point where objectives were reduced to a 1year supply; then, in June 1945, it was reduced to a 3month supply.

As WWII wound down, sentiment increased for a continued national defense stockpile after hostilities ceased. The military wanted a stockpile to avoid critical materials shortages in the future, whereas materials' producers anticipated adverse effects from the dumping of Government materials stocks, which included an estimated 400,000 tons of zinc metal and zinc concentrate, into the postwar market. In July 1946, legislative pressure to assist domestic basic industries in adjusting to a peacetime economy aided passage of the Strategic and Critical Stock Piling Act of 1946 (P. L. 79520). Wartime zinc stocks that had not been sold to domestic industries to assist conversion and reconstruction efforts were transferred or eventually smelted to metal for addition to the national strategic stockpile. The zinc industry in the 194648 period was able to maintain relatively high production rates owing to stockpile purchases and to the smelting of Government concentrates.

KOREAN WAR DECADE

The advent of the Korean War (19501953) resulted in continued zinc purchases for the stockpile. Zinc prices and imports increased, and even with Government allocation limits in place, there were no serious shortages of zinc during the war. In the period 19541960 the government continued to accumulate zinc for the stockpile via direct purchases, the Defense Production Act (DPA) of 1950 (P.L. 81744), and barter agreements (P.L. 82480 and the Office of Defense Mobilization authorization of May 1956). Zinc additions from DPA programs were small, but zinc metal obtained through barter of perishable surplus agricultural products totaled 324,000 tons from 1954 through 1960. Zinc was removed from the barter list in 1960. Stockpile materials obtained under the barter programs were not placed in the national stockpile but were placed in a "supplemental" stockpile. that could only be released by Congress.

POST 1960'S ERA

The year, 1959, marked the

end of significant acquisitions for the Government's zinc stockpiles. Thereafter, almost all zinc stockpile transactions involved sales in excess of goals. Small adjustments in the stockpile inventory after 1975 were generally the result of the return of excess metal withdrawn from Government stocks in earlier years by various Government agencies. Stockpile goals or objectives often were reviewed and modified based on assumptions governing military and industrial preparedness and length of a potential U.S. war emergency. Stockpile goals were generally based on 5year and 3year wars but in the late 1950's, the war-period was reduced to 1 year based on the assumption that nuclear war would be short and decisive and would eliminate protracted conventional conflict. In 1958, the stockpile goal for zinc was reduced from 1,250,000 tons to 178,000 tons and in 1963, to zero. The assessment that no stockpile of zinc was required was, in part, based on the assumption that North American sources could supply the necessary zinc requirements in the event of a nuclear or conventional war. The reduction in goals essentially made about 1.5 million tons of zinc in the Nation's stockpiles available for reentry into the U.S. and world economies. In July 1964, P. L. 88374 authorized the release of 67,500 tons of zinc to domestic primary producers and 7,500 tons to independent, zinc-based alloyers. In April 1965, P.L. 899 authorized the stockpile release of 150,000 tons of zinc for disposal to industry and 50,000 tons for use by the Government, including the Agency for International Development, the U.S. Mint, and the Department of Defense. Another 200,000 tons of zinc was authorized for release in November 1965 (P. L. 89322). Sales from the above public laws continued through 1971.

A new zinc stockpile goal, 560,000 tons, was established in 1969, but despite earlier disposals, large stocks in excess of this goal remained. In April 1972, P. L. 92283 authorized the disposal of 515,000 tons of zinc from the national stockpile for domestic consumption only. A total of 440,000 tons was to be released through domestic primary producers and 75,000 tons was for off-the-shelf sale. In April 1973, the zinc stockpile goal was reduced to 202,700 tons, making available even more excess zinc. In December 1973, P. L. 93 212 authorized the release of an additional 357,300 tons of zinc, with 150,000 tons to be sold directly to consumers; the remaining 207,300 tons was to be sold through participating primary zinc producers.

All zinc sales ceased in June 1976 when a new stockpile goal, incorporating the stocks of the previous goal and unsold excess stocks, was established. In October 1976 a new stockpile goal of 1,313,000 tons of zinc was approved. In 1979 the Strategic and Critical Stockpile Revision Act of 1979 (P. L. 9641) created the NDS by combining the national and supplemental stockpiles, limited stockpile materials use only for the common defense of the United

States, and prohibited stockpile sales for economic purposes. The last mentioned provision was added to ensure that the stockpile could not be used as a means of controlling commodity prices as an anti inflation mechanism or be used to produce revenue for budgetary purposes, as both were perceived to have been important underlying factors in previous stockpile disposal legislation. In May 1980, the stockpile contained 383,000 tons of zinc and a new stockpile goal of 1,425,000 tons was established for zinc. Other than a small drop in the zinc inventory, the zinc stockpile situation of 1980 was unchanged until sales resumed following passage of disposal legislation in 1992. A summary of stockpile goals, acquisitions and disposals, by year since 1946, appears in Table 4.

During the period of stockpile sales, the Government attempted to aid zinc producers by price stabilization payments, import quotas, tariffs, and marketing of stockpiled metal through primary producers. The Lead-Zinc Small Producers Stabilization Act of 1961 (P. L. 87347) was passed to encouraged zinc production by authorizing prorated payments for small mine producers for the difference between 14.5 cents and the monthly average price of zinc metal, if it fell below 14.5 cents. This program expired in 1969. Import quotas were imposed on imported concentrates and metal from 1958 through 1965. In August 1971, the President froze the price of zinc, but in doing so increased tariffs in the form of surtaxes on imported zinc products temporarily. The surcharge was dropped in December of that same year.

Government stockpile sales and price controls were, to some extent, viewed as detrimental to the health of the domestic zinc-producing sector. Domestic zinc smelting companies maintained their market shares during the stockpile disposal period. However, the sales and prospects of future sales discouraged domestic investments in new technology and facilities. The sales also tended to hold down zinc prices, thereby limiting industry profit, which might have encouraged additional investment in the industry. Price controls held U.S. zinc prices below world prices from August 1971 through the end of 1973. As a result, domestic smelters were adversely affected because they could not effectively compete for foreign concentrates on world markets. U.S. zinc metal production declined after 1969, even though domestic zinc metal consumption was the highest ever in the early 1970's. In the 19711975 period, seven smelters permanently closed for a number of reasons, including high costs, obsolescence, and environmental concerns. However, price controls and stockpile sales undoubtedly contributed to their demise as well. The Government also was adversely affected by price controls and the price-depressing effect of sales, in that it received less value for its metal. This was especially true during the second half of 1973 when the average LME zinc price was more than twice the U.S. controlled price and the EPP price was about 50% higher.

SUMMARY

During WWI the United States was self-sufficient in zinc production and capable of supplying its allies with needed refined zinc products. At the start of WWII, the United States continued to be self-sufficient, but became more import dependent beginning in 1940. From that time through the end of the war, the U.S. mining sector was not capable of supplying the greatly expanded refining sector with sufficient zinc feed material. As a consequence, the United States became 30% dependent on imported zinc concentrates during WWII, but remained essentially self-sufficient in refined zinc production.

During the Korean and Vietnam Wars a somewhat similar situation prevailed. But in these later wars, U.S. stockpiles contained large quantities of zinc that were available if needed. In the 19401974 period, the U.S. zinc smelting sector, though dependent upon imports for almost 40% of its concentrate supply, provided about 90% of U.S. requirements of refined zinc products.

The peak year for domestic slab zinc production was 1969; thereafter production declined, owing to the permanent closure of 11 of 14 smelters that had operated in the United States in 1968. Two of the remaining three smelters closed for a period of time, but reopened after extensive modification. One new "greenfields" smelter opened in 1978, resulting in the four operating primary smelters at the present time. The U.S. primary zinc smelter capacity however, had dropped from 1,318,000 tons in 1968 to 410,000 tons in 1993.

Mine production followed a similar but less dramatic decline, although in the past few years the development of a large zinc mine in Alaska reversed the trend and restored U.S. mine output to 1960's levels (about 600,000 tons per year). Unfortunately, the Alaskan mine, which accounts for about one half of U.S. zinc mine output, present problems from a strategic point of view. The mine is isolated, depends on sea transport for supplies and shipments, and has port facilities that are ice-bound for 9 months a year. Since its opening, all concentrate shipments have been exported, but even if they had been redirected to the United States, there is no excess domestic smelter capacity to process the Alaskan concentrate. The largest domestic primary smelter processes large amounts of secondary material, which effectively lowers the primary processing capacity of the United States to about 300,000 tons, about the same as the mine production of the "Lower 48 States." In today's environmentally conscious climate with permitting difficulties and the long-term threats of Superfund, it is doubtful that any new zinc smelters will be built in the United States in the foreseeable future. As a result of the decline in U.S. zinc smelter capacity, a major shift in the zinc import pattern occurred in the early 1970's. Concentrate imports fell sharply, whereas slab zinc imports rose dramatically. With the above change, the U.S. net import reliance (dependence) for refined zinc products jumped to about 60%. Determining import dependence on a concentrate basis, the United States dependence is about 25%. Although domestic mine output has improved since the opening of the Red Dog Mine, much of this new production from a strategic standpoint is essentially unsecured and unavailable for reasons stated earlier. In this uncertain world, when considering whether or not to have some zinc in the NDS, one should consider the old adage: "Keep (at least some of) your powder dry."

Table 4: Zinc National Defense Stockpile, Inventories, Receipts, Releases, and Goals, 1946-1994 (short tons)

Year	Inventory as of Dec 31	Net Increases Releases ( )	Goal or Objective
1946	69,223	69,223	0
1947	93,381	24,158	1,500,000
1948	490,595	397,214	1,500,000
1949	594,657	104,062	1,500,000
1950	644,146	49,489	780,000
1951	649,163	1/ 5,017	740,000
1952	2/ 661,714	12,551	740,000
1953	700,320	38,606	740,000
1954	824,463	124,143	1,100,000
1955	966,551	142,088	1,100,000
1956	1,147,710	181,159	1,250,000
1957	1,462,023	314,313	1,250,000
1958	1,548,235	86,212	178,000
1959	1,583,564	35,329	178,000
1960	1,578,719	(4,845)	178,000
1961	1,579,616	897	178,000
1962	1,579,907	291	178,000
1963	1,580,941	1,034	0
1964	1,505,234	(75,707)	0
1965	1,312,868	(192,366)	0
1966	1,212,368	(100,500)	0
1967	1,198,122	(14,246)	0
1968	1,160,606	(37,516)	0
1969	1,142,185	(18,421)	560,000
1970	1,141,490	(695)	560,000
1971	1,137,937	(3,553)	560,000
1972	949,583	(188,354)	560,000
1973	677,009	(272,574)	202,000
1974	391,600	(285,409)	202,000
1975	385,714	(5,886)	202,000
1976	385,192	(522)	1,313,000
1977	383,415	(1,777)	1,313,000
1978	381,259	(2,156)	1,313,000
1979	381,051	(208)	1,313,000
1980	380,303	(748)	1,425,000
1981	378,320	17	1,425,000
1982	378,317	(3)	1,425,000
1983	378,316	(1)	1,425,000
1984	378,316	--	1,425,000
1985	378,316	--	1,425,000
1986	378,316	--	1,425,000
1987	378,316	--	1,425,000
1988	378,316	--	1,425,000
1989	378,316	--	1,425,000
1990	378,760	444	1,425,000
1991	378,768	8	1,425,000
1992	375,873	(2,895)	0
1993	359,220	(16,653)	0
1994	316,363	(42,857)	0

1/ Authorized release of 15,000 tons; actual release 90
2/ Does not include Zn in DPA inventory
  (data are not available)

ZINC BASICS

"'Found your problem.' He put the remains of both props on the concrete dock.

"God, what did we hit."

…"Water, Doc, just water."

"Did you have the boat surveyed before you bought it?"

"Sure…[the surveyor said] there was something wrong with the sinks…I had a plumber check…but they were fine."

"Sinks?"

"That's what he told me over the phone."

"Zinks," Kelly said laughing. "Not sinks…What destroyed your props was electrolysis. Galvanic reaction. It's caused by having more than one kind of metal in saltwater….The surveyor meant for you to replace the zinc anodes on the strut.'"

Without Remorse (1989)
Tom Clancy

SPECIFICATIONS AND STANDARDS

Zinc is a bluish-white metal with a melting point of 419.6 degrees C. and a boiling point of 907 degrees C. The melting point permits low-temperature casting, and the boiling point is an important factor in purifying the metal by distillation and in producing metal dust and zinc oxide in fine particles, a feature desirable for many applications. The properties of being chemically active and alloying readily with other metals are utilized industrially in preparing numerous useful zinc-containing compounds and alloys. The relatively high position of zinc in the electromotive series accounts for its large use in protecting iron and steel products from corrosion. Common zinc dissolves readily in most acids but less so with increasing purity of the zinc. This increasing resistance is due to the fact that the overvoltage of hydrogen on zinc is higher than on most other metals. Zinc is malleable and can be rolled if of high purity, otherwise heating to 100-150 degrees C. is required.

Zinc produced from ores is termed "primary" zinc, and when produced from scrap or residues, it is called "secondary", "redistilled", or in some cases, "remelt" zinc. Zinc metal is further categorized as "electrolytic" or "distilled" according to the reduction method used. Commercial zinc metal is cast into various sizes and shapes and as such, is termed "slab zinc"; the older, historical term, "spelter", is now rarely used as a term for slab zinc in the United States.

The purity of zinc plays an important role in the use and consumption of the metal. Until the latter half of the Nineteenth century, the quality or purity of zinc metal was largely dependent on the ores used in the smelters. The desirability of purer zinc metals was recognized in applications such as munitions brass, batteries, and rolled products; however redistillation was difficult and expensive. The above created premium markets for "Bergenpoint" and "Bertha" spelter produced from the high quality calamine ores produced in Pennsylvania and Virginia, respectively. By the end of the Nineteenth century, zinc and other metals were increasingly required to meet quality and purity standards for industrial and military needs; this, in turn, led to implementation of standards or specifications for commercial zinc metal, dust, and oxide.

In 1911 the American Society for Testing Materials (ASTM) adopted standard specifications for four primary or virgin spelters—high grade, intermediate, brass special, and prime western. In 1917 ASTM added another grade, selected. During WWI, when so much zinc was bought on Government specification, it became the custom to use the Government's letter-grading system, for example, "grade A", instead of "high grade". Grade B was the equivalent to intermediate; grade C, to brass special and later also to select; and grade D, to prime western. After the war the market tended to revert back to the more familiar ASTM grading terms; however the letter system continued to be used and was implemented again during World War II for Government purchases. After the war, as in WWI, the industry again reverted back to ASTM terminology.

ASTM has approved as many as six commercial grades of zinc metal at one time. In 1977 Brass Special and Select were dropped and currently ASTM lists only three; Special High Grade (SHG), High Grade (HG), and Prime Western (PW). (The above grades are considered to be proper names and when written are capitalized).

Over time the specifications in a few cases have slightly changed; the most recent impurity specifications of the various U.S. metal grades are shown in Table 5. In recent times, two other grades of nonspecified slab zinc, Continuous Galvanizing Grade (CGG) and Controlled Lead Grade (CLG), have gained acceptance for galvanizing purposes. These grades are prepared to customer specifications. CGG contains up to 0.35% lead and some aluminum, whereas CLG contains less than 0.18% lead and no aluminum.

Table 5: Commercial Zinc Metal Grades in the United States

	Zinc Min. %	Lead Max.%	Iron Max.%	Cadmium Max.%	Other
Special High Grade 1/	99.990	0.003	0.003	0.003	Tin not to exceed 0.001%.
High Grade	99.900	0.030	0.020	0.020	
Intermediate 2/	99.500	0.200	0.030	0.400	
Brass special 2/	99.000	0.600	0.030	0.500	Aluminum not to exceed 0.0005% for rolled zinc or brass.
Selected (Select) 3/	98.750	0.800	0.040	0.750	
Prime Western	98.000	1.400	0.050	0.200	

Source: ASTM
1/ First specified for die casting in 1930.
2/ Dropped as a specified grade in 1977.
3/ Dropped as a specified grade in 1968.

The zinc used in hot-dip galvanizing historically has been PW, HG, or special grades, all of which contain more than 97% zinc. In the 1980's, two "galvanizing" alloys having superior coating qualities, gained widespread acceptance. They are known by their trademark names as Galfan and Galvalume. Galfan is a 95% zinc and 5% aluminum-mischmetal alloy and is used mainly as a coating for sheet, wire, and tube. Galvalume is a 45% zinc and 55% aluminum (with some silicon) alloy used mainly for sheet applications. The typical coating weight of zinc in traditional galvanizing of sheet is anywhere from 0.1 to 2.0 or more ounces per square foot. Designations such as G-60 and G-90 indicate coating weight; G-60, for example, indicates a coating weight of not less than a total of 0.6 ounce of zinc per square foot of sheet; the above weight refers to the amount of zinc on both sides of the sheet, not one side. If the coating is an alloy, it is designated by an "A", A-60, for example.

The zinc used in electrogalvanizing is normally SHG or HG metal, although in some processes, high-purity zinc oxide and zinc sulfate are used. Zinc calcine (roasted zinc sulfide concentrate) also has been used as the source of zinc in electrolytic wire coating processes. In recent years a number of electrogalvanized coatings of various compositions, mainly zinc with iron, manganese, and/or nickel, have greatly increased their share of the automotive sheet market at the expense of the traditional products and coatings used. The main advantage of electrogalvanized coatings in products such as exposed automobile skins is that it provides a superior corrosion resistant surface that can be readily painted in the quality necessary. Prior to this, exterior automobile skins were much more vulnerable to corrosion, being only protected by primers, paints, epoxy coatings, and in some cases by expensive coating processes such as Zincrometal. The coating weights on electrogalvanized steel sheet typically range from 20 to 120 grams per square meter. A designation, such as 70/50, shows that the sheet is coated on both sides but the two sides have different coating weights.

Zinc-base alloys are widely used as die casting and gravity casting metals, but only since the early 1930's were superior alloys developed for widespread commercial use. Prior to this time the use of zinc in casting was limited to simple items —souvenirs, statuettes, hat molds, printing type—owing to mechanical weakness caused by inter-granular corrosion especially in aqueous environments, and in the case of gravity-type castings, by the growth of large crystals due to slow cooling. Because of the poor performance of the early zinc casting alloys, zinc gained an adverse reputation with metallurgists, such that the War Department during WW11, specifically banned the use of zinc die-castings in certain types of military equipment. Despite this, two discoveries made in the 1920's led to the development of suitable zinc-base, die-cast alloys. One was the discovery of the beneficial effect of small additions of magnesium in reducing inter-granular growth and corrosion in high-strength zinc alloy, and the other was the discovery that tin, lead, and cadmium impurities were responsible for most of the disadvantages suffered by zinc alloys. These two discoveries plus the recognition that high purity zinc was required to make strong, durable die castings gradually resulted in the expansion of zinc-base alloy uses. Zinc die-casting alloys, commonly known commercially as Zamak alloys or die-cast alloys No. 3, 5, and 7, are any of several zinc-base alloys in which aluminum and copper are the principal alloying components. Zinc-base alloys for sand and permanent-mold foundry use were used by the Germans during WWII as a substitute for bronze, but only became commercially acceptable and available in the domestic market in 1956. However, so far, their use has been limited. They are generically identified by their approximate aluminum contents, 8%, 12%, and 27%, and commercially designated as ZA-8, ZA-12 and ZA-27; ZA, a trademark, stands for zinc-aluminum. A new series of zinc-base, die-cast alloys with high copper and aluminum contents, developed in 1993 for structural applications at elevated temperatures, are designated by trademark as ACuZinc5, 10, etc.

Zinc dust is made by zinc distillation of slab zinc or secondary metal, dross, etc. The zinc vapor enters a cooling chamber with a non-oxidizing atmosphere and condenses out as fine particles, the size of which can be controlled to some extent by the rate of distillation and condensation. The principle grades of zinc dust have metallic contents of 95% to 97%, with the remainder in oxide form. Materials of lower metallic content are generally not used as pigment but are suitable for other uses. Dust with an average particle size of 7 to 9 microns is classed as standard grade, while an average particle size between 4.5 to 7 micron is called superfine and dust of finer size is ultrafine. The term "blue powder" generally refers to the zinc distillation product from pyrometallurgical zinc processes and owes its name to its bluish-gray appearance caused by surface oxidation on the particles.. Blue powder was the earliest used form of zinc dust, but, for the most part, it has been returned to the furnace for redistillation and recovery as slab zinc.

Zinc powder is produced by atomizing a stream of molten zinc metal by a high-pressure blast of air or other gas, followed by solidification in a cooling chamber. The resultant particles are considerable coarser than distilled zinc dust, and are sized by sieving before use. Zinc powder is also made by atomizing the stream of molten zinc by two converging high pressure streams of water. Particles produced in this fashion are more jagged and have less surface oxidation and, thereby, have greater surface area for cementation or chemical reactions. Leafed or flaked zinc powder can be made by milling the spherical particles.

Zinc vapor burns in air to form zinc oxide, the purity, composition, particle size and shape of which can be controlled by the production method, composition of the ore or metal, and combustion conditions. Zinc oxide is produced commercially by three processes, from metal by the indirect or French process, from ore and oxidic zinc secondary materials by the direct or American process, and from chemical solutions or dissolved drosses and fume by wet or chemical processes. Depending on use, zinc oxide may or may not receive some form of treatment, such as milling, screening, washing, blending, densifying, heating, etc.

The oldest method is the French process, in which the oxide is made by burning zinc vapor produced by melting and vaporizing metal in a retort. This oxide, of high purity, is made directly from high-purity metal. The highest purity French-process oxide, sometimes known as "pharmaceutical oxide", is used for salves, lotions and cosmetics. The other French-process oxides sometimes are marketed by "Seal" grades, which indicate the degree of purity, brightness, density, etc. For pigments, White Seal is the brightest and the least dense; Green Seal is less bright and Red Seal is the least bright. These oxides and other standard grades of French-process zinc oxides are used in most applications.

American-process zinc oxide results from the reduction of zinc ore or concentrate and/or other oxidic zinciferous materials by a carbonaceous fuel to produce a zinc vapor followed by its oxidation. American-process oxide is less pure than French-process oxide but has a wide variety of useful chemical and physical properties owing to the fact that its crystal structure is affected by the higher impurity levels and its surface by absorption of furnace gases. American-process zinc oxide is sometimes heat treated to improve its properties, including its purity, surface condition, crystal size and density. Because of a wide variation in crystal size and shape and chemical characteristics of these oxides, there have been few industry-wide specifications other than minimum zinc oxide content and certain maximum impurity levels. As a result each producer of American-process oxide has tended to find his own market niche. In recent years the production of American-process zinc oxide has fallen drastically, such that very little is now produced in the United States or in the world, but, because this oxide has certain desirable properties, some producers treat and/or adulterate French-process oxide to obtain those properties.

Wet-process zinc oxides are the least produced of the three types. They sometimes, are known as "secondary zinc oxide" because they are produced secondarily, as from solutions obtained as a result of chemical processes or by dissolving galvanizers dross or brass fume. The zinc solution is purified, followed by precipitation as a normal or basic carbonate. The precipitate is calcined to obtain zinc oxide and ground to fine size. The product can be of high purity but generally is coarser grained than other oxides; these oxides have been mainly used in rubber compounding and in production of other zinc compounds.

Lithopone, a manufactured white pigment composed of co-precipitated barium sulfate (70%) and zinc sulfide (30%), is produced by mixing purified solutions of barium sulfide and zinc sulfate. The precipitate is filtered, dried, calcined, and milled to a fine state. The use of lithopone in the United States began at the start of the 20th century, rose to about 200,000 tons by 1929, stabilized at about 125,000 tons per year through the late 1940's, and declined rapidly to only a few thousand tons by the 1980's. Lithopone found wide use in paints and as a pigment in linoleum, floor coverings, coated fabrics and textiles, paper and rubber.

GEOLOGY AND RESOURCES

Zinc is an ubiquitous element occurring in all rocks in the earth's crust and is a trace constituent in the oceans and the atmosphere. Zinc is the twenty-third most abundant element in the earth's crust and is estimated to have an average crustal abundance of about 80 parts per million or 0.008%. Based on this average, the amount of zinc in a cubic mile of crustal rock would total a surprising one million tons. Zinc is much less abundant than iron, aluminum, and manganese but more abundant than either copper or lead.

Many minerals contain zinc as a major component; however, only a few are economically important. Sphalerite, zinc sulfide, is by far, the most important zinc mineral and has been the source for most of the world's production. Sphalerite oxidizes readily above the water table; the solubilized zinc reacts, for the most part, with carbonate and silica forming the principal components of calamine: smithsonite, zinc carbonate; hemimorphite, a hydrous zinc silicate; and less commonly, hydrozincite, the basic zinc carbonate. These three minerals, and others that form under special conditions, tend to form near-surface deposits over or in zinc sulfide ore bodies and in earlier years, were the first zinc ores to be mined. They were the major ores of zinc in ancient times for making brass and for the production of most zinc metal through the mid-1800's. The only other important ore sources of zinc, other than the four listed above are willemite, zinc silicate; franklinite, a magnetic zinc-manganese-iron oxide; and zincite, zinc oxide. The latter three minerals occur elsewhere but have been found in abundance at only one place in the world, the famous Franklin and Sterling Hill deposits in New Jersey where zinc ores were mined for more the 120 years before ore depletion in 1986.

Sphalerite is commonly associated with cadmium, copper, lead, and silver minerals and to a lesser degree, with tin minerals and gold. Sphalerite nearly always contains some iron and cadmium and commonly gallium, germanium, indium, mercury, and thallium, all of which can substitute for zinc atoms in the mineral's crystal structure. Sphalerite has been, in fact, the source for most of the world's cadmium, indium, and germanium production. Calamine deposits generally contain associated cadmium but, not infrequently, the zinc (and cadmium) calamine minerals tend to form separate bodies from any oxidized lead deposits, despite close association of galena and sphalerite in the primary ore.

Zinc minerals and associated metallic minerals are widely scattered over the earth's land masses but only in rather restricted localities and only in favorable geologic environs are they concentrated in sufficient quantity to be classed as ore deposits. Exploitable primary zinc deposits can be divided into any number of deposit types based on the criterion used; many deposits that may have originally been one type or another were later altered by igneous intrusion, hydrothermal action, or metamorphism complicating categorization. Further complicating simple classification is the fact that some deposits and mining districts have representatives of two or more general deposit types. Most fall into seven general categories based on interpretation of their geologic history, morphology, or observed occurrence. These are: (1) volcanic-hosted submarine exhalative massive sulfide deposits; (2) sediment-hosted submarine exhalative deposits; (3) stratibound carbonate-hosted deposits; (4) stratibound sandstone-hosted deposits; (5) vein deposits; (6) metamorphic deposits; and (7) other deposits, including diatremes. Oxidized zinc deposits, excluding the unique hypogene ores in New Jersey, can form in or near any of the primary deposits listed above by supergene enrichment under the right conditions.

Deposit types (1) and (2) are

thought to be syngenetic, being initially related to distribution of zinc sulfide mineralization into concurrently accumulating volcanic and sedimentary material. The mineralization is injected into sea water from submarine volcanic sources and from very hot hydrothermal systems that enter deep ocean water at plate boundaries. The recent recognition of these types of deposits has led to reinterpretation of some older, well-known deposits. The Red Dog deposit in Alaska is thought to be of this type, as it is a stratiform, stratibound deposit of volcanogenic origin occurring in relatively unaltered, Upper Paleozoic marine black chert and shale units and Lower Permian argillite.

Deposit types (3) and, less common, type (4) are epigenetic in origin, being deposited generally long after the host rock was formed. Stratibound zinc and zinc-lead deposits in carbonate rocks are common and have been the source of much of the world's output of zinc. In the United States and much of the world, carbonate-hosted deposits are referred to as Mississippi Valley Type (MVT), owing to the fact that large deposits of this type are common and widely distributed in the Mississippi River drainage basin. These deposits typically occur in areas measured in many square miles in shallow marine carbonate rocks characterized by faulting, well developed jointing, karst features, collapse breccias, and/or stromatilite or bioherm structures. Ore minerals occur as replacements and as fillings. The ore is relatively simple, mineralogically, and tends, for the most part, to be low grade. There are strong indications that these types of deposits may have formed at favorable sites through the action of huge, low-temperature hydrothermal systems resulting from uplift and mountain building. Typical of deposits of this type are those in the Tri-State, Tennessee, Virginia and Washington.

Veins, type (5), as used here, is a general category based on the morphological character of the deposit, rather than genetic origin. Veins have comprised a relatively small portion of the world's zinc production, but in the United States, veins have been major contributors to its overall zinc output. The principal zinc-producing veins in the United States have been those occurring in or adjacent to igneous rocks, as at Butte, MT, and those associated with regional metamorphism, as in northern Idaho. Vein deposits fill fissures and replace wall rock and tend to be tabular structures, narrow in width compared to depth and length. Vein ores are generally high-grade but contain complex suites of minerals. Other prominent zinc-producing veins in the United States include those occurring in Colorado and Utah.

Metamorphic zinc deposits, type (6), are those that occur in metamorphic rocks, mainly those that are or were formerly carbonate rocks, and that may have obscure pre-metamorphic histories. Host rocks are commonly referred to as marbles, skarns, tactite, schists, quartzite, etc., which, for the most part, result from contact metamorphism or pyrometasomatism and regional metamorphism. Zinc ore may occur as tabular bodies that are parallel to the "bedding" of the host rock, as veins and replacement bodies, or as irregular or pipe-like bodies that are roughly conformable to nearby igneous rocks. Deposits of this type are found in New Mexico, New York, Colorado, Utah, California, and New Jersey. Zinc production from these deposits, other than those in New York, New Jersey, and the Central district in New Mexico, has been small.

Diatremes, type (7), are the throats and vents of volcanoes that were mineralized in the dying phases of volcanic activity. They are mined mainly for gold and silver and typically contain only minor disseminated metal-sulfide mineralization. They have been very minor zinc production sources because most deposits of this type have been mined only for their precious metals. One mine of this type was opened in Montana in the 1980's and has been a significant producer of zinc despite average zinc grades of less than 0.6%.

MINE DISCOVERY

Most of the important early zinc-mining areas in the United States were discovered as outcrops and, with the exception of the zinc deposits in New Jersey, Pennsylvania, and Tennessee, were developed first for their lead, silver, copper, and gold resources. Only after a number of years were the zinc resources in many of these districts deemed to be economic and exploited. With few exceptions, the zinc-bearing mining districts discovered in the 19th century have, by far, been the major sources of zinc in the United States until only a few years ago (see Table 3, page 105).

In recent years, surface outcrops of base-metal sulfides in the form of gossans have aided in the finding of several major zinc deposits in Alaska. However, modern prospecting and exploration for zinc and other base metals generally involves identification of geologically favorable areas to investigate, based on general geology, ore formation theory, previous mineral history, and broad surveys by space or aerial mapping; identification of specific target areas in the region for further exploration as indicated by surface mapping and by geophysical and geochemical means; and follow-up in targeted areas by detailed geologic mapping, sampling and drilling.

Airborne magnetic surveys were instrumental in finding zinc in Missouri and Wisconsin, and geochemistry was widely used to pinpoint favorable sites for zinc deposits in Tennessee, Virginia and elsewhere on the basis of anomalous quantities of zinc in soils and vegetation. In the mid 1960's, exploration for zinc deposits having no surface indications, began in central Tennessee based largely on an intuitive concept that the paleophysiographic setting of the region's Lower Ordovician carbonate rocks would be ideal for the localization of MVT ore deposits. The exploration was carried out despite the fact that promising ore zones were totally obscure and were expected, if present, to be a 1,000 or so feet below the surface. A random-walk exploration program was instituted consisting largely of wide-spaced drilling. On the 79th hole at a depth of almost 1,400 feet, and 38 miles from first hole, the Elmwood orebody was struck, thereby opening up a hitherto unknown zinc district.

After initial discovery, exploration continues with close-spaced drilling, shaft sinking, drifting, detailed sampling, and ore testing to prove out a workable ore body of sufficient extent and value to justify expenditures for development. Development is the driving of openings to and into a proved ore body for the purpose of economic extract-ion of that body. In some cases deposit exploration and mine development are carried out simultaneously. Many variations of the above steps are possible depending on factors such as the type and character of deposit, ore mineralization, geological setting, environmental considerations, and so forth.

FLOW OF ZINC IN THE UNITED STATES

The principal relationships of zinc production, consumption, recycling, and disposal in the U.S. economy are shown in Figure 5. These relationships have essentially remained unchanged since the early years of the U.S. zinc industry. However, the relative importance and technology of the various sectors may have changed significantly. American-process zinc oxide from ore, for example, is no longer manufactured in the United States.

MINING

Mining involves the opening up of orebodies, the fragmenting and removal of the ore, and maintenance and support of haulageways and areas being mined, as well as those facilities, etc. to insure mine safety. The principal methods in zinc mining are "room and pillar", "shrinkage stoping", "cut and fill", and "open pitting". "Square-set" stoping was commonly employed in western U.S. zinc mines and, in many other mines, where necessary for support or where pillar removal was desired. Many of the zinc mines using square-set stoping as a primary system were closed down by the early 1950's. Any of several different mining methods may be employed depending on the physical characteristic of the ore and enclosing rocks, and the size, shape, and structural nature of the deposit.

Room and pillar mining involves the extraction of ore from a flat or gently dipping ore body by excavation of rooms supported by pillars of waste rock and/or ore. Zinc and zinc-lead stratiform bodies are the types of deposits most often mined by this method. Examples are to be found in the zinc mines of Tennessee and New York and the lead-zinc mines of Missouri.

Figure 5: Zinc Flow Diagram

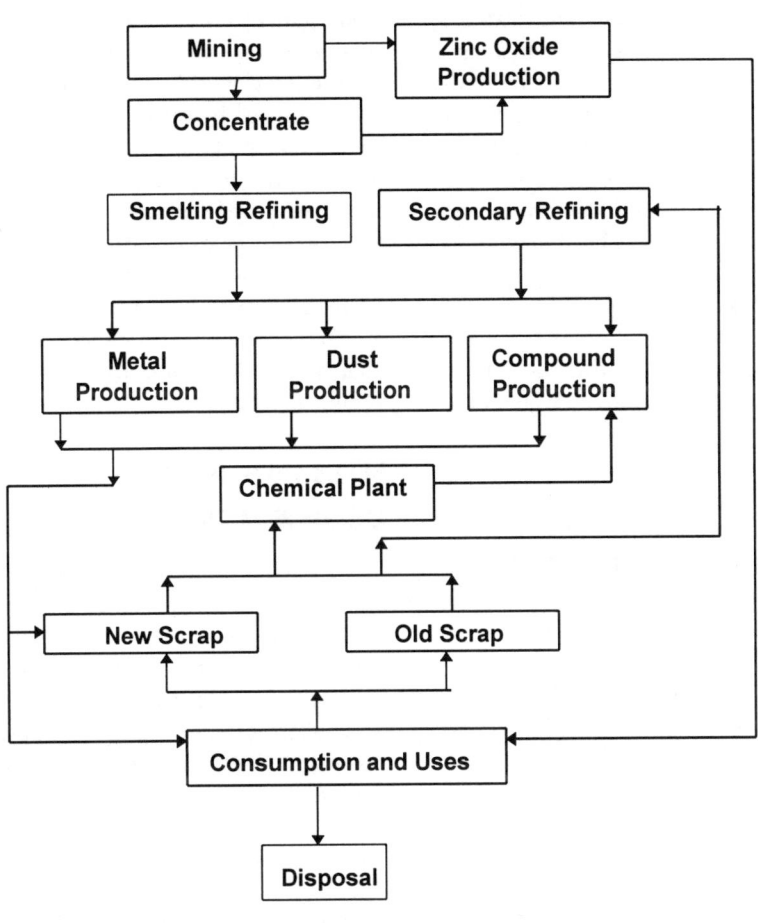

In shrinkage stoping, ore is mined in successive flat or inclined cuts or slices progressing upward from a haulage level. After each slice, enough broken ore is drawn off through chutes or drawpoints to provide space and a work surface for the miner to make the next cut. Shrinkage stoping is generally employed in mining vein deposits but is sometimes used in conjunction with room and pillar mining of thick ore sections in Tennessee.

Cut and fill mining is generally employed in mining vein and steeply dipping deposits. It involves the complete extraction of the broken ore from the stope after each cut and is followed by the filling of the stope with tailings, rock, and/or sand to within working distance to the back before the next cut is taken.

Mines in Idaho and New Jersey, all now closed, used this method. The ore at the Sterling mine was so high in zinc grade that the sand fill was cemented to form a hard surface so that very little of the broken ore would be lost in loose fill. A variation of this method is the underhand cut and fill approach. The ore is mined by progressive downward cuts of the ore which after breaking is largely removed and the opening is then filled completely with cemented sand. The next cut then is taken beneath the above sand-filled workings. This method is employed where dangerous conditions such as rock bursts, are a problem.

Open pit mining, the working of an ore body by surface methods, was employed in the early years of the domestic zinc industry, mainly in extraction of near-surface calamine ore and in mining of surface ores at Franklin, NJ. Since those early years, very little surface zinc mining had taken place in the United States until recently. In the late 1980's two zinc-producing, open-pit mines, one in Alaska and the other in Montana, came on-stream. The Montana mine is a moderate-sized producer of zinc, but the Alaska mine is both large and rich in zinc, such that by itself, it has resulted in a doubling of U.S. zinc production and, to some extent, has restored domestic zinc mine output to the relatively lofty levels attained in the mid 1960's.

The cycle of operations in mining consists of drilling, blasting, and removal of the ore. Drilling is generally done with compressed air percussion drills that, often, are mounted on self-propelled, rubber-tired jumbos. Following blasting, broken ore is loaded by diesel or electric, mechanical shovels or loaders and moved by trucks and/or load-haul-dump units to bins near the shaft for hoisting. Track haulage is employed largely to move ore, waste rock, and supplies on main haulageways only. However, track systems, for the most part, have been replaced in zinc operations by rubber-tired, high-speed, high-capacity, maneuverable machines.

MILLING AND CONCENTRATING

Zinc ores, with few exceptions, are treated to some form of beneficiation to concentrate the zinc minerals and/or to increase the zinc content before the material is sent to a smelter or zinc plant. Simple washing to remove clay and sand, jigging, and hand sorting were commonly used to upgrade calamine ore. This was sometimes followed by calcining to further upgrade the zinc content by driving off carbon dioxide and water. Coarse sulfide ore chunks were often passed along a picking belt where high-grade pieces were hand-selected and stockpiled for direct shipment; this practice continued to be used at some operations up to WWII.

Most beneficiation processes, however, involve crushing and grinding to liberate the individual mineral constituents from each other. Because it is desirable to prevent overgrinding or sliming owing to costs and poorer recover, grinding and ore classification (sizing) often is carried out in several stages rather than in just one stage. Bull jigs were used prior to WWII at some operations to sharply reduce the amount of material containing little or no ore minerals from entering the grinding and flotation circuits. In Tennessee, some zinc mills have installed heavy-media separation devices, which eliminates as much as half of the ore fed to the mill from further processing with only insignificant loss of zinc.

Before flotation was available, the crushed ore was washed, sized and subjected to some form of gravity separation to separate the heavy ore minerals from the lighter gangue minerals and with further gravity processing, to separate the sphalerite from the galena or other ore minerals. This process worked well for ores containing easily liberated minerals of relatively large grain size and widely disparate specific gravities, but less so with fine-grained ores and complex ores because fine grinding was required, particle liberation incomplete, and presence of minerals having specific gravities close to those of sphalerite. The former situation was typical of the Tri-State ore and led to its early dominance as the principal zinc district in the United States. The latter situation applied to many of the western mines and some in Wisconsin, and delayed zinc production until satisfactory mineral separations could be made. The crushing and grinding of ore in itself, caused problems with ore recovery even in the Tri-State, because grinding created many small particles or slimes of the ore minerals that could not be recovered by tables, buddles, and jigs. In the 1870 through the 1890's, sphalerite recovery from ores mined in the Tri-State district was only about 60%, owing in large part to unrecovered sphalerite slimes. The Wifley table improved zinc recoveries in the late 1890's, but high recovery of sphalerite in fines and slimes was possible only after the development of flotation.

The unique oxidized (primary) ores in New Jersey were initially crushed and jigged, but the three major ore minerals could not be readily separated from each other resulting in a low-grade zinc concentrate, suitable mainly for oxide production. With the invention of the Wetherill magnetic separator in 1896, good concentrates of both franklinite and willemite-zincite could be made; the former was used for oxide and spiegel production and the latter was either exported to Europe or shipped to the Bertha smelter in Virginia for high-purity slab zinc production. These ores were not readily amenable to recovery by flotation and this method was never used, but it was of little consequence because zinc recoveries using gravity and magnetic methods were excellent. After the Franklin Mine closed in 1954, the Sterling ore was not beneficiated but was merely crushed and shipped directly to

Palmerton, PA. for American-process zinc oxide production.

Zinc production in Wisconsin until the early 1900's was hampered by high levels of marcasite in the zinc concentrate. Wetherill-type magnetic separators were tried, but, for the most part, were unsuccessful. High-grade zinc concentrates were made after it was found that the marcasite could readily be extracted if made more magnetically susceptible by subjecting the concentrate to a slight roast. The early zinc concentrates at Leadville, CO. were similarly plagued, only with pyrite not marcasite; this pyrite problem was solved by the same method as that used in Wisconsin.

Flotation is based on the fact that certain reagents can modify the surface conditions of a mineral particle to be non-wettable or wettable; the former will attach itself to a rising air bubble, whereas the latter will not and will sink. The rising particles accumulate in a surface froth and can be collected. Reagents that increase wettability are known as depressants, and those that decrease wettability are known as collectors. Flotation is carried out in cells of various size and construction. As time progressed, cells became larger with increased capacity; and recently column cells, large cylinders on the order of 25 to 40 feet in height have become important in zinc flotation. They have a large throughput, improved recovery, and proportionally, are less costly to operate than standard flotation cells.

A typical scheme for the flotation of a mixed sulfide ore is as follows: (1) flotation of lead and copper sulfides and depression of the zinc and iron sulfides; (2) separate the lead and copper into two concentrates by flotation; (3) activate and float the sphalerite from the pyrite and gangue; and (4) float the pyrite, if desired, or discard with the gangue. The tailings are sent to thickeners and subsequently discharged into a tailings dam, or in some cases, pumped back into the mine for sand-filling of mine workings. Carbonate tailings are sometimes sold for agricultural uses.

Many zinc and zinc-lead mills are highly automated and require few operating personnel. Most have instrumentation, sensors, and on-stream X-ray analyzers to provide central control of most mill functions ranging from head feed rates and flotation reagent control to tailings disposal.

High-grade concentrates, 62% or better, are produced from straight zinc ores, such as those in Tennessee. Lower concentrate grades are produced from lead-zinc ores and from complex ores. Mixed bulk concentrates containing 20% to 30% each of lead and zinc, are produced but they require special smelting processes such as the Imperial Smelting or Kivcet processes, which are capable of recovering both metals economically. Lead smelter slag is also a source of zinc refinery feed; zinc is extracted from the slag by the slag-fuming process.

SMELTING AND REFINING

Three major smelting technologies have been employed in the United States to produce primary metal. These are known as the horizontal retort, the vertical retort, and the electrolytic zinc processes. From 1860 through 1914, all domestic smelters used horizontal retort processes. The first electrolytic plant became commercial in 1915; then came the externally-fired vertical retort and the electrothermic vertical retort in 1930 and 1936, respectively. Before WWII, the horizontal retort process was recognized as an economic anachronism because of its disadvantages relative to continuous processes and its high labor and fuel costs. Modernization during WWII, national defense requirements and the Korean War prolonged the use of the process in many plants, but soon, thereafter, horizontal retort plants began to close. The process tended to survive principally in the mid-continent gas-belt and elsewhere by reason of available high-grade concentrates, cheap fuel, and proximity to suitable retort clay until the late 1960's, when environmental problems, mine closings and associated losses of concentrate sources, high costs, and foreign competition, resulted in the closing of all U.S. horizontal retort smelters by 1976. For somewhat similar reasons, all of the vertical retorts, but one, were closed by 1981 and all, but three, of the electrolytic plants by 1984. Historically, most of the zinc metal produced in the United States has come from retort-type smelters; not until 1975, did electrolytic metal production exceed distilled output. Since that time most primary zinc has been produced electrolytically.

Reduction of the concentrate to zinc metal is accomplished by electrolytic deposition from a sulfate solution or by distillation in retorts or furnaces. For either method, the concentrate is roasted to eliminate most of the sulfur and produce an impure zinc oxide known as calcine. At a few foreign zinc plants, pressure leaching is employed instead of roasting. The pressure leach process yields zinc-sulfate in solution directly and extractable elemental sulfur, thereby circumventing the roast and acid leach requirements of the typical plant. During the roasting process available iron in the concentrate reacts with zinc to form zinc-iron ferrite. This is of no consequence in pyrometallurgical processing but can result in significant reduction in zinc recovery at electrolytic zinc plants because the ferrite in insoluble in the acid leach employed in most processes and remains in the leach residue. This zinc is lost unless an additional process involving hot strong acid is used to dissolve the ferrite. The Tainton process employed at Bunker Hill to leach calcine directly or several separate, in-plant, processes known by the iron residue product produced—"jarosite", "goethite", and "hematite"—are effective in extracting zinc from ferrite.

The electrolytic process has

changed little in its essential features since it was developed in 1914. The calcine is leached with return sulfuric acid from the electrowinning cells. The resultant zinc sulfate solution is purified and piped to electrolytic cells, where the zinc is electrodeposited on aluminum cathode sheets. The cathodes are lifted from the cells at intervals and stripped of the zinc, which is melted and cast into slabs. The quality of of the zinc is SHG or HG. The electrolysis regenerates the acid, which is reused in succeeding leaching cycles. The lead and silver compounds in the calcine are not dissolved in the leaching process and remain in the leach residue, which if rich enough is shipped to a lead plant for extraction of those metals. The potential ease with which byproducts, including gold and copper, could be recovered by the electrolytic process versus retort processes was a driving force in the development of the electrolytic process in both Montana and British Columbia because their ores were complex and the zinc concentrates produced were rich in byproduct metals.

There are three basic types of distillation retort plants: batch horizontal retorts, continuous vertical retorts externally heated by fuel, and continuous vertical retorts internally heated electrothermally. Fine-sized roaster calcine is suitable for the electrolytic process, but a coarse feed is required for thermal reduction processes, owing mainly to the need to maintain space for good heat transfer and to provide easy pathways for the zinc vapor and other gasses to escape from the charge. Agglomeration of particles for retort feed is accomplished by sintering, pelletizing, and briquetting. The coarse-grained sphalerite concentrates produced by gravity and magnetic means were suitable feed for the early retort process, but when the fine-grained flotation concentrates came on the scene, some form of agglomeration was required. Because the vertical retort processes operate with heavy, stacked charges, high-strength briquettes or sinter and coking coal are required.

Distillation processes are based on the reduction of zinc oxide to elemental zinc by carbon fuels. Horizontal retorting is a batch reduction-distillation process carried out in banks of tube-shaped, ceramic retorts that are externally heated by natural or producer gas, oil, or coal. Retorts range from 4 to 5 feet in length with an opening of 8 to 9 inches in diameter; walls are about 1 inch thick. Furnaces generally contain 500 to 1,000 retorts. The charge for a retort generally consists of 65% to 70% calcine, blue powder, and residues, and the remainder, carbonaceous reductant and, sometimes, some salt. Salt is added to enhance the coalescence of blue powder into liquid metal in the condenser, which is a 2-foot long conical, tapered ceramic tube that is attached to the open end of the retort. It extends through the furnace wall and is cooler than the retort, thus providing a suitable vessel in which the zinc vapor will condensed and from which the metal can be extracted. At elevated temperatures, the zinc oxide in the retort is reduced to elemental zinc, which instantly vaporizes and is forced by rising gasses into the cooler condenser where it condenses as liquid metal. A prolong may or may not be attached to the condenser; it is used to capture escaping blue powder and cadmium, which is expelled in the early phases of the firing cycle. Zinc metal is tapped at intervals throughout the smelting cycle. After the completion of smelting, the retort is opened, cleaned of residue, and recharged.

Vertical retorting is a continuous operation that can be carried out in two different types of furnaces. Fuel-fired vertical retort processes are accomplished in tall vertical retorts that are heated externally in large vertical muffle furnaces. Retorts are long-lived and range from 25 to 35 feet in height and have cross-sections of 5 to 7 feet by 1 foot. The retorts are top charged with briquetted or sintered calcine and coked coal. The charge moves downward from removal of spent charge at the bottom of the retort. As it drops in the retort, the temperature increases such that reduction occurs and the zinc vapor rises through the charge and is condensed in a conventional surface or splash-type condenser.

The other type of vertical retorting is carried out in a large shaft furnace using internal electrical (electrothermic) heating instead of external fuel heating. The shaft furnace can range up to 40 feet in height and have an inside diameter of 5 to 8 feet. The furnace is heated by electrical resistance of the charge, sintered ore and coke, via electrical current supplied by large graphite electrodes that protrude into furnace. The charging, reduction and collection of the zinc is essentially the same as with the other vertical method described above.

There are a number of other processes that are used elsewhere that have not been used commercially in the United States. The most widely used of this group is the Imperial Smelting Process (ISP), which is a blast furnace process developed in the 1950's, and is suitable for processing bulk zinc-lead concentrate. A solid sintered calcine and heated coke are fed into the top of the furnace and pre-heated, hot (about 1,000 degrees C.) air is blown in at the bottom. Reduction occurs, the hot gasses and zinc vapor exit near the top of the furnace and molten slag and lead are tapped at the bottom. The zinc is captured in lead in a splash condenser. Other processes, such as Kivcet, QSL, and Sirosmelt, are somewhat different from each other but each is continuous, involves melted metals and slags, can process bulk concentrates and complex ores, and results in volatilized zinc. Zinc produced by these methods is generally recovered in oxide form and may require additional refining.

RECYCLING

Recycled zinc, which accounts for about 30% of U.S. zinc consumption, is derived from about

70% new scrap and 30% old scrap. New scrap consists mostly of drosses, skims, furnace dusts, and residues from galvanizing and die-casting operations, brass mills, and chemical plants, and clippings from the processing, stamping, and trimming of galvanized and brass sheet and rolled zinc. Old scrap is derived from end-use items and products that are obsolete, broken, used up, etc. Old scrap, historically, has consisted almost entirely of die-castings from scrapped automobiles; brass products; and rolled zinc from roofing, gutters and engraving plates. In large part, the types of old scrap that were processed were limited due to ease of collection and to the narrow but desirable range in composition of the collected scrap. The types of old scrap recycled remained as above until the 1980's, when environmental mandates resulted in large amount of zinc being recovered from electric-arc-furnace dusts generated in recycling mainly old and new galvanized steel scrap at steelmaking mini-mills. Since then, small additions to old scrap recycle were made by recovery of zinc from burning rubber tires for energy; despite the need to dispose of billions of old tires in the United States, burning, though environmentally safe, is unpopular and likely to remain so in the near future, thereby, limiting recovery of zinc from this source.

In the early years of the zinc industry, very little old scrap was recycled because collection was difficult and processing was not cost effective; even in the 1990's, only about 10% of the total zinc consumption of the United States is derived from old scrap. This is due to the many varied uses of zinc, many of which are dissipative or could not be collected easily or recycled economically because the zinc is merely a coating or a small part of the item being scrapped. Rapidly dissipated zinc products include fertilizers, animal and human trace element supplements, oils, inks, fluxes, fungicides, pharmaceuticals, etc. Other products are dissipative but over a longer period; this group includes losses to the immediate environment by corrosion of galvanized surfaces, rubber tire abrasion, flaking paint, and sacrificial anode consumption. Most other zinc items, with the possible exception of large steel products and the traditional recycled zinc materials, are land filled or merely tossed away.

Because of wide differences in the character and content of zinc-bearing scrap, zinc recycling processes vary widely. Metals can be separated from other materials by magnetics, sink-float, and hand sorting. In mixed nonferrous shredder scrap, zinc can be selectively melted from higher melting point metals like copper and aluminum in a sweat furnace. Zinc in galvanized steel is recovered in the furnace dusts when the steel is melted in a steelmaking process. The zinc recovered in the above processes may require additional processing before a suitable zinc product is made. Brass scrap is largely just remelted, alloyed to specs, and recast. New scrap is often remelted or redistilled to make metal, dust, or oxide, or chemically treated to manufacture zinc chemicals, such as sulfates and chlorides. The zinc products produced from scrap (see Figure 6) are generally of the same quality as those produced from primary materials; however, up through WWII, it was not unusual for the Government or industry to specify the use of virgin zinc only in contracts. The ASTM specifications for primary slab zinc do, however, require that the zinc be made from ore or other material by distillation or electrolysis, and not by sweating or remelting of scrap.

CONSUMPTION AND USES

Zinc is found in all sectors of the economy, but its role or presence is often not obvious because zinc tends to lose its identity in many end products, such as alloys, metal coatings, rubber, paint, batteries, coinage, etc. Even zinc die-castings are often described by the public as being "pot metal" rather than zinc alloy. Zinc coatings on steel are described as being galvanized, which is proper usage, but the public tends to associate the product with steel, not zinc.

Zinc is used extensively by the military, industry and the general public for construction, transportation, electrical, machinery, and chemical purposes. Zinc-coated (galvanized) steel sheet, structural shapes, fencing, storage tanks, fasteners (nails, screws, clips, etc.), wire and wire rope are used in the construction of houses, industrial plants, culverts, roads, bridges, transmission towers, barns and silos.

> "Zinc is a metal possessing many valuable properties. Granting that old uses pass away there are new ones that may be developed. The zinc industry of the world has been largely blind in looking in that direction... It is not so much a matter for scientific research, whereof much has been done, as it is of sales promotion."
>
> *W.R. Ingalls (1931)*

Zinc sacrificial anodes are used for corrosion protection of ship hulls, offshore drilling rigs, water heaters, and buried steel tanks and pipes. Brass has a myriad of uses, including ammunition, tubes, valves, fittings, electrical connections, heat exchangers, scientific instruments, bearings, and architectural products. Zinc die-castings are common in everyday society in the form of handles, grills, bezels, brackets, locks, hinges, and housings and complex parts for appliances, vehicles, business machines, scientific instruments, sporting equipment, etc. Rolled zinc is or has been used for engraving plates, battery cans, glass-jar tops, coinage, decorative trim, flashing, gutters, roofing, boiler plate, wash boards and linings in ice boxes and refrigerators. Zinc dusts and powders are used in primers and paints; in alkaline dry cell batteries; in tear-gas formulations; in the making of artificial adrenaline; in the sheradizing process; in match-head compositions and fireworks; for precipitating gold from solution; and by electrolytic zinc smelters to purify zinc solutions before electrolysis. Other uses of zinc metal include the production of zinc chemicals, primarily zinc oxide, and use as an alloying component in light metal (aluminum and magnesium) alloys and bronzes.

A breakdown of the basic-use areas of zinc metal (galvanizing, die-casting, brass and bronze, rolled zinc, and other uses) for the 1900-1994 period are shown in Figure 7. Galvanizing has been the largest consumer of zinc metal, and overall, has accounted for about one-half of the domestic consumption. The most significant change in metal consumption was the rapid rise in the use of zinc-base alloys following their development in the late-1920's. Spurred on by use in the automobile industry, zinc die casting uses increased many-fold reaching their highest levels in the 1960's and early 1970's. The energy crisis in the early 1970's led to extensive down-sizing and weight reduction programs in the automotive industry, and a substantial reduction in the use of zinc-base alloys. Other significant changes in zinc usage over time were the relative increases in brass production during the two World Wars, the rise in rolled zinc consumption, beginning in 1982, when the zinc penny was first produced, and the huge increase in electrogalvanizing in the late 1980's, mainly to meet the requirements of the automobile industry for corrosion resistant steel body parts.

About one-fourth of the zinc consumed in the United States is in the form of zinc chemicals (compounds). They are spread throughout the industrial and consumer sectors performing numerous functions, either as a component of a product or as a part a process but not present in the final product. An example of the former is zinc chloride where it is used as a wood preservative or as a battery electrolyte; the latter would be represented by zinc sulfate used in rayon production to promote coagulation and crenelation of viscose rayon filaments when they are injected into a sulfuric acid-zinc sulfate bath from a spinneret. Some chemicals, such as zinc oxide in rubber production, for example, may perform both functions: it's a necessary chemical to accelerate the rubber-making process and a useful component in rubber itself.

Zinc oxide is the largest-used zinc chemical, finding a myriad of uses, including skin lotions, face powders, golf ball covers, sun screen creams, rubber compounding, photocopying, paint pigments, adhesives, white surgical tape, ceramics and glass, phosphors, plastic stabilizers, animal and plant nutrients, lubricants, and ferrites for radio, television, recording and telecommunication equipment. Zinc oxide is also a principal starter material for the production of other zinc chemicals.

> "In those days it was generally believed that the cost of prepared paints depended largely on the price of linseed oil and zinc oxide. Conventions were therefore held about the middle of, November, after the size of the flax seed crop for the year was known. At these meetings also the New Jersey Zinc Co. announced its prices for the coming year...If the price was to be higher, the company was objurgated. If it was to fall, the company was praised as the big brother of the industry. There prevailed for many years a feeling that it was the duty of the company to hold the price of zinc, by a comfortable margin, under the price of dry lead, and, as a matter of fact, that was the usual practice."
>
> *Paint Industry Reminisces*
> *G.B. Heckel (1931)*

Zinc sulfate, the most widely-used zinc salt, is used in water treatment, mineral flotation, electrogalvanizing, fire proofing, medicine, plant and animal nutrition, and as a starter chemical for the production of zinc stearate, zinc bacitracin, and many other chemicals.

Zinc chloride is the second-most important zinc salt. It also has varied uses, including dental preparations; textiles where it can act as a mordant and resist in dyeing and printing fabrics, a sizing and weighting agent for cotton goods, and as a fireproofing agent in woolen goods; adhesives; welding fluxes; embalming fluids; pharmaceuticals; and wood preservatives.

Many other zinc compounds are used in a myriad of ways. Applications include medicines that range from anti-bacterial agents to insulin; dental cement; personal items such as toothpaste, cosmetics, and mouthwashes; varnishes, inks, and dyes; oil and lubrication greases; pigments and phosphors; food additives and vitamins; catalysts and conditioning agents; and insecticides and fungicides. (Sodium zincates, for example, are used in athlete's foot preparations.)

> **...studies showed a 42% reduction in the duration of the common cold when the zinc (gluconate) lozenge was taken within 48 hours of the first sign of a cold.**
>
> *Metals World (1992)*

Figure 6. U.S. Secondary Zinc Recovery and Form, 1907-1994.

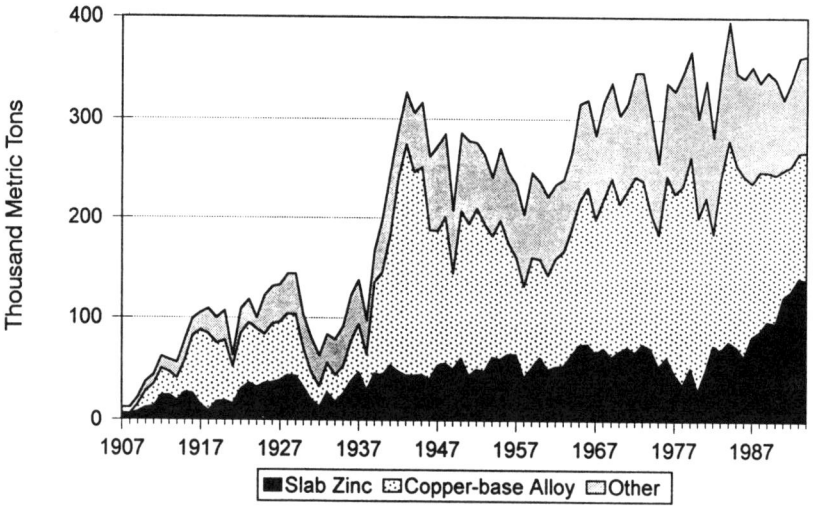

Figure 7. Slab Zinc Consumption, 1900-1994.

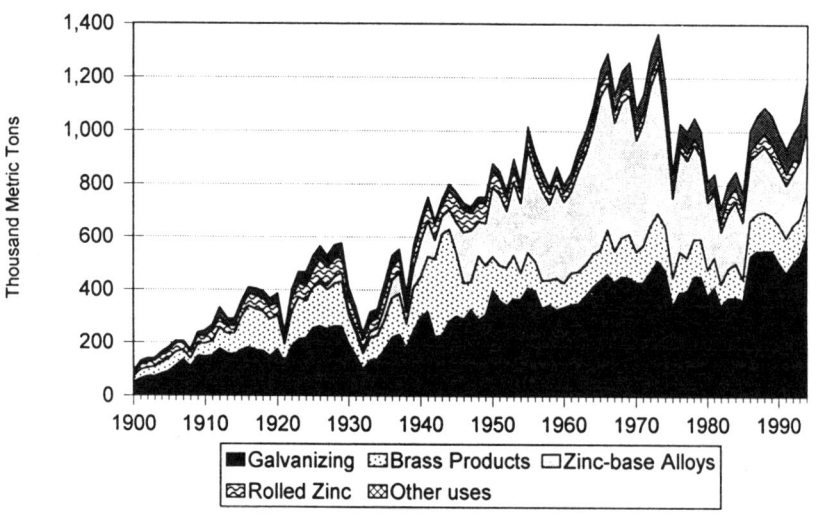

Source: USBM IC 7450, 1900-1945 and Minerals Yearbook, 1946-1994.

ZINC MARKETING

U.S. TARIFF HISTORY

In Eighteenth Century colonial America, the only zinc-containing material of any commercial importance was brass, virtually all of which was imported from England. Zinc ores were not mined in colonial America, but a small brass industry developed, producing commercial brass mainly by remelting scrap but also by the cementation process using imported calamine. As a result of the latter process, the first basic zinc material mentioned in a tariff act of the United States was "lapis calaminaris" (calamine) in 1792, where it was placed on the free list. Calamine continued to be duty free until 1842 when it was unenumerated and, as a result, fell into a catch-all crude materials category having a duty of 20% ad valorem. In 1857 the calamine duty was reduced to 15% ad valorem and, in 1861, was again made duty free, where it was so classified until the tariff act of 1909. Up to this time, calamine had been the only zinc ore mentioned in the tariff schedule; thereafter, however, all zinc ores were classified together in phrases, such as "zinc-bearing ore of all kinds (initially with but later without) including calamine" and "zinc ores and concentrates".

Metallic zinc was first mentioned in the tariff act of 1816, where "zinc, teutengue or spelter" was admitted duty free. Free status continued up to 1846 when spelter was made dutiable at 5% ad valorem. Zinc sheet was also made dutiable in 1846 at 15% ad valorem. In 1861 the spelter duty was raised to 1 cent per pound and sheet to 1.5 cents per pound. Interestingly, this rise coincided with the start of spelter production in the United States. The duty rates for spelter and sheet zinc tended to rise in subsequent tariff acts, and in 1897 the duty rates were 1.5 and 2 cents, respectively.

Zinc dust, until the tariff act of 1909, was free of duty. Although it had a number of uses at that time, it was sometimes imported in the category "articles in a crude state used in dyeing and tanning". It was apparently placed in this category as a convenience because a paste of zinc dust, known as "indigo auxiliary", was widely used to discharge locally the color in dyeing cotton goods, giving a white pattern on the background of dye. Up to 1909, for the most part in U.S. smelting practice, domestically produced dust was charged back into the retorts to make metal because spelter was considerably more valuable owing to protection by a high tariff, whereas the dust was not. The domestic zinc-dust market was largely in the hands of German producers, but it was partially captured by domestic producers in 1909 when the duty on zinc dust was set at 1.375 cents per pound. Despite the tariff, the Germans were highly competitive in the U.S. market until World War I broke out in 1914, Virtually all European imports ceased at that time, and by the next year, there were seven U.S. dust producers compared with only two in mid-1914.

Up to 1900 imports of zinc ores were insignificant and inconsequential to the domestic industry; however beginning in 1902, the industry began to take notice of large imports of blende from British Columbia. By 1905 large imports of Mexican calamine and blende by Kansas smelters had created enough concern among Missouri mining companies that they demanded that duties be placed on imported zinc ores. As a result, one of the more interesting tariff cases related to zinc ores and ores in general arose. In 1906 the Secretary of the Treasury issued the following:

> "You are hereby instructed...to classify ores chiefly valuable for the zinc they contain as metallic mineral substances in a crude state under paragraph 183 of the existing tariff act (1897) at a rate of 20% ad valorem. You will admit calamine, (defined as) silicate of zinc, free under paragraph 514."

In response, the importing smelters brought a test case in the name of The Kansas City Smelting Company. Their position was that both the carbonate and the silicate of zinc should be duty free under the traditional meaning of calamine, and that blende was duty free as a "crude mineral, not advanced". The case was argued before the Board of U.S. General Appraisers, who agreed with the importing smelters, thus reversing Treasury's interpretation. An editorial in the February 6, 1907, *Engineering and Mining Journal* summed up the reversal thusly:

> "The decision as to calamine was a foregone conclusion inasmuch as anything different would have violated the accepted commercial and metallurgical practice of a century, as was done ignorantly in the Treasury order of February 10, 1906. The fact that mineralogists had a more recent and more restricted meaning for this term, or rather some mineralogists... could not rationally be allowed to supersede the preemption of the term by metallurgist, who had never abandoned it. The surprising thing is that the Government should have considered it had any ground for contest on this point."

> "Commercially...the most important part of the recent decision...relates to blende. It is to be admitted that when the Dingley Law (Tariff Act of 1897) was enacted, no one thought that in a few years the United States would be importing zinc blende and when such importations were begun there was ground for the belief that blende fell under the catch-all paragraph, providing that 'metallic mineral substances, in a crude state, not else where specified', should be dutiable at 20% ad valorem as the Secretary

of the Treasury ordered. Now the Board of General Appraisers...holds that the phase 'metallic mineral substances' applies only to that class of mineral substances in which the metal appears in a free state, such as the ores of gold, silver, and copper. Inasmuch as the only ores of the above character which are of commercial importance are those of gold, silver, and copper, and they are specifically on the free list, this clause in the Dingley Act under the recent decision becomes meaningless."

In April 1908, the U.S. Circuit Court upheld the Board's decision, but this earlier controversy became a moot point with passage of the Tariff Act of 1909 with its catch-all term for zinc ores and imposition of tariffs on all ores containing more than 10% zinc. The Tariff Act of 1909 also provided for the establishment of bonded warehouses for the smelting and refining in bond of ores and crude metals but provided for the withdrawal (use in the United States) of lead only. Cancellation of the bond for the zinc duty would occur only upon exportation of 90% of the zinc contained in the bonded ore. In 1913 provision was made for smelting in bond with withdrawal applicable to all metals, not just lead, with payment of duty.

The Tariff Act of 1913 replaced the 1909 content-based, variable duty rates on zinc ores with a single 10% ad valorem rate on zinc in all zinc-bearing ores. This also led to considerable controversy, in that duties would be paid on zinc that could not be recovered. In a series of Treasury rulings, duties were rescinded for zinc in ores where it was not commercially recoverable, but in mixed ores where zinc was recoverable, the duty was appraised at the cost of spelter in the United States, less an allowance for the cost of transportation, duty, profit, smelting charges, and overhead expenses. Ores containing zinc but no other recoverable metals were appraised at the price of the ore.

In 1922 the ad valorem rate on zinc in ores was dropped in favor of content-based, variable duty rates as in 1909. This in turn was replaced in 1930 by a single 1.5 cents per pound rate for all zinc in ores, except pyrite ores containing less than 3% zinc and lead and copper ores in which zinc is not recovered. This relatively high tariff protected the domestic industry from imports throughout most of the 1930's, in that the tariff accounted for 30% or more of the U.S. spelter price during that period. In the 1930 tariff act, the rate of duty on the zinc content of zinc-bearing ores was equal to about 86% of the zinc metal rate (1.5 cents versus 1.75 cents). This differential was intended to approximately compensate for metallurgical losses incurred in converting ore to metal. As time progressed, improvements in smelting resulted in metal recoveries higher than 86%, thereby providing smelters using imported ores with a small but increasing element of protection and profit they did not have earlier.

Based on comparisons of

the U.S. and London zinc prices, the effectiveness of U.S. ore and spelter tariffs in protecting the domestic zinc industry was mixed. From 1901 to 1914 the tariff generally assured higher than world prices for domestic producers of zinc concentrates although the United States was a net exporter of zinc ore in that period. This paradoxical situation arose because only a limited domestic market existed for the large quantities of willemite produced at the Franklin Mine in New Jersey. Because willemite could be treated economically in Europe, this trade was carried out irrespective of the tariff, and as such, was of no consequence to the protected domestic ore or metal market. In the 1914-1928 period, the United States was a relatively large exporter of zinc metal resulting in the tariff being relatively ineffective. Thereafter, with the onset of world depression and slow world growth to the start of WWII, U.S. exports of zinc virtually disappeared owing to excess world production of zinc in that period.

The zinc tariff protected the domestic zinc producers and prevented the dumping of excess world zinc production on the U.S. market, while at the same time keeping the domestic zinc prices well above London prices. However, during 1936 and 1937 the price differential between the U.S. and London markets actually exceeded the import duty and resulted in a substantial jump in zinc metal imports in those years.

"...the (zinc) industry remains remarkably international. The United States is the only major country which is effectively isolated in trade..."

W. Y. Elliott, et.al., 1937

In 1939 the signing of the Canadian Trade Agreement, which reduced the duty on imported Canadian zinc ores and metal by 20%, squeezed the profitability of the depression-weakened domestic zinc producers because the reduction could not be regained by higher prices but had to be absorbed as a cost of business. World War II intervened, however, restoring good zinc prices. By 1941 zinc demand had outstripped the production capacity of domestic mines and large imports of ores and concentrate and metal began, most of which entered duty free during the war for the account of the Government. In late 1942, to insure adequate materials for the war effort, the United States and Mexico concluded the Mexican Trade Agreement, which reduced the duty on imported zinc concentrate and metal from Mexico by 50%. After the war, the rates provided for in the Mexican Trade Agreement were retained for all countries having "most-favored nations" status.

With few adjustments, the duty rates on zinc concentrates and metal since World War II through 1979 changed only slightly in absolute terms, but in terms of protection for the domestic industry, they declined substantially as a percentage of zinc prices. To aid the industry and defense efforts, the Government, rather than increase tariffs, instituted a number of programs: the stockpiling of zinc, DMEA/ OME programs, a small miner's lead/zinc price stabilization program from 1962 to 1969 and imposed import quotas on imported zinc in the 1958-1965 period. To increase production, tariffs were suspended on scrap during the Korean War and again, with "zinc in ores and concentrates", from 1975 through June 1978.

In 1963 a revised method for computing duties on concentrates was introduced. The duty was raised from 0.6 cent per pound to 0.67 cent but the latter allowed deductions for processing losses. The duty arising from either the old or the new basis, however was about the same.

In 1979 the Government approved new tariff rates stemming from the Tokyo round of multilateral trade negotiations. Beginning January 1, 1980 progressively lower rates were phased in over the next 8 years on most zinc materials, except alloys, which was unchanged. In 1982 the duty on imported scrap was again suspended.

In 1989 the U.S. tariff structure changed dramatically; the Trade Schedule of the United States (TSUS) was replaced by the international Harmonized Trade Schedule (HTS), and the United States and Canada implemented the Free Trade Agreement (FTA). The HTS resulted in definition and category changes and in some cases, rate reductions.

The FTA was set up to eliminate or gradually reduce the tariffs between the two countries on virtually all materials and products over 10 years. A similar agreement, known as the North American Free Trade Agreement (NAFTA) between Mexico, Canada, and the United States was approved in 1994; its purpose is the same as that of the FTA. Also in 1994 the Uruguay round of multilateral trade negotiations was approved; its purpose is to reduce trade barriers worldwide. The Uruguay negotiations created a new organization, the World Trade Organization (WHO), to replace the former world trade group, the General Agreement on Tariffs and Trade (GATT), in administering the new international agreement.

CARTELS AND SYNDICATES

Syndicates and cartels to control zinc prices and production were formed a number of times from the mid 1800's up to the late 1970's. The abnormally high prices for spelter in the United States in 1875 and 1876, which were not exceeded until World War I, for example, were due largely to the manipulations of Midwest zinc producers who attempted to control the supply of zinc in the spring of 1875.

Zinc and bismuth have the distinction of being the world's first metals cartels.

W. Y. Elliott, et. al., (1937)

Interestingly, the high prices so stimulated production that by mid-1876 prices began to fall causing disbandment of supply control measures. Because of poor prices in 1879 and 1882, European syndicates were organized to control zinc supply and prices, but these efforts had only temporary effects on the market as zinc prices continued to fall.

In 1885 the first significant international zinc cartel was organized in an attempt to raise zinc prices. All of the Silesian, Belgium, Rhenish, French, and English zinc works joined a spelter combine to restrict zinc output. U.S. zinc producers were not participants in the cartel because the United States was a net importer of zinc and not a factor in the world's zinc trade. At that time the cartel members produced about 85% of the world's spelter output with most of the remainder produced in the United States. Production cuts were instituted and prices improved. In 1889 the actions of the cartel proved successful enough that they agreed to extend production restrictions to the end of 1894. The market continued to be satisfactory; however by the end of 1890, new production, encouraged to a large extent by favorable zinc prices, resulted in price decreases. This result plus economic depression in Europe and the United States in the 1893-1895 period ended the cartel's effectiveness.

In 1909 the European producers again organized to control zinc production, forming a cartel known as "The Zinc Convention", also known as "The Spelter Convention". The purpose of the Convention was to fix beforehand, within certain limits, the production of spelter by its members, and if the market showed that consumption was not keeping pace with the production, further to curtail output. After the first year, the clause restricting production was modified to allow unrestricted production if prices exceeded an agreed price or if stocks fell below an agreed amount. The Convention proved to be reasonably successful in meeting its objectives and in 1913 was extended to 1916. However in 1914, World War I intervened and the cartel collapsed. The German members of the syndicate tried to continued on after the war but the effort was soon abandoned; in 1922 most of the German zinc industry was awarded to Poland.

The next significant attempt to control zinc production was in 1928 when concern over depressingly low zinc prices led to the creation of the "European Zinc Cartel". Production cuts were instituted and spelter exports to Europe by Canadian and Australian producers, although not members of the Cartel, were voluntarily curtailed. Problems within the Cartel, including the relatively poorer economic position of integrated smelters versus custom smelters in a period of low zinc prices, the more severe economic burden born by electrolytic plants versus retort smelters in cutting back production, and the failure to bring all European electrolytic production fully into the Cartel, led to the Cartel's demise at the end of 1929. Efforts to revive the cartel languished until July 1931 when the extremely low world zinc price brought all the members of the 1928-1929 cartel and virtually all other world zinc producers of importance, except those in the United States, into a new "European Zinc Cartel". An initial production cutback of 45% of capacity was imposed, and one year later this was raised to 55%. However internal problems plagued the Cartel's existence almost from the start. The British abandoned the "gold standard" and placed a tariff on non-British Empire zinc. The Germans wanted increased tariff protection for their industry, sought higher production quotas, and in 1933, established a policy to become self-sufficient in zinc production. All of the cartel participants were impacted by the low zinc prices, leading some, especially those with devalued currencies, to cheat on their agreement by increasing zinc production and sales despite the possibility of severe penalties. The problems became overwhelming and at the close of 1934 the Cartel was dissolved. A number of efforts to revive the cartel failed, although in 1938 a short-lived "International Sheet Zinc Cartel" involving 10 countries was formed, but it and other cartel proposals were swept away by the start of war in Europe.

The most recent cartel, known in some circles as the "Zinc Club", was set up in 1964 and disbanded in May 1976. Unlike previous zinc cartels, the Club was a secret group with meetings and actions never officially acknowledged. Membership not only included Australian and Canadian zinc companies but also virtually all the zinc producers and smelters in Western Europe. No U.S. companies were members; however others such as those in Japan and Peru, although not members, cooperated in the Club's efforts. It should be noted that both EC and OECD rules banned cartels, but the governments of most Club-member countries, although aware of the cartels existence, took minimal action until initiation of a U.S. Justice Department investigation in April 1976 on the affect of the cartel on U.S. zinc prices.

The Club was set up to stabilize international zinc prices, provide support for their newly introduced "European Producer Price" pricing basis for zinc and limit the importance of the LME in setting world zinc prices. To carry this out, they set agreed-upon price limits, withheld metal, and purchased metal and metal futures on the LME when necessary to support their price limits. After the Club's collapse, the European Producer Price mechanism continued but frequent price changes, deep discounting from the set price and dissatisfaction with the process of setting concentrate prices, led to the demise of the EPP in 1988.

The most recent concerted effort to keep zinc prices artificially high occurred in 1992. From March through September, a small number of traders and producers was able to obtain and maintain large options positions at the LME. By rolling the options over when called, they, in effect "squeezed" the market by creating supply tightness by withholding metal from the market. By the end of September, however, in the face of weak world demand and a continuous and relentless rise in world stocks, option support collapsed resulting in a sharp drop in zinc prices.

ZINC METAL PRICING

U.S. zinc metal price averages before the 1860's are generally unavailable. Consumption up to that time was small, as the only significant use of zinc metal was as an alloying component in brass. Calamine, zinc dust, zinc oxide, and rolled zinc were also commercial products but prices are also lacking. Imports of zinc metal and dust from Europe provided virtually all domestic zinc metal needs until 1860 when the first commercial production of zinc metal began in the United States.

Consumption grew slowly, but in the early 1870's two major events occurred that firmly established the U.S. zinc industry and led to published U.S. zinc prices. The first was the development of the vast Joplin, MO zinc-mining district, which provided easily mined and smelted domestic ore. The second was the phenomenal growth in zinc metal use for brass, rolled zinc products and galvanizing in response to rapid industrialization and to Western expansion. Because of its growing importance, zinc prices were published in *Iron Age* and the *Engineering and Mining Journal (E/MJ)*, and later by *American Metal Market* (*AMM*). In 1907, *AMM* began its annual *Metal Statistics* series. The initial zinc prices published were for common western spelter in New York City and were established by inquiry of producers and consumers.

By the turn of the century, U.S. smelting activity had grown rapidly, mainly developing in Kansas, Missouri, and Illinois because of closeness to Joplin and Wisconsin zinc ores and adequate coal and/or low-cost natural gas for smelting fuel.. Because of these nearby activities, St. Louis, MO. and, later, East St. Louis, IL. emerged as a second focal point of U.S. zinc pricing. By 1909, the St. Louis price became the principal U.S. zinc price, although New York prices remained important and were concurrently quoted. Generally, New York prices were higher to account for shipping charges from Midwest smelters. In 1953 an attempt was made to overturn the traditional East St. Louis basing point for pricing slab zinc when a series of new prices on a delivered basis was announced by the American Smelting & Refining Co.(later known as ASARCO). Deliveries west of the Continental Divide were priced one-quarter cent per pound higher than those east of the Divide. This effort was short-lived, but in 1971, the same company successfully introduced its delivered-basis pricing system, thereby eliminating both East St. Louis and New York zinc pricing and establishing what became known as the U.S. producers' price, essentially a narrow range of quoted prices cited by domestic zinc metal producers. This also resulted in *Metals Week*, an offshoot from *E/MJ*, becoming the medium and barometer for zinc metal pricing in the United States. The *Metals Week* daily weighted price, an average based on sales, became the basis for most other transactions. However, producers' pricing, per se, was strongly influenced by outside forces soon after its beginning. The U.S. price freeze dictated zinc prices in the 1971-1973 period, but from the mid 1970's the *Metal's Week* averages and the LME price became more prominent in U.S. zinc pricing owing largely to the decline of the domestic smelting sector and the resulting increase in imported metal.

The base grade of metal through most of the U.S. price series is (see Table 6) western spelter or Prime Western (PW) zinc. Higher grades of zinc metal have generally sold at a premium to the PW price. Historically, this developed because PW is the standard product of pyrometallurgical smelters, which accounted for all U.S. production through 1915 and which, until the early 1960's, was the dominant domestic production method. PW producers charged more for High Grade (HG) or Special High Grade (SHG) metal because of the additional refinement steps required. Electrolytic smelters, on the other hand, alloyed the higher purity metal produced electrolytically in order to make PW, a lesser valued product. In the 1970's, the industry wanted to change the price basis to resolve this dilemma because most metal was then produced electrolytically. In September 1980, the base price of zinc was changed from PW to HG and *Metals Week* introduced its weighted average price based on daily sales of HG. PW, ironically, was then priced the same or higher than HG. Producers' pricing continued through 1990, but because of discounting and other factors, the *Metals Week* weighted average was believed to reflect the price of zinc more accurately, and is also used in Table 6.

Outside the United States, the world zinc prices have essentially been set by "cartels" and those quoted by the LME, which introduced its first zinc contract in 1915. Prior to 1915, European zinc producer cartels set world zinc prices. Cartels formed after WWI were generally weak and the LME prices were more influential up to the start of WWII. In August 1939 the LME ceased operations because of the war and did not resume dealings on zinc until January 1953. In 1964, a group of non-U.S. zinc producers, concerned that the volatility and high prices on the LME might adversely affect zinc prices and consumption, established the European Producer Price (EPP) in 1964. This was an attempt to control as well as stabilize the then "free world" zinc market by maintaining zinc prices unaffected by short-term fluctuations and by adjusting smelter output to maintain price goals. As a result, the impact of the LME on zinc prices declined after 1964; however, it remained important as the only medium to reflect day-to-day changes in the zinc market and, as such, served as the barometer for world price trends. Most zinc outside the North American market was sold on an EPP basis; however, its questionable character as a cartel and continuing dissatisfaction with the EPP pricing system, mainly as it related to the settlement price of zinc concentrates and/or the determination of smelter treatment charges, led to its demise in 1988 and to reemergence of LME zinc quotations as the principal basis for world zinc pricing.

Essentially, two zinc metal markets, a European and World market and a U.S. market, existed from the late 1800's through the early 1950's. The United States was largely self-sufficient in metal production throughout most of the period, and metal trade (except during and immediately after the World Wars) was small, due partly to severe tariff barriers. Cartel action to control production and prices was international in scope but excluded U.S. producers. After WWII, foreign producers added virtually all of the world's new smelter capacity and played an increasingly important role in determining U.S. zinc prices. During this period, U.S. production waned and became less significant as many smelters closed because of environmental constraints and technical obsolescence. U.S. zinc smelter capacity fell from more than 1 million metric tons in the 1960's to 325,000 tons in 1988. U.S. production, which accounted for about 30% of world output of zinc metal in the 1960's, accounted for only 5% in the 1980's and supplied only one-third of domestic demand. As a result of these changes, LME and EPP zinc prices became the dominant factor in the determination of U.S. zinc prices. By 1990, the MW weighted average had become a minor factor in U.S. pricing of zinc and in 1991 was dropped in favor of an LME zinc pricing basis. The U.S. zinc price since is tracked by *Platt's Metals Week,* formerly *Metals Week.*

The U.S. zinc price series is characterized by extended periods of moderate fluctuation interspersed by shorter periods of significantly higher or lower prices mainly owing to the influence of major world events, such as wars and economic depressions. Throughout the price series, short-term economic events have caused sharp 1-year changes in the average annual price.

The early years of WWI produced the most dramatic rise in U.S. prices, when average prices reached levels unsurpassed until 1951 and were the highest ever in 1987 dollar terms. With the outbreak of war, European zinc production, which accounted for about 60% of the world production in 1914, was sharply curtailed. Some of the major battlefields were near the major smelting districts in continental Europe. This, coupled with limited zinc production capacity in The United Kingdom, resulted in a scramble for zinc from the United States, the only large source outside

Europe. British and French buyers drove up prices, initially bidding against each other, but later commissioning one buyer for both governments, resulting in lower prices. The higher grade metals, used largely for brass production for munitions, were in tremendous demand. In early June 1915, the prices for the highest grade metals were reported at more than 40 cents per pound or more than $5 per pound in 1987 dollars.

The most lengthy stable period of zinc prices is the 18-year period of 1953-1971 and can be attributed partially to U.S. Government policies pertaining to stockpile programs, import quotas, and tariffs. The stockpile program, begun in 1946, initially absorbed the production from excess capacity that had built up during WWII. New and higher stockpile goals for zinc determined during and after the Korean War (1950-1953) resulted in the additional acquisition of 836,000 tons of excess world zinc output in the 1950-1959 period. Although the stockpile buildup helped to stabilize prices in the 1950's the large quantity of zinc in the stockpile, which reached 1.4 million tons in 1959, overhung the world market and, together with stockpile releases, acted as a suppressant on increased U.S. and world prices in the 1960's. More than 230,000 tons of zinc was released from the stockpile in the 1964-1965 period, a time of strong domestic economic growth and high zinc consumption. This dampened price increases in a time of tight supply. From 1967-1971, government programs played less of a role in stabilizing prices. Adjustments in the zinc supply by both foreign and domestic producers tended to maintain a reasonable supply-demand balance. The President, concerned by increasing inflation, initiated his economic stabilization program during the latter half of 1971, which froze the price of zinc at 17 cents per pound. Price controls were abolished at the end of 1973, resulting in a sharp rise in domestic zinc prices to world levels in 1974. Other periods of relative price stability were those associated with price controls imposed during wartime.

In 1987 dollar terms, Table 7 shows the decades that included major conflicts between nations tended to have high average prices despite price controls. Those decades characterized by extended periods of economic growth tended to have low average prices. The 1910-1919 period was one of especially high prices owing to the war, which not only caused demand to increase, but also disrupted more than one-half of world smelter production. The decades of low average prices experienced economic downturns in their early years followed by long periods of relative economic stability and low inflation rates. The 1990's decade, so far, appears to be following this latter trend.

MARKETING OF ORES, CONCENTRATES, AND ZINC PRODUCTS

Zinc-bearing ores and concentrates tend to be sold on terms negotiated individually between the miner and the smelterer. Concentrate transactions involving traders have sharply declined in recent years owing in large part to the requirements of financial institutions that long-term financing of mine projects include purchasing or supply contracts to ensure loan repayment. Ore traders, however, remain important as the buyers and sellers of last resort for ores and concentrates, transactions involving complex trades, arbitrage, and transactions involving hedging strategies.

The price of zinc is a major factor in valuing concentrate but terms can vary considerably between different parties owing to differences in the character and composition of ore and concentrate and the feed desired by the smelterer. Not all zinc ores or concentrates are suitable feed for all facilities, not only for technical reasons but also for a variety of environmental reasons. The terms for purchase are typically set out in a "smelter schedule", which generally consists of a list of premiums, penalties for undesirable contaminants, deductions before payment for other extractable metals such as gold or silver, treatment charges, and the price basis for metal payment calculations. Payments for contained metals are based on current spot prices if the material is bought outright or, the monthly average price, average of week of arrival, etc., if delivered at a later date.

After WWII, numerous custom smelters were constructed in Japan, Europe, and elsewhere to meet their internal industrial needs. As a result, a large, more diverse, world trade in zinc concentrates developed. To insure adequate smelter feed deliveries for the following year, the individual smelting companies and producers of concentrate began in the fall of the year, annual negotiating sessions, sometimes collectively referred to as "the mating season", to determine contract treatment charges (TC) for the coming year or, in some cases, two or three years. The TC basically represents smelter operating costs, including labor, fuel, flux, power, and supplies at some fixed price basis to treat a ton of concentrate. As such, it comprises only a part of the smelter schedule contract. Adjustments to the TC are primarily based on fluctuations in the zinc content of the concentrate and zinc price at the time of delivery, although a contract may call for further adjustment for changes in labor and supply costs. The negotiated level of TCs is often dependent on the relative supply of concentrates. A stronger negotiating position is gained by the miner when the concentrate supplies are limited, whereas the smelterer gains the upper hand when concentrates are plentiful. In recent decades, these contracts have increasingly been negotiated in terms of U.S. dollars and, as a result, both miners and smelterers have on occasion gained or lost financially due to currency fluctuations.

In the early years of the U.S. zinc industry, the marketing of ores and concentrates took place at the mine site or the local mill. The miners sold directly to brokers or agents for the smelting companies, who bought the ore or concentrate f.o.b. mine and who, in turn, arranged and paid for shipment. The Tri-State ores were sold similarly but were also somewhat unique owing to the early leasing and mining customs adopted for the district. Most leases required mine operators to have their own gravity mills and stipulated that none of the ore and concentrate could be moved off the lease area until its weight and royalty could be assessed.

Tri-State ores (concentrates) generally were sold by weight up to the mid 1890's and thereafter by assay. In 1910 the Zinc Ore Producers' Association adopted the Joplin Ore Scale, a sliding-scale method based on the zinc metal price.

Custom mills sprang up in many mining districts to serve groups of mines. Some custom mills operated on a toll basis, storing an individual's ore until enough was on hand to batch process it. Other mills took in ores from many operations, processed commingled ore, and sold the concentrate to brokers and ore buyers. Payment to the miners was based on their ore assays with adjustment for penalty elements and total concentrate produced.

Prior to 1950, there were a number or custom milling operations in many of the mining districts of the United States to accommodate the milling requirements of domestic mining operations without mills. However, since that time most custom mills in the United States have closed and many have been dismantled.

"The producer of spelter from ore he owns is desirous of getting maximum price and is adverse to selling at a low price. The custom smelter on the other hand is more indifferent to price; his profit being derived from the margin between purchase and sale and his desire to keep his plant running at near capacity."

World Survey of the Zinc Industry, Ingalls, 1931

Custom ore sometimes is purchased by mines that operate their own mill if they have excess capacity and if the ore is compatible with their own ore and process. However, such arrangements have become rare. The trend in recent years is for mining companies with limited capital that open or reopen zinc-producing mines with no standing mill, to lease a nearby closed mill or arrange, if possible, for tolling the ore at a nearby facility.

Secondary-zinc materials, such as ash, dross, sludge, etc., are sold by type of material and zinc content; the price per pound or unit of material is typically related to some percentage of the zinc metal price. An agreement, similar to a smelter schedule for ore and concentrate but generally much simpler, is used in scrap purchases. In recent years minimum standards for the character and content of most types of zinc scrap and secondary materials have been established and accepted by U.S. recycling industries. This has simplified the classification and standards of the scrap sold and made redress easier in cases where the secondary material sold is not of the quality implied at the time of sale. International sales and shipments of secondary-zinc materials that are classified as being hazardous, became subject to the rules and regulations of the Basel Convention and/or OECD agreement on scrap materials in recent years. Under these agreements, hazardous zinc scrap could not be shipped to any country that did not recycle or dispose of the material in an environmental sound manner, or at least to the standards imposed for the recycle or disposal of such material in the exporting country.

Commercial zinc metal is traded or sold mainly in the grades produced at primary zinc plants or special zinc alloys as requested by die casters and galvanizers. The sizes and shapes of the metal sold are varied, but generally are in the form of slabs, blocks, and waffle ingots. Slabs weigh about 25 kilograms, whereas blocks range up to 2 tons. Each slab or block is identified as to grade and producer by a registered brand cast in the metal. Zinc metal is also commercially available in the form of plate, sheet, strip, wire, dust, powder and flake.

All world zinc metal pricing is now based on the LME price. The zinc price in any one area or country, however, is typically composed of the LME basis price plus the tariff, insurance, and delivery costs; local shortages or surpluses in supply, taxes, etc., may cause additional price fluctuations. Because the LME basis metal is SHG, other grades of slab zinc are sold at prices slightly below that of SHG zinc, whereas special alloy compositions typically are higher priced. In the past, producer pricing and quotes at various pricing points, such as East St. Louis and New York were major factors in U.S. sales prices. Spot sales generally stipulated a market price at the time of delivery, as published in some specific trade journal, usually the *Engineering and Mining Journal*, in the United States. Prices for zinc dust, powder, oxide and various chemicals are influenced by the slab zinc prices, but other factors such as production costs, volume of sale, physical and chemical quality, etc., also influence the price of these materials.

The LME is a terminal market for zinc, as well as a market for hedging purposes. Activity on the LME is not restricted to the periods of floor trading, but operates as a 24-hour market because the members (ring dealers, associate brokers, and clearing brokers) can write client contracts at any time. Floor and kerb trading periods are held in late morning and in the afternoon (in London); the important difference between the two sessions is the fact that no official prices are announced in the afternoon. LME warehouses for acceptance or delivery of approved brands of SHG metal are scattered throughout the world; in the early 1990's, six approved LME warehouses were established in the United States.

Zinc oxide is the most widely marketed zinc chemical. It is purchased according to production method, purity, brightness, crystal size and shape, and customer requirements. For example, acicular and plate-like forms of zinc oxide and pigment brightness are favored for paint use, whereas nodular forms are preferred in rubber compounding. Special treatment, size, or quality may be necessary for photoconductive or pharmaceutical uses. Other zinc chemicals such as zinc sulfate and zinc chloride, are marketed in high-purity and technical grade in both solid and liquid form. Solid chemicals are sold in bags, drums and carload lots and liquids are sold in drums and tanker-load volumes. Prices for zinc oxide typically are based on the zinc metal price but only partially so for the sulfate and chloride chemicals. The prices for the latter two chemicals can vary substantially depending on quality, quantity, and other factors.

Table 6: Average Annual Zinc Prices in the United States, 1850-1994

Year	Average Annual Price	Constant 1978 Dollars	Year	Average Annual Price	Constant 1987 Dollars
1850	NA	NA	1889	0.050	NA
1851	NA	NA	1890	0.055	NA
1852	NA	NA	1891	0.050	NA
1853	0.055	NA	1892	0.046	NA
1854	NA	NA	1893	0.040	NA
1855	NA	NA	1894	0.035	NA
1856	NA	NA	1895	0.036	NA
1857	NA	NA	1896	0.039	NA
1858	NA	NA	1897	0.041	NA
1859	NA	NA	1898	0.046	NA
1860	NA	NA	1899	0.058	NA
1861	NA	NA	1900	0.044	NA
1862	NA	NA	1901	0.041	NA
1863	NA	NA	1902	0.048	NA
1864	0.139	NA	1903	0.054	NA
1865	NA	NA	1904	0.051	NA
1866	NA	NA	1905	0.059	NA
1867	NA	NA	1906	0.061	NA
1868	NA	NA	1907	0.058	NA
1869	NA	NA	1908	0.046	NA
1870	NA	NA	1909	0.054	0.772
1871	NA	NA	1910	0.054	0.760
1872	NA	NA	1911	0.056	0.803
1873	NA	NA	1912	0.068	0.932
1874	NA	NA	1913	0.055	0.767
1875	0.070	NA	1914	0.051	0.680
1876	0.072	NA	1915	0.142	1.820
1877	0.060	NA	1916	0.136	1.545
1878	0.049	NA	1917	0.089	0.832
1879	0.052	NA	1918	0.080	0.696
1880	0.055	NA	1919	0.070	0.515
1881	0.052	NA	1920	0.078	0.484
1882	0.053	NA	1921	0.047	0.356
1883	0.045	NA	1922	0.057	0.467
1884	0.044	NA	1923	0.066	0.527
1885	0.043	NA	1924	0.063	0.504
1886	0.044	NA	1925	0.076	0.606
1887	0.046	NA	1926	0.073	0.588
1888	0.049	NA	1927	0.062	0.512

Table 6: Average Annual Zinc Prices in the United States, 1850-1994 (cont.)

Year	Average Annual Price	Constant 1987 Dollars	Year	Average Annual Price	Constant 1987 Dollars
1928	0.060	0.484	1962	0.116	0.486
1929	0.065	0.524	1963	0.120	0.437
1930	0.046	0.380	1964	0.136	0.486
1931	0.036	0.327	1965	0.145	0.505
1932	0.029	0.296	1966	0.145	0.489
1933	0.040	0.421	1967	0.139	0.457
1934	0.042	0.404	1968	0.135	0.423
1935	0.043	0.405	1969	0.147	0.435
1936	0.049	0.462	1970	0.153	0.428
1937	0.055	0.586	1971	0.161	0.427
1938	0.046	0.418	1972	0.178	0.450
1939	0.051	0.472	1973	0.207	0.493
1940	0.063	0.573	1974	0.360	0.785
1941	0.075	0.641	1975	0.390	0.772
1942	0.083	0.664	1976	0.370	0.692
1943	0.083	0.648	1977	0.344	0.602
1944	0.083	0.638	1978	0.310	0.505
1945	0.083	0.624	1979	0.373	0.560
1946	0.087	0.527	1980	0.374	0.512
1947	0.105	0.559	1981	0.446	0.558
1948	0.136	0.677	1982	0.385	0.450
1949	0.122	0.610	1983	0.414	0.468
1950	0.139	0.685	1984	0.486	0.530
1951	0.180	0.844	1985	0.404	0.428
1952	0.162	0.747	1986	0.380	0.391
1953	0.109	0.496	1987	0.419	0.419
1954	0.107	0.479	1988	0.602	0.584
1955	0.123	0.533	1989	0.820	0.762
1956	0.135	0.564	1990	0.746	0.665
1957	0.114	0.462	1991	0.528	0.449
1958	0.103	0.409	1992	0.584	0.485
1959	0.115	0.446	1993	0.462	0.374
1960	0.130	0.494	1994	0.493	0.387
1961	0.116	0.437			

NA -- Not Available

Sources: Reported by C.E. Siebenthal, published in Mineral Resources of the United States 1916-Part 1 (1853 and 1864) ;New York price, Ingalls, W. R., Lead and Zinc in the United States, McGraw-Hill Book Co. Inc. NY, 1908, p. 342 (1875-1904); St. Louis or E. St. Louis prices, American Metal Market/ Metal Statistics and E/MJ Metal and Mineral Markets (1905-14 and 1919-70); St. Louis price, calculated from total receipts from sales by Producers (1915-18); and Metals Week (1971-94).

Table 7: Average Zinc Prices by Decade, 1900-1994

Decade	Average Annual Price	Based on Constant 1987 Dollars
1875-1899	4.9	NA
1900-1909	5.16	NA
1910-1919	8.01	93.5
1920-1929	6.47	50.52
1930-1939	4.47	41.71
1940-1949	9.2	61.61
1950-1959	12.87	56.65
1960-1969	13.29	46.49
1970-1979	28.46	57.14
1980-1989	47.3	51.02
1990-1994	56.26	38.47
Averages:		
1900-1994	17.1	--
1910-1994	18.6	56.46
1920-1994	20.00	51.25

NA -- Not Available

BIBLIOGRAPHY

AIME, 1936. Metallurgy of Lead and Zinc. Am. Inst. of Min. and Met. Eng., New York, vol. 121, 748pp.

Alenius, E.M.J., 1930. Methods and Costs of Stripping and Mining at the United Verde Open-Pit Mine, Jerome, Arizona. BuMines IC 6248, pp. 1-3.

American Cyanamid Company, 1942. The Central Mill of the Eagle-Picher Mining and Smelting Company, Cardin, Oklahoma. Ore Dressing Notes, No. 11, pp. 14-17.

American Cyanamid Company, 1942. The Mascot Mill of the American Zinc Co. of Tennessee. Ore Dressing Notes, No. 11, pp. 12-13.

Anderson, G.M., 1991. Organic Maturation and Ore Precipitation in Southeast Missouri. Economic. Geology, vol. 86, no. 5, pp. 909-926.

Bateman, A.M., 1951. Economic Mineral Deposits (2nd ed.). John Wiley & Sons Inc., New York, 916pp.

Berg, J.E., 1930. Mining methods at the Page Mine of the Federal Mining and Smelting Co., Page, Idaho. BuMines IC 6372, pp. 1-8.

Blackburn, W.H., 1931. Milling Methods and Costs at the Concentrator of the Treadwell Yukon Co., Ltd., at Tybo, Nev. BuMines IC 6430, pp.1-13.

Blake, W.P., 1894. Notes on the Structure of the Franklinite and Zinc-Ore Beds of Sussex County, New Jersey. AIME Trans. 24, pp. 521-524.

Blake, W.P., 1894. The Zinc-Ore Deposits of Southwestern New Mexico. AIME Trans 24, pp. 187-195.

Boyle, J.R., and L. Williams, 1965. Mining Methods and Practices at the Young Mine, American Zinc Co. of Tennessee, Jefferson County, Tenn. BuMines IC 8269, 27pp.

Branner, G.C., 1927. Outlines of Arkansas' Mineral Resources. Bureau of Mines, Manufactures and Agriculture and State Geological Survey, Little Rock, AR., pp. 248-266.

Brichta, L.C., 1960. Catalog of Recorded Exploration Drilling and Mine Workings, Tri-State Zinc-Lead District - Missouri, Kansas, and Oklahoma. BuMines IC 7993, 13 pp.

Brown, H.E., 1957. Zinc Oxide Rediscovered. The New Jersey Zinc Company, New York, 99pp.

Brown, H.E., 1990. Zinc Chemicals: Applications. International Lead Zinc Research Organization, Inc., Research Triangle Park, NC. 373 pp.

Brown, J.S., 1951. A Graphic Statistical History of the Joplin or Tri-State Lead-Zinc District. Mining Engineering, September, pp. 785-790.

Buck, W.K., 1986. Intergovernmental Mineral Commodity Arrangements. Ctr. for Resource Studies, Queen's University, Kingston, Ontario, Canada. 70pp.

Bureau of Mines. Materials Survey: Zinc. Compiled for the Materials Office, National Security Resources Board, March 1951, 616pp.

Bureau of Mines. Mineral Resources of the United States 1924-1931 and Minerals Yearbook 1932-1992.

Bureau of the Census, 1971. Colonial Times to 1970. In Part 1, Historical Statistics of the United States. Department of Commerce, pp 603-4.

Callahan, W.H., 1968. Geology of the Friedensville Zinc Mine, Lehigh County, Pennsylvania. Ore Deposits of the United States, 1933-1967, ed. by J.D.Ridge, Vol. 1, AIME, New York, pp. 95-108.

Callahan, W.H., 1977. The History of the Discovery of the Zinc Deposit at Elmwood, Tennessee, Concept and Consequence. Economic Geology, vol.72, no. 7, pp. 1382-1392.

Carrillo, F.V., M.H. Hibpshman and R.D. Rosenkranz. 1974. Recovery of Secondary Copper and Zinc in the United States. BuMines IC 8622, 58pp.

Carus, H.D., 1959, Historical Background, in ZINC, , ed. by CH. Mathewson, Amer. Chem. Soc. Monograph No. 142, Reinhold Publ. Corp., New York, pp. 1-8.

Chance, H.M., 1890. The Rush Creek, Arkansas, Zinc-District. AIME Trans., 18, pp. 505-508.

Cigan, J.M., T.S. Mackey and T.S. O'Keefe (eds.) 1980. Lead-Zinc-Tin '80. Proc. World Conf. on Metallurgy and Environmental Control, Las Vegas, NV., Feb. 24-28, The Met. Soc. of AIME, 1045pp.

Cole, W.A., 1957. Mining and Milling Methods and Costs, Tri-State Zinc, Inc., Jo Daviess County, MO. BuMines IC 7780, 19pp.

Constantopoulos, J., 1994. Oxygen Isotope Geochemistry of the Coeur D'Alene Mining District, Idaho. Economic Geology. vol. 89, pp. 944-951.

Cotterill, C.H., and J.M. Cigan (eds.). 1970. AIME World Symposium on Mining & Metallurgy of Lead and Zinc, New York. Vol. I, 1017pp. ; vol. II, 1090pp.

Coy, H.A., 1930. Mining Methods and Costs, American Zinc Co. of Tennessee, Mascot, Tenn. BuMines IC 6239, pp. 1-8.

Crabtree, E.H., 1930. Milling Practice at the White Bird Concentrator, Canam Metals Corporation, Picher, Okla. BuMines IC 6353, pp. 1-7.

Craddock, P.T., Freestone, I.C., Gurjar, L.K., Hegde, K.T.M. and V.H. Sonawane, 5. Early Zinc Production in India. Mining Mag., January 1985, pp. 45-52.

Dawkins, F.M., 1950, Zinc and Spelter. Zinc Development Association. Alden Press, Oxford, 35 pp.

Drinker, H.S., 1871-1873. Abstract of a Paper on the Mines and Works of the Lehigh Zinc Company. AIME Trans. 1, pp. 67-75.

Dunn, P.J., and B.T. Kozykowski, 1991. The Resurrection of Sterling Hill. The Mineralogical Record, vol. 22, pp. 367-376.

Dunning C.H., with E.H. Peplow Jr., 1966. Rock to Riches. Hicks Publishing Corp., Pasadena, CA., 406pp.

Engineering and Mining Journal. 1892-1903. Zinc and Cadmium.In The Mineral Industry, volumes I-XII.

Fay, A.H., 1920. A Glossary of the Mining and Mineral Industry. BuMines B. 95, 754 pp.

Foreman, C. H., 1930. Mining Methods and Costs at the Hecla and Star Mines, Burke, Idaho. BuMines IC 6232, pp. 1-6.

Frondel, C., and J.L. Baum, 1974. Structure and Mineralogy of the Franklin zinc-Iron-Manganese Deposit, New Jersey. Economic Geology, vol. 69, no. 2, pp. 157-180.

Hague, J.M., Baum, J.L., Herrmann, L.A., and Pickering, R.J., 1956. Geology and Structure of the Franklin-Sterling Hill Area, NJ. Geology Soc. America Bulletin, v. 67, pp. 435-473.

Handy, R.S., 1930. Milling Methods and Costs at the Northern Idaho Mills of the Bunker Hill and Sullivan Mining and Concentrating Co. BuMines IC 6314, pp. 1-51.

Hartnagel, C.A., and J.G. Broughton, 1951. The Mining and Quarry Industries of New York State, 1937 to 1948. New York State Museum, Bulletin no. 343, pp. 120-124.

Hayes, A.L., and R.E. Daugherty, 1964. The Jefferson City Flotation Mill. In Milling Methods in the Americas, ed. by N. Arbiter, Gordon & beach Science Publishers, New York, pp. 415-448.

Haynes, F.M., and S.E. Kesler, 1994. Relation of Mineralization to Wall-Rock Alteration and Brecciation, Mascot-Jefferson City Mississippi Valley-Type District, Tennessee. Economic Geology, vol. 89, pp. 51-66.

Healy, J.F., 1978. Mining and Metallurgy in the Greek and Roman World. Thames and Hudson, London, pp. 139-214.

Henrich, C., 1892-1893. Zinc-Blende Mines and Mining Near Webb City, Mo. AIME Trans. 21, pp. 3-25.

Heyl, A.V., 1963 Oxidized Zinc Deposits of the United States, Part 2. Utah. USGS Bulletin 1135-B, 104pp.

Heyl, A.V., 1964. Oxidized Zinc Deposits of the United States. Part 3. Colorado. USGS Bulletin 1135-C, 98pp.

Heyl, A.V., and C.N. Bozion. 1962. Oxidized Zinc Deposits of the United States. Part 1. General Geology. USGS Bulletin 1135-A, 52pp.

Hill, J.M., 1929. Historical Summary of Gold, Silver, Copper, Lead, and Zinc Produced in California, 1848 to 1926, 22 pp.

Hoffman, H.O., 1922. Metallurgy of Zinc and Cadmium. First edition, McGraw Hill Book Co., New York. 341p.

Holmes, R.W., and M.B. Kennedy, 1983. Mines and Minerals of the Great American Rift (Colorado-New Mexico). Van Nostrand Reinhold Company, Inc., pp. 1-266.

Horvick, E.W., 1961. Zinc and Zinc Alloys. In Metals Handbook. American Society for Metals, Novelty, OH., 8th Ed., vol. 1, pp. 1157-1172.

Ingalls, W.R., 1908. Lead and Zinc in the United States. Hill Publishing Co., New York, 367pp.

Ingalls, W.R., 1931. World Survey of the Zinc Industry. Min. and Met. Soc. of America, New York, 128pp.

International Lead and Zinc Study Group, 1984. Secondary Lead and Zinc. Special Meeting of the Scrap Sub-Committee 7th-9th September, 1983, Final Report, Washington, D.C., 325 pp.

Jackson, C.F., and J.H. Hedges, 1939. Metal-Mining Practice. BuMines B. 419, 512pp.

Jackson, C.F., J.B. Knaebel and C.A. Wright, 1935. Lead and Zinc Mining and Milling in the United States. Current Practices and Costs. BuMines B. 381, 204pp.

Jagnaux, R., 1891. Histoire De La Chimie. Baudry, Paris, II, p. 207.

Jolly, J.H., 1991. Zinc. In Metal Prices in the United States through 1991. BuMines Special Publication, pp. 191-195.

Jolly, J.H., 1993. Materials Flow of Zinc in the United States 1850-1990. Resources, Conservation and Recycling, No. 9, pp. 1-30.

Jolly, J.H., 1993. Zinc. In Chemical Industry Applications of Industrial Minerals and Metals. Special Publication of the U.S. Bureau of Mines, pp. 151-156.

Jolly, J.H., 1993. Zinc. In Recycled Metals in the United States. Special Publication of the U.S. Bureau of Mines, pp. 69-72.

Jolly, J.L., and A.V. Heyl, 1968. Mercury and Other Trace Elements in Sphalerite and Wallrocks from Central Kentucky, Tennessee and Appalachian Zinc Districts. USGS Bulletin 1252-F, 29pp.

Keener, O.W., 1930. Methods and Costs of Mining at Hartley-Grantham Mine, Tri-State zinc and Lead District. BuMInes IC 6286, pp. 1-4.

Kegler, V.L., 1931. Mining Methods of the Ducktown Chemical and Iron Co., Mary Mine, Isabella, Tenn. BuMines IC 6397, pp. 1-9.

Kilgore, C.C., S.J. Arbelbide and A.A. Soya, 1983. Lead and Zinc Availability – Domestic. BuMines IC 8962, 30pp.

Kisvarsany, O., S.K. Grant, P.W. Pratt and J.W. Koening (eds.), 1983. International Conference on Mississippi Valley Type Lead-Zinc Deposits. Proc. Conf. held October 1982, Univ. of Missouri-Rolla, MO., 603pp.

Kline, H.D., W.A. Calhoun and B.M. Reynolds, 1961. Mining and Milling Methods and Costs, Madison Mine, National Lead Co., St Louis Smelting and Refining Division, Madison County, Mo. BuMines IC 8028, 26 pp.

Koch, Jr, G.S., and J.H. Schuememeyer, 1979. Zinc Availability in the United States; A Statistical Analysis. Final Report (prepared for the USBM., Contract No. GO188015). Univ. of Georgia, March 1979, 164pp.

Li, Chiaoping, 1948. The Chemical Arts of China. Journal of Chemical Education, Easton, PA.

Lunt, H.F., 1920. Historical Sketch of Ore Treatment in Colorado. Ch. in Investigation of the Low-Grade and Complex Ores in Colorado, Bulletin 10, Colorado Bureau of Mines, pp. 287-295.

Lyon, D.A., and C.C. Ralston, 1919. Recovery of Zinc from Low-Grade and Complex Ores. BuMines B. 168, 145pp.

Mackey, T.S., and R.D. Prengaman (eds.), 1990. Proc. World Symposium on Metallurgy and Environmental Control of Lead and Zinc. Anaheim, CA., Feb. 18-21, 1990. TMS, 1086pp.

Macone, A.J., and V.A. Cammarota, Jr., 1982. Overview of the International Lead and Zinc Study Group. Proceedings of the Lead Industries Assoc./Zinc Institute, Inc., March 1982, St. Louis, MO, pp. 173-180.

Magnuson, R.G., 1968. Coeur D'Alene Diary. The First Ten Years of Hardrock Mining in North Idaho. Metropolitan Press, Portland, OR., 319pp.

Maher, S.W., 1958. The Zinc Industry of Tennessee. State of Tennessee, Div. of Geology, Information Circular No. 6, 28 pp.

Maier, G., G.L. Oldright and G.W. Kuerner, 1924. Possibilities in the Hydrometallurgy of Oxidized Zinc Ores. Utah Engineering Experiment Station Research Investigations for the year 1922-23. Univ. of Utah Bulletin No. 14, v. 14, no. 7, January 1924, pp 5-36.

Maley, T.S., 1983. Handbook of Mineral Law, 3rd Ed. Mineral Land Publications, Boise, ID., 711 pp.

Mathewson, C.H., (ed.), 1959. Zinc. The Science and Technology of the Metal, Its Alloys and Compounds. Am. Chem. Soc. Monograph Series, Reinhold Publishing Corp., New York, 721 pp.

Matson, J.T., and C. Hoag, 1930. Mining Practice at the Pecos Mine of the American Metal Co. of New Mexico. BuMines IC 6368, 18 pp.

Mavrogenes, J.A., R.D. Hagni and P.R. Dingess, 1992. Mineralogy, Paragenesis, and Mineral Zoning of the West Fork Mine, Viburnum Trend, Southeast Missouri. Economic Geology, Vol. 87, pp. 113-124.

McGilvra, D.B., and A.J. Healy, 1930. Methods of Mining at the Black Rock Mine, Butte & Superior Mining Co., Butte District, . BuMines IC 6370, pp. 1-3.

Mining Engineering, 1953. The New Jersey Zinc Story. AIME, December, 52 pp.

Missouri Geological Survey and Water Resources, 1969. Zinc. In Missouri Minerals - Resources, Production, and Forecasts. Special Publication Number 1, pp. 122-127.

Morgan, S. W.K., 1985. Zinc and Its Alloys and Compounds. Ellis Horwood Ltd., West Sussex, England, 235p.

Nason, F.L., 1894. The Franklinite-Deposits of Mine Hill, Sussex County, New Jersey. AIME Trans. 24, pp. 121-130.

National Research Council, 1979. Zinc. Report of NRC Committee on Medical and Biologic Effects of the Environmental Pollutants. University Park Press, Baltimore, MD., 471pp.

NcMahon, A.D., C.H. Cotterill, J.T. Dunham and W.L. Rice, 1974. The U.S. Zinc Industry: A Historical Perspective. BuMines IC 8629, 76pp.

Netzeband, W.F., 1958. Mining Methods and Costs at the Rialto Mine, Nellie B Division, American Zinc, Lead & Smelting Co., Ottawa County, Okla. BuMines IC 5823, 23 pp.

New York Division of Commerce, 1943. Zinc, Lead, and Pyrite. In The Expanding Mineral Industry of the Adirondacks. State of New York, pp. 53-62.

New York State Division of Commerce, 1950. Zinc and Lead.In The Mineral Industries of New York State, pp. 40-48.

Norris, J.D., 1968. A History of the American Zinc Company. State Historical Soc. of WS. Worzalla Publishing Co., Stevens Point, WS., 244pp.

North, R.M. The Metallic Mineral Resources of New Mexico. California Mining Journal, February, pp. 3-5 and 82.

Nriagu, J.O., (ed.), 1980. Zinc in the Environment. Part 1., Ecological Cycling., and Part 2., Health Effects. The Environmental Science and Technology Series. John Wiley & Sons, Inc., New York, 900pp.

Oder, C.R.L., and J.E. Ricketts, 1961. Geology of the Mascot-Jefferson City Zinc District, Tennessee. State of Tennessee, Div. of Geology, Report of Investigations No. 12, 29 pp.

Pack, J.W., 1874-1875. Process of Spelter Production, as Practiced at the Carondelet, Missouri, with Comparisons. AIME Trans. 3, pp. 125-130.

Paliwal, H.V., L.K. Gurjar, and P.T. Craddock, 1986. Zinc and Brass in Ancient India. CIM Bulletin, January, pp. 75-78.

Partington, F. R., 1935. Origins and Development of Applied Chemistry. p. 83.

Payne, H.M., 1928. The Undeveloped Mineral Resources of the South. American Mining Congress, Washington, D.C., pp. 186-193.

Pehrson, E.W., 1929. Summarized Data of Zinc Production. BuMines Economic Paper 2, 47pp.

Platt, J.C., 1876-1877. The Franklinite and Zinc Litigation Concerning the Deposits of Mine Hill, at Franklin Furnace, Sussex County, N.J. AIME Trans. 5, pp. 580-584.

Plumlee, G.S., et. al., 1994. Chemical Reaction Path Modeling of Ore Deposition in Mississippi Valley-Type Pb-Zn Deposits of the Ozark Region, U.S. Mid-continent. Economic Geology, vol. 89, pp. 1361-1418.

Rauch, D.O., and B.C. Mariacher (eds.), 1970. AIME World Symposium on the Mining and Metallurgy of Lead and Zinc. Two Volumes. AIME, New York. Port City Press, Inc., Baltimore, MD., 2017pp.

Raymond, R.W., 1879-1880. Note on the Zinc Deposits of Southern Missouri. AIME Trans. 8, pp. 165-167.

Reeder, E.C., 1930. Methods and Cost of Mining Fluorspar at Rosiclare, Illinois. BuMines IC 6294, pp. 1-3.

Richard, F.W., 1930. Mining Methods and Costs at the Ground Hog Unit, Asarco Mining Co., Vanadium, New Mexico. BuMines IC 6377, 13 pp.

Rickard, T.A., 1932. Man and Metals. Whittlesey House, McGraw Hill, New York, pp. 154-161.

Rogers, A.F., 1937. Introduction to the Study of Minerals. Mcgraw-Hill Book Co. Inc., New York, 626 pp.

Ruhl, O., S.A. Allen and S.P. Holt, 1949. Zinc-Lead ore Reserves of the Tri-State District Missouri-Kansas-Oklahoma. BuMines IR 4490, 59 pp.

Salsbury, M.H., W.H. Kerns, F.B. Fulkerson, and G.C. Branner, 1964. Marketing Ores and Concentrates of Gold, Silver, Copper, Lead and Zinc in the United States. BuMines IC 8206, 150 pp.

Schultze, J.F., 1983. Gordonsville Concentrator Replaces Elmwood Production for Jersey Miniere Zinc Co. E&MJ, January, no. 1, pp. 73-76.

Seymour, C.M., 1930. History of Zinc in Tennessee. Mining Cong. Journal, vol. 16, no. 11, pp.821-822 and 833.

Smith, D.A., 1977. Colorado Mining. Univ. of New Mexico Press, Albuquerque, NM., pp. 100 and 101.

St, Joe Resources Company, Mining Division, 1982. St. Joe - Mining in Northern New York. 12 pp.

Strachan, C.B., 1930. Milling Methods of the American Zinc Co. of Tennessee, Mascot, Tennessee. BuMines IC 6379, pp. 1-12.

Tennessee Division of Geology, 1969. Papers on the Stratigraphy and Mine Geology of the Kingsport and Mascot Formations (Lower Ordovician) of East Tennessee. Report of Investigations No. 23, 90 pp.

The New Jersey Zinc Company, 1948. The First Hundred Years of The New Jersey Zinc Company, 1848-1948. Marbridge Printing Co., Inc., New York, 70 pp.

Thompson, C.J.S., 1990 (reprint, originally published in 1932). The Lure and Romance of Alchemy. Bell Publishing Co., New York, 249 pp.

Thompson, R. C., 1936. A Dictionary of Assyrian Chemistry and Geology. Clarendon Press, Oxford, pp. 76 and 78.

Thrush, P.W. (ed.), 1968. A Dictionary of Mining, Mineral and Related Terms. U.S. Bureau of Mines Special Publication, 1269 pp.

U.S. Geological Survey. Mineral Resources of the United States. Annual Volumes, 1883-1923.

Van Hise, C.R., and H.F. Bain, 1902. Lead and Zinc Deposits of the Mississippi Valley, United States. Inst. of Mining Engineers, Trans. 23, pp. 376-434.

Vanderburg, W.O., 1931. Milling Methods at the Hughesville Concentrator of the St. Joseph Lead Co., Hughesville, Mont. BuMines IC 6447, pp.1-14.

Wedow, H., Jr., T.H. Killsgaard, A.V. Heyl and R.B. Hall, 1973. Zinc. Ch. in United States Mineral Resources. USGS Prof. Paper 820, pp. 697-711.

Wethered, C.E., and L.J. Coady, 1930. Mining Methods at the Morning Mine of the Federal Mining and Smelting Company. Mullan, Idaho. BuMines IC 6238, pp. 1-5.

Winslow, A., 1894. Lead- and Zinc-Deposits of Missouri. AIME Trans. 24, pp. 634-682.

Young, H.I., 1930. History of the American Zinc, Lead, and Smelting Company. Mining Cong. J. 16:ll, November 1930, pp. 813-817.

Zeigler, W.L., 1932. Milling Methods and Costs at the Lead Concentrator of the Hecla Mining Co., Gem, Idaho. BuMines IC 6600, 17 pp.

Zinc Development Association, 1973. Zinc Chemicals. 243 pp.

STATISTICS

Table A1: U.S. and World Zinc Mine and Metal Production, 1800-1994

Year	Mine Production		Smelter Production	
	World	United States	World	United States
1800-1810	25	---	5	---
1811-1820	70	---	13	---
1821-1830	130	---	89	---
1831-1840	210	---	157	---
1841-1850	400	---	333	---
1851-1860	500	3	416	1
1861-1870	800	28	667	25
1871	160	7	139	6
1872	165	8	142	7
1873	170	8	151	8
1874	190	14	165	12
1875	200	17	175	15
1876	210	19	182	15
1877	230	17	207	14
1878	250	20	218	17
1879	250	23	220	19
1880	265	28	233	23
1881	290	35	262	27
1882	310	38	278	31
1883	300	42	287	33
1884	325	44	298	34
1885	325	47	298	36
1886	325	52	298	38
1887	330	58	306	45
1888	350	65	321	51
1889	370	66	335	53
1890	390	75	345	57
1891	425	90	364	73
1892	440	99	377	79
1893	425	89	380	71
1894	430	82	385	68
1895	465	96	415	81
1896	475	88	425	73
1897	490	109	445	91
1898	525	129	470	104
1899	550	146	491	117
1900	540	151	479	116
1901	570	168	511	128
1902	635	193	547	142
1903	650	180	574	145
1904	690	202	629	169
1905	720	213	660	184
1906	790	201	704	204
1907	810	236	738	233
1908	800	213	723	197
1909	860	277	775	240
1910	910	297	810	256
1911	1,000	313	894	273
1912	1,080	344	963	331

Table A1: U.S. and World Zinc Mine and Metal Production, 1800-1994 (cont.)

Year	Mine Production		Smelter Production	
	World	United States	World	United States
1913	1,120	369	1,015	338
1914	940	377	876	339
1915	925	533	837	471
1916	1,070	638	973	632
1917	1,101	647	1,095	623
1918	935	577	823	479
1919	806	498	659	441
1920	855	533	722	440
1921	511	233	472	198
1922	804	428	711	351
1923	980	554	946	499
1924	1,463	579	1,002	501
1925	1,354	645	1,128	555
1926	1,603	703	1,218	598
1927	1,604	652	1,306	576
1928	1,591	631	1,401	591
1929	1,751	657	1,451	610
1930	1,567	540	1,394	483
1931	1,045	372	997	284
1932	904	259	781	201
1933	1,134	349	983	306
1934	1,353	398	1,168	348
1935	1,451	470	1,332	408
1936	1,612	522	1,464	485
1937	1,751	568	1,623	552
1938	1,755	468	1,566	433
1939	1,789	529	1,650	506
1940	1,927	603	1,620	657
1941	2,072	680	1,749	800
1942	2,094	697	1,800	857
1943	2,018	675	1,840	899
1944	2,059	652	1,622	833
1945	1,616	557	1,302	738
1946	1,583	521	1,392	701
1947	1,769	578	1,599	782
1948	1,858	572	1,706	771
1949	1,910	538	1,825	789
1950	2,150	565	1,969	826
1951	2,359	618	2,141	844
1952	2,585	604	2,232	871
1953	2,667	497	2,359	879
1954	2,658	429	2,449	789
1955	2,903	467	2,658	934
1956	3,112	492	2,812	958
1957	3,148	482	2,903	960
1958	3,057	374	2,731	751
1959	3,121	386	2,858	777

Table A1: U.S. and World Zinc Mine and Metal Production, 1800-1994 (cont.)

Year	Mine Production		Smelter Production	
	World	United States	World	United States
1960	3,338	395	3,025	788
1961	3,488	421	3,248	818
1962	3,565	459	3,406	851
1963	3,661	480	3,487	864
1964	4,028	522	3,693	930
1965	4,302	554	3,949	978
1966	4,483	519	4,081	1,005
1967	4,835	498	4,126	918
1968	4,975	480	4,628	999
1969	5,342	502	4,973	1,008
1970	5,464	485	4,827	866
1971	5,515	456	4,744	769
1972	5,436	434	5,131	641
1973	5,709	434	5,332	605
1974	5,781	453	5,609	575
1975	5,850	426	5,013	450
1976	5,690	440	5,675	515
1977	5,906	408	5,812	454
1978	5,878	303	5,882	441
1979	5,998	267	6,260	526
1980	5,954	317	6,049	370
1981	5,919	312	6,081	397
1982	6,125	303	5,866	302
1983	6,283	275	6,249	305
1984	6,523	253	6,527	331
1985	6,758	227	6,798	334
1986	6,842	203	6,699	316
1987	7,188	216	7,014	344
1988	6,774	244	7,109	330
1989	6,808	276	7,245	358
1990	7,150	515	7,178	358
1991	7,270	518	7,311	376
1992	7,260	523	7,136	399
1993	6,960	488	7,400	381
1994	6,810	570	7,360	356
TOTALS	302,128	43,522	285,382	52,864

Sources: Siebenthal, 1919 (U.S. production before 1882); Ingalls, 1900; Pehrson 1929; Mineral Resources of the United States; and the U.S. Bureau of Mines Minerals Yearbook.

Table A2: U.S. Historical Salient Zinc Statistics (metric tons)

Year	Slab Zinc Production Primary	Slab Zinc Production Secondary	Imports for Consumption Mine 1/ Production Recoverable	Imports for Consumption Slab Zinc	Imports for Consumption Ore Zinc 2/ Content	Exports Slab Zinc	Exports Zinc 3/ Content	Slab Zinc 4/	Consumption Consumed as Ore 5/	All Classes 6/	Net Import Reliance (percent) 7/
1859	45	NA	245	NA	NA	NA	NA	NA	NA	NA	NA
1860	725	NA	925	NA	NA	NA	NA	NA	NA	NA	NA
1861	1,350	NA	1,650	NA	NA	NA	NA	NA	NA	NA	NA
1862	1,350	NA	1,600	NA	NA	NA	NA	NA	NA	NA	NA
1863	1,550	NA	1,800	NA	NA	NA	NA	NA	NA	NA	NA
1864	1,630	NA	1,900	NA	NA	43	NA	NA	NA	NA	NA
1865	1,900	NA	2,200	NA	NA	83	671	NA	NA	NA	NA
1866	1,800	NA	2,000	NA	NA	63	450	NA	NA	NA	NA
1867	2,900	NA	3,300	NA	NA	141	203	NA	NA	NA	NA
1868	3,350	NA	3,800	NA	NA	463	166	7,700	NA	7,700	NA
1869	3,900	NA	4,500	NA	NA	NA	378	8,731	NA	8,731	NA
1870	4,900	NA	5,400	NA	NA	49	NA	13,660	NA	13,660	NA
1871	6,250	NA	7,000	NA	NA	34	693	13,362	NA	13,362	NA
1872	7,070	NA	8,200	NA	NA	28	436	14,746	NA	14,746	NA
1873	6,661	NA	8,000	NA	NA	33	167	17,251	NA	17,252	NA
1874	11,900	NA	13,500	NA	NA	19	10	16,627	NA	16,627	NA
1875	14,363	NA	17,200	NA	NA	17	115	16,240	NA	16,240	NA
1876	15,400	NA	18,500	NA	NA	61	139	19,376	NA	19,376	NA
1877	14,150	NA	16,900	NA	NA	644	461	17,860	NA	17,860	NA
1878	17,800	NA	19,700	NA	NA	1,154	291	14,684	NA	14,684	NA
1879	19,300	NA	23,000	NA	NA	967	728	17,791	NA	17,791	NA
1880	21,082	NA	28,400	NA	NA	620	483	19,481	NA	19,481	NA
1881	27,450	NA	34,800	NA	NA	676	590	27,646	7,335	34,981	NA
1882	30,631	NA	38,000	NA	NA	675	516	29,308	7,257	36,565	NA
1883	33,451	NA	42,000	NA	NA	386	494	40,306	7,263	47,563	NA
1884	34,967	NA	44,000	NA	NA	57	138	42,315	8,709	51,024	NA
1885	36,911	NA	47,000	NA	NA	46	216	38,004	9,435	47,439	NA
1886	38,683	NA	52,000	NA	NA	349	310	39,294	10,886	50,180	NA
1887	45,667	NA	58,000	NA	NA	164	396	40,445	13,063	53,508	NA
1888	50,714	NA	65,000	NA	NA	28	54	49,217	13,063	62,280	NA
1889	53,397	NA	66,000	NA	NA	399	206	52,555	14,515	67,070	NA
1890	57,772	NA	75,000	NA	NA	1,494	1,213	54,388	12,316	66,704	NA
1891	73,367	NA	90,000	NA	NA	1,948	3,508	57,538	17,200	57,538	NA
1892	79,161	NA	99,000	NA	NA	5,667	171	71,327	17,200	88,527	NA
1893	71,515	NA	89,000	NA	NA	3,301	41	73,465	19,958	93,423	NA
							44	69,058	17,461	86,519	NA

Table A2: U.S. Historical Salient Zinc Statistics (cont.)
(metric tons)

Year	Slab Zinc Production Primary	Slab Zinc Production Secondary	Imports for Consumption Mine 1/ Production Recoverable	Imports for Consumption Slab Zinc	Exports Ore Zinc 2/ Content	Exports Slab Zinc	Exports Zinc 3/ Content	Consumption Slab Zinc 4/	Consumption Consumed as Ore 5/	Consumption All Classes 6/	Net Import Reliance (percent) 7/
1894	68,336	NA	82,000	NA	NA	1,636	--	64,028	14,506	78,534	NA
1895	81,361	NA	96,000	NA	NA	1,388	21	79,557	15,030	94,587	NA
1896	73,934	NA	88,000	NA	NA	9,189	1,882	63,609	14,515	78,124	NA
1897	90,700	NA	109,000	NA	NA	12,923	3,746	80,541	18,144	98,685	NA
1898	104,688	NA	129,000	NA	NA	9,524	4,771	98,156	23,950	122,107	NA
1899	117,073	NA	146,000	NA	NA	6,127	11,429	113,021	29,135	142,156	NA
1900	112,388	NA	150,546	NA	NA	20,322	19,079	90,173	35,446	125,619	NA
1901	127,752	NA	167,646	NA	NA	3,075	20,029	128,529	35,198	163,727	NA
1902	142,362	NA	192,922	406	NA	2,937	25,280	138,511	40,591	179,102	NA
1903	144,441	NA	180,194	183	NA	1,380	17,876	140,052	48,308	188,360	NA
1904	169,378	NA	201,951	309	727	9,205	16,289	164,120	49,343	213,463	NA
1905	184,929	NA	213,003	388	7,301	5,004	14,037	181,834	53,724	235,558	NA
1906	203,908	NA	201,147	926	17,816	4,237	12,574	200,289	58,915	259,204	NA
1907	226,669	6,369	235,824	1,550	33,168	511	9,234	205,903	59,944	272,538	NA
1908	190,893	6,495	212,758	704	17,614	1,119	11,842	194,289	43,549	244,104	NA
1909	232,022	8,412	277,075	8,545	37,356	303	5,650	245,602	49,114	310,094	NA
1910	244,200	11,597	297,295	897	23,078	3,388	8,941	223,062	45,896	296,733	NA
1911	259,932	12,740	313,215	293	14,943	13,023	8,292	254,065	41,841	330,140	NA
1912	307,360	23,645	343,656	9,724	15,936	6,808	10,591	308,752	58,089	408,701	NA
1913	314,499	23,579	368,694	4,686	12,244	12,234	8,034	267,955	64,067	361,587	NA
1914	320,281	18,638	377,082	177	11,006	67,594	5,040	272,140	64,064	364,877	NA
1915	444,084	27,001	533,057	57	52,316	119,213	377	330,991	72,187	459,036	NA
1916	605,506	26,105	638,039	19	134,397	137,211	71	416,685	81,897	577,969	NA
1917	607,427	15,272	647,159	16	65,747	199,639	1,197	375,251	88,517	564,368	NA
1918	469,856	8,997	577,052	10	22,506	96,751	56	334,458	93,239	537,589	NA
1919	422,515	17,915	497,905	29	15,430	132,718	--	293,895	90,909	475,194	NA
1920	420,469	19,387	532,993	--	23,265	103,661	--	293,060	106,122	497,362	NA
1921	181,891	15,942	232,820	5,986	12,963	4,495	--	184,703	51,917	295,881	NA
1922	321,395	29,926	428,220	36	NA	31,571	1,538	338,462	86,104	513,990	NA
1923	463,058	35,774	554,009	1	788	29,885	2,544	405,071	106,633	609,178	NA
1924	469,322	32,192	578,763	10	5,489	69,165	334	406,652	99,207	602,735	NA

Table A2: U.S. Historical Salient Zinc Statistics (cont.)
(metric tons)

Year	Slab Zinc Production		Mine 1/ Production Recoverable	Imports for Consumption		Exports			Slab Zinc 4/	Consumption		Net Import Reliance (percent) 7/
	Primary	Secondary		Slab Zinc	Ore Zinc 2/ Content	Slab Zinc	Zinc 3/ Content			Consumed as Ore 5/	All Classes 6/	
1925	519,768	35,544	644,870	--	18,566	69,264	62,551		453,680	100,698	647,160	NA
1926	561,023	37,012	702,672	--	12,508	38,936	86,411		505,326	117,934	723,867	NA
1927	537,522	38,813	651,850	35	5,572	41,454	42,380		468,446	110,677	679,990	NA
1928	546,652	44,149	630,648	--	3,951	22,942	4,099		568,351	112,491	788,199	NA
1929	567,396	42,953	657,236	205	1,705	13,073	64		575,427	125,192	808,953	NA
1930	451,819	31,614	540,161	255	23,264	4,203	--		409,050	95,254	582,693	NA
1931	264,894	19,618	352,234	249	706	583	12		335,658	68,039	470,783	NA
1932	187,922	13,352	258,757	281	1,727	5,870	--		234,961	49,895	339,731	NA
1933	278,671	27,294	348,613	1,715	1,935	1,039	734		317,787	65,317	453,193	NA
1934	329,843	17,868	398,006	1,565	12,952	4,631	3,285		326,496	68,946	458,948	NA
1935	381,593	25,991	469,834	4,032	9,544	1,467	418		429,099	78,018	581,967	NA
1936	446,455	38,291	522,152	10,578	156	34	222		527,982	87,090	709,871	NA
1937	505,215	46,769	568,226	33,755	3,035	226	285		553,383	101,605	758,104	NA
1938	404,914	28,679	468,742	6,559	4,409	--	122		381,925	62,596	522,944	NA
1939	460,157	45,748	529,621	28,086	30,393	4,096	275		567,898	77,111	798,679	NA
1940	612,599	44,377	603,340	9,204	40,494	71,750	406		665,018	87,090	907,883	NA
1941	745,724	53,980	679,595	36,549	140,171	81,020	--		750,637	122,470	1,086,368	NA
1942	809,093	48,248	696,741	32,978	256,885	121,507	--		660,584	104,326	1,031,676	NA
1943	854,849	43,740	675,123	50,943	468,694	88,395	1		740,968	104,054	1,157,226	NA
1944	788,618	44,486	651,941	57,721	376,485	19,578	--		806,148	128,820	1,223,677	NA
1945	693,598	44,672	557,336	87,779	300,762	7,060	--		773,204	118,841	1,191,969	NA
1946	660,668	40,384	521,478	94,406	151,396	42,841	81		735,039	121,563	1,101,486	24
1947	728,011	54,016	578,428	65,374	176,740	96,769	1,274		713,374	132,449	1,087,273	25
1948	714,648	56,536	571,506	83,910	121,394	59,454	3,218		741,837	120,656	1,114,419	8/
1949	739,158	49,932	538,145	113,910	99,369	53,260	2,654		645,771	79,832	900,135	1
1950	765,181	60,754	565,516	140,915	215,514	11,718	1,034		877,369	121,956	1,247,865	39
1951	799,804	44,141	617,964	79,871	179,618	33,121	2,803		847,284	121,422	1,203,002	24
1952	820,530	49,996	604,186	102,560	491,979	52,357	3,057		773,632	99,134	1,099,189	39
1953	831,077	47,967	496,620	206,524	407,990	16,301	2,679		894,418	107,269	1,217,795	47
1954	727,948	61,700	429,526	145,277	436,282	22,674	--		802,223	90,035	1,071,106	49
1955	874,076	59,912	466,902	176,955	348,947	16,392	--		1,015,877	107,170	1,332,727	43

Table A2: U.S. Historical Salient Zinc Statistics (cont.)
(metric tons)

Year	Slab Zinc Production Primary	Slab Zinc Production Secondary	Imports for Consumption Mine 1/ Production Recoverable	Imports for Consumption Slab Zinc	Exports Ore 2/ Zinc Content	Exports Slab Zinc	Exports Zinc 3/ Content	Consumption Slab Zinc 4/	Consumption Consumed as Ore 5/	Consumption All Classes 6/	Net Import Reliance (percent) 7/
1956	892,316	65,433	492,003	222,012	419,463	7,995	775	915,159	102,864	1,200,226	45
1957	894,299	65,764	482,387	243,873	616,356	9,784	6	848,780	100,072	1,117,283	47
1958	708,735	42,279	373,765	168,458	487,792	1,881	—	787,733	86,126	1,036,155	56
1959	724,538	52,452	385,829	149,197	384,769	10,550	1	867,448	98,039	1,159,724	53
1960	725,309	62,352	395,013	109,701	347,396	68,170	12	796,403	80,082	1,051,371	46
1961	768,200	50,110	421,288	113,567	324,457	45,409	1,515	844,782	88,225	1,095,398	46
1962	797,774	53,415	458,574	123,373	351,372	32,751	123	936,053	92,154	1,209,560	47
1963	809,739	54,706	480,181	120,050	337,399	30,711	15	1,002,542	94,987	1,282,956	49
1964	865,531	64,951	521,503	121,670	282,529	24,064	35	1,095,215	96,114	1,393,210	44
1965	902,107	75,858	554,429	139,667	365,537	5,388	NA	1,228,412	111,486	1,580,377	51
1966	929,924	75,535	519,416	254,290	359,585	1,276	NA	1,291,528	114,937	1,651,088	54
1967	851,692	66,683	498,419	201,397	391,286	15,249	NA	1,134,592	103,692	1,456,814	52
1968	926,137	72,452	480,305	276,407	439,806	29,947	NA	1,225,295	112,590	1,583,362	58
1969	944,014	64,005	501,786	294,616	512,772	8,435	NA	1,256,796	114,951	1,645,785	59
1970	796,337	69,995	484,560	235,988	408,932	261	NA	1,076,784	113,199	1,425,728	53
1971	695,297	73,412	455,899	294,159	423,989	12,107	NA	1,137,664	108,185	1,497,485	60
1972	574,411	66,876	433,924	468,691	157,907	8,923	NA	1,286,705	107,325	1,672,870	61
1973	529,323	75,466	434,408	535,920	139,864	13,214	NA	1,364,350	117,617	1,752,613	65
1974	503,658	71,246	453,477	493,333	121,321	17,293	NA	1,168,178	115,315	1,517,732	60
1975	397,394	52,513	425,792	340,124	388,769	6,257	NA	839,445	75,053	1,117,484	63
1976	452,554	62,192	439,543	630,612	141,342	3,187	NA	1,028,876	91,844	1,394,268	59
1977	408,364	45,913	407,889	503,621	109,277	215	NA	999,505	86,940	1,367,704	58
1978	406,698	34,774	302,669	622,470	106,315	723	10,973	1,050,585	89,959	1,441,810	67
1979	472,481	53,212	267,341	524,130	87,499	279	20,095	1,000,606	79,710	1,394,314	63
1980	340,456	29,396	317,103	410,163	182,370	302	54,457	811,146	58,986	1,142,409	60
1981	346,563	50,192	312,418	612,007	245,710	323	54,232	840,875	60,643	1,189,369	65
1982	228,176	74,288	303,160	456,233	66,809	341	77,289	795,000	35,515	1,038,600	58
1983	235,694	69,390	275,294	617,679	63,156	427	60,168	933,000	36,912	1,246,300	65
1984	253,432	78,113	252,768	639,228	86,172	760	30,579	980,000	45,487	1,344,000	68
1985	261,209	72,563	226,545	610,900	90,186	1,011	23,264	961,000	39,886	1,257,000	70
1986	253,369	62,912	202,983	665,126	75,786	1,938	3,269	999,000	19,236	1,274,000	73

Table A2: U.S. Historical Salient Zinc Statistics (cont.)
(metric tons)

Year	Slab Zinc Production		Imports for Consumption		Exports				Consumption			Net Import Reliance (percent) 7/
	Primary	Secondary	Mine 1/ Production Recoverable	Slab Zinc	Ore Zinc 2/ Content	Slab Zinc	Zinc 3/ Content	Slab Zinc 4/	Consumed as Ore 5/	All Classes 6/		
1987	261,345	82,589	216,327	705,985	46,464	1,082	16,921	1,052,000	2,536	1,324,000		69
1988	241,294	88,492	244,314	749,130	62,966	482	33,590	1,089,000	2,412	1,340,000		70
1989	260,305	97,904	275,883	711,554	40,974	5,532	78,877	1,060,000	2,107	1,311,000		61
1990	262,704	95,708	515,355	631,742	46,684	1,238	220,446	991,000	2,178	1,240,000		40
1991	253,276	122,457	517,804	549,137	45,419	1,253	381,818	933,000	w	1,165,000		24
1992	271,867	127,623	523,430	644,482	44,523	5,886	307,118	1,035,000	w	1,276,000		30
1993	240,282	141,472	488,283	723,563	33,093	8,756	311,278	1,148,000	w	1,354,000		26
1994	216,600	139,000	570,000	793,000	27,400	6,680	389,000	1,190,000	w	1,420,000		25

Source: U.S. Bureau of Mines
NA Not Available. w Withheld to protect proprietary data.
1/ Estimated for 1859-1899.
2/ Imports ore: 1904-1909 were estimated (Minerals Yearbook, 1926) and for 1910-1936, imports are the zinc content of General Imports.
3/ Exports ore: 1897-1915, the zinc content estimated at 50% of ore shipped; 1907-1915, virtually all ore shipped was willemite concentrate (50% zinc).
4/ Data through 1981 are reported consumption; 1982 forward data are apparent consumption.
5/ 1880-1900 data are zinc oxide from ore; 1901-1907 data are for zinc and zinc-lead oxide from ore; 1908-1918 data are pigments from ore; 1919-1924 data are pigments and salts from ore; 1925-1956 data are zinc oxide from ore; 1957-1988 includes ore.
6/ Apparent consumption of slab zinc and ores, concentrates, and secondary materials used to make dust and chemicals.
7/ Import reliance is expressed as a percentage of apparent consumption.
8/ Net exporter.

The U.S. Zinc Industry

Table A3: Mine Production Of Recoverable Zinc In The United States, By State, 1906-1994.
(Metric tons)

State	1907	1908	1909	1910	1911	1912	1913	1914	1915	1916	1917	1918	1919	1920	1921	1922	1923
Alaska	---	---	---	---	---	---	---	---	---	---	---	---	---	---	---	---	---
Arizona	103	308	2,712	2,488	2,069	3,973	4,276	4,442	8,264	8,926	9,477	1,030	779	661	---	95	236
Arkansas	430	549	463	902	602	679	679	552	2,911	6,182	6,070	863	171	298	14	122	134
California	105	---	---	---	1,274	1,971	480	177	5,939	6,920	4,931	2,523	214	529	384	1,366	---
Colorado	23,944	13,667	23,228	34,967	42,913	59,975	54,134	43,896	47,443	60,911	54,574	40,431	16,883	22,131	1,070	10,550	24,563
Idaho	3,169	17	613	2,542	3,783	6,308	10,512	19,056	31,821	39,238	36,221	20,485	7,255	12,670	15	1,864	12,679
Illinois	669	1,558	1,962	3,220	3,827	3,688	2,028	4,365	5,020	3,088	3,871	3,440	6,158	4,282	2,201	2,834	1,148
Iowa	132	358	32	87	---	---	---	---	---	19	---	---	---	---	---	---	---
Kansas	16,122	12,809	10,383	12,001	9,319	9,646	9,152	10,237	13,032	11,293	18,370	27,394	43,215	55,405	33,560	51,006	91,598
Kentucky	---	---	51	5	143	445	297	209	693	994	681	286	33	8	200	129	245
Maine	---	---	---	---	---	---	---	---	---	---	---	---	---	---	---	---	---
Missouri	105,916	97,435	118,081	116,654	111,144	123,877	113,365	96,156	123,649	141,485	120,580	50,795	28,613	22,234	9,838	14,670	16,570
Montana	111	744	4,246	14,351	19,872	12,210	40,222	50,612	84,888	103,991	84,486	94,918	76,550	83,614	10,558	54,009	64,165
Nevada	983	506	1,367	1,228	1,609	6,043	6,541	5,888	11,057	14,716	10,119	7,586	4,084	4,852	32	1,188	6,426
New Hampshire	---	---	---	---	---	---	14	5	22	33	41	13	---	---	---	---	---
New Jersey	62,931	66,931	72,085	59,339	57,788	69,454	72,179	75,358	107,217	101,489	109,808	92,821	83,929	71,224	51,208	66,821	68,245
New Mexico	340	1,622	5,936	8,205	4,644	6,153	7,495	8,348	11,523	16,588	13,698	10,909	3,445	4,542	103	2,039	7,482
New York	---	---	---	---	73	129	---	---	2,227	4,089	4,710	3,426	4,645	5,129	1,426	4,369	7,678
North Carolina	---	---	---	---	---	---	9	---	---	---	---	---	---	---	---	---	---
Oklahoma	1,503	4,109	7,081	5,801	4,672	5,233	10,581	12,693	12,985	26,085	77,868	146,421	161,851	199,333	110,107	190,220	219,921
Oregon	---	---	---	---	---	---	---	---	---	---	---	---	---	---	---	---	---
Pennsylvania	---	---	---	---	---	---	---	---	---	---	---	---	---	---	---	---	---
South Dakota	---	---	---	---	---	---	---	---	---	---	---	---	---	---	---	---	---
Tennessee	99	312	540	876	1,013	1,988	5,065	9,457	14,933	23,975	25,853	19,115	21,089	17,433	8,792	14,123	14,424
Texas	20	16	---	---	---	108	296	98	14	105	10	---	---	---	---	---	---
Utah	2,473	662	4,472	7,424	8,092	7,742	8,554	7,253	11,019	13,414	9,655	8,346	2,010	3,700	32	2,322	5,139
Virginia	---	640	53	720	937	226	2,467	158	1,149	2,065	1,545	863	---	---	---	---	---
Washington	---	---	---	---	9	---	---	---	111	768	542	17	---	193	204	557	1,372
Wisconsin	16,774	16,516	21,003	23,521	26,962	29,982	27,315	28,225	37,560	51,531	54,197	45,372	36,981	24,753	3,075	9,935	11,985
Other	---	---	---	---	---	---	---	---	---	---	---	---	---	---	---	---	---
Totals	235,824	218,759	274,308	294,331	300,745	349,830	375,661	377,185	533,477	637,905	647,307	577,054	497,905	532,991	232,819	428,219	554,010
Employment, mine and mill	---	---	---	---	---	---	---	---	---	---	---	---	---	---	---	---	---

146

Table A3: Mine Production Of Recoverable Zinc In The United States, By State, 1906-1994.
(Metric tons)

State	1924	1925	1926	1927	1928	1929	1930	1931	1932	1933	1934	1935	1936	1937	1938
Alaska	--	--	--	--	--	--	--	--	--	--	--	--	--	--	--
Arizona	--	3,326	5,872	1,029	580	1,115	739	--	--	5	821	3,027	3,256	4,560	5,274
Arkansas	3	39	79	116	78	8	--	--	--	10	62	139	165	219	138
California	1,388	5,212	9,269	3,657	--	--	--	73	--	132	327	146	7	18	--
Colorado	25,731	27,951	29,483	32,536	32,415	26,699	32,894	14,685	99	1,166	700	1,090	1,172	3,853	4,130
Idaho	6,958	14,169	23,865	24,293	28,361	41,434	34,155	17,753	9,300	19,022	22,497	28,272	44,453	49,169	39,943
Illinois	2,279	2,471	2,338	473	15	28	8	--	--	--	--	--	--	--	--
Iowa	--	--	--	--	--	--	--	--	--	--	--	--	--	--	--
Kansas	95,610	107,754	114,584	99,271	97,296	99,654	67,407	35,426	23,838	37,146	34,710	49,088	71,683	72,847	66,246
Kentucky	446	389	1,667	780	83	--	--	--	42	207	113	115	216	245	292
Maine	--	--	--	--	--	--	--	--	--	--	--	--	--	--	--
Missouri	11,721	13,421	23,603	16,998	11,770	9,994	9,808	2,908	894	4,574	6,404	6,589	16,972	18,688	9,277
Montana	58,276	52,306	66,860	72,784	74,142	61,848	23,969	6,121	1,993	18,800	27,870	49,696	45,102	35,533	8,023
Nevada	4,990	6,723	4,907	2,878	3,082	7,675	13,230	9,463	115	5,794	12,646	14,094	12,226	12,915	8,114
New Hampshire	--	--	--	--	--	--	--	--	--	--	--	--	--	--	--
New Jersey	76,539	89,261	73,145	86,813	90,601	94,111	88,565	85,534	73,899	68,152	69,448	77,754	81,540	91,996	77,872
New Mexico	9,417	8,388	10,933	27,036	28,307	31,257	29,724	25,280	23,218	28,054	24,060	20,072	18,750	21,706	25,615
New York	4,231	4,679	4,573	3,402	10,212	9,299	20,385	21,863	15,235	16,087	21,036	21,518	24,440	29,656	27,121
North Carolina	--	--	--	--	--	--	--	--	--	--	--	--	--	--	--
Oklahoma	244,157	257,070	247,269	187,434	163,522	174,218	123,516	70,880	57,549	82,613	97,769	117,719	117,186	123,101	102,443
Oregon	--	--	--	--	--	--	5	--	5	5	34	--	55	22	--
Pennsylvania	--	--	--	--	--	--	--	--	--	--	--	--	--	--	--
South Dakota	--	--	--	--	--	--	--	--	--	--	--	--	--	--	--
Tennessee	13,042	14,747	10,975	9,435	29,862	36,794	43,678	34,756	16,796	29,728	43,284	44,300	40,747	50,127	51,497
Texas	--	--	--	--	--	--	--	--	--	--	--	--	--	--	--
Utah	8,420	23,864	43,173	44,990	42,573	46,729	40,365	33,829	26,913	26,984	25,581	28,220	32,833	43,546	30,534
Virginia	2,416	2,478	5,289	7,552	b	b	b	b	2,037	3,056	b	b	b	b	b
Washington	413	552	474	581	39	961	319	4,512	2,037	3,056	1,747	1	3,994	3,734	10,344
Wisconsin	12,725	18,352	24,313	29,793	16,708	15,409	11,392	9,152	6,824	7,076	8,897	8,095	7,372	6,294	1,881
Other	--	--	--	--	--	--	--	--	--	--	--	--	--	--	--
Totals	578,762	653,152	702,671	651,851	629,646	657,233	540,159	372,235	258,757	348,611	398,006	469,935	522,169	568,229	468,744
Employment, mine and mill	--	--	--	--	--	--	--	--	--	--	--	--	--	--	--

Table A3: Mine Production Of Recoverable Zinc In The United States, By State, 1906-1994.
(Metric tons)

State	1939	1940	1941	1942	1943	1944	1945	1946	1947	1948	1949	1950	1951	1952
Alaska	---	---	---	---	---	---	---	---	---	---	---	---	1	---
Arizona	6,088	14,021	14,962	16,803	17,851	26,378	36,492	39,612	49,572	49,422	64,100	54,867	48,080	42,767
Arkansas	112	399	187	164	87	17	275	77	16	20	2	5	45	24
California	5	72	399	556	1,684	7,670	9,002	6,239	4,912	28	1	7	8,711	8,545
Colorado	1,660	4,590	14,263	29,225	40,001	36,247	32,453	32,792	35,149	4,831	6,540	6,850	50,543	48,265
Idaho	43,136	64,048	71,744	79,157	78,859	82,891	75,716	64,870	75,359	40,972	43,275	41,527	70,870	67,419
Illinois	303	4,371	8,344	8,516	5,308	6,588	7,539	7,981	9,138	78,260	69,450	79,732	19,755	17,070
Iowa	---	---	---	---	---	---	---	---	---	11,775	16,472	24,478	---	---
Kansas	62,569	51,739	64,776	50,688	51,659	57,790	43,902	43,275	37,645	32,275	26,701	24,654	26,221	23,117
Kentucky	825	1,159	387	369	845	309	165	285	461	580	848	663	3,136	2,976
Maine	---	---	---	---	---	---	---	---	---	---	---	---	---	---
Missouri	13,695	11,524	19,896	33,016	27,590	33,227	20,117	20,170	15,489	5,863	5,362	7,429	10,411	12,688
Montana	31,569	47,706	55,075	49,637	34,116	32,774	15,788	15,213	41,439	53,610	49,165	61,396	77,611	74,557
Nevada	5,650	10,735	13,725	9,251	12,380	18,778	19,465	20,547	15,395	18,405	18,546	19,601	15,824	13,932
New Hampshire	---	---	---	---	---	---	---	---	---	---	---	---	---	---
New Jersey	80,482	82,922	85,077	85,312	84,245	72,836	73,838	58,472	69,736	69,247	46,252	49,921	57,077	53,696
New Mexico	26,631	27,499	34,348	42,149	53,999	46,019	36,555	32,752	40,010	37,650	26,622	26,547	41,203	46,244
New York	32,871	32,374	34,878	41,555	41,730	32,242	22,660	29,497	30,950	31,358	34,449	34,764	36,334	29,607
North Carolina	---	---	---	---	---	---	---	---	---	---	---	---	---	---
Oklahoma	127,350	147,812	151,139	132,912	103,496	82,961	62,868	63,552	46,323	39,754	39,946	42,401	48,489	49,819
Oregon	---	---	---	---	---	---	1	---	1	---	5	19	2	1
Pennsylvania	---	---	---	---	42	51	---	---	17	26	---	---	---	---
South Dakota	---	---	---	104	---	---	---	---	---	---	---	---	---	---
Tennessee	51,006	31,566	32,813	39,890	37,889	37,041	30,685	22,329	28,315	26,784	27,023	32,047	35,053	34,491
Texas	---	---	---	---	---	---	---	40	20	---	---	---	22	2
Utah	31,321	39,724	38,146	41,316	42,543	35,375	30,509	25,666	39,619	37,639	36,895	28,738	31,132	29,889
Virginia	b	15,356	20,786	14,507	16,876	17,842	14,583	15,336	15,230	14,408	11,944	11,245	6,651	12,164
Washington	9,191	10,487	12,991	13,062	11,070	10,799	10,608	10,277	12,519	11,465	9,743	13,433	16,501	18,236
Wisconsin	5,356	5,234	5,659	8,551	13,052	14,106	14,117	12,951	11,089	7,134	4,804	5,191	14,292	18,677
Other	---	---	---	---	---	---	---	---	---	---	---	---	---	---
Totals	529,620	603,338	679,595	696,740	675,122	651,941	557,338	521,933	578,427	571,506	538,145	565,515	617,964	604,186
Employment, mine and mill	---	---	---	---	---	---	---	---	---	---	---	---	---	---

Table A3: Mine Production Of Recoverable Zinc In The United States, By State, 1906-1994.
(Metric tons)

State	1953	1954	1955	1956	1957	1958	1959	1960	1961	1962	1963	1964	1965	1966	1967
Alaska	---	---	---	---	---	---	---	---	---	---	---	---	---	---	---
Arizona	24,975	19,469	22,684	23,206	30,758	25,884	33,861	32,487	26,839	29,835	23,060	22,398	19,738	14,501	13,000
Arkansas	---	---	---	---	---	---	44	45	34	191					
California	4,681	11,284	6,202	7,302	2,693	46	71	422	276	292	92	130	204	304	400
Colorado	34,300	31,888	32,069	36,511	42,638	33,686	32,103	28,375	38,689	39,327	43,644	48,180	48,870	49,734	47,575
Idaho	65,456	55,817	48,366	44,961	52,463	45,110	50,529	33,385	52,884	57,030	57,395	53,794	52,648	55,335	51,281
Illinois	13,205	13,088	19,686	21,808	20,126	22,625	24,326	26,807	24,308	24,869	18,449	12,519	16,614	13,782	18,521
Iowa															
Kansas	14,075	17,336	25,048	26,004	14,387	4,911	923	1,921	2,219	3,577	3,182	4,232	5,904	4,326	4,323
Kentucky	444	415	---	378	759	1,141	611	788	1,041	1,063	1,325	1,872	5,129	5,975	5,731
Maine															
Missouri	9,055	4,726	4,061	3,973	2,677	328	83	2,559	5,304	2,533	291	1,362	3,912	3,600	6,740
Montana	72,821	55,295	62,222	63,975	45,831	30,153	25,263	11,386	9,310	34,181	29,884	26,362	30,650	26,417	3,031
Nevada	5,273	939	2,422	6,793	4,801	83	197	381	411	255	518	528	3,500	5,286	2,753
New Hampshire															
New Jersey	41,458	33,943	10,562	4,234	11,367	551			102	13,888	29,699	29,870	34,742	22,895	23,624
New Mexico	12,132	5	13,859	31,761	29,647	8,196	4,203	12,492	20,774	19,972	11,737	27,064	33,076	26,577	19,396
New York	46,746	48,261	48,095	53,625	58,658	48,094	39,430	60,204	49,680	48,674	48,530	55,115	63,394	66,636	64,006
North Carolina											12				
Oklahoma	30,312	39,164	37,687	24,961	13,563	4,778	952	2,116	2,856	9,084	12,016	11,030	11,535	10,194	9,680
Oregon									2		2	w	w		
Pennsylvania						9,808	15,166	12,445	21,254	22,052	24,847	27,900	25,070	25,474	31,812
South Dakota															
Tennessee	34,895	27,511	36,483	41,751	52,063	53,642	81,585	82,911	74,148	64,907	86,951	105,182	111,028	93,546	102,571
Texas															
Utah	26,475	30,872	39,513	38,441	37,055	40,807	31,954	32,183	33,783	31,128	32,821	28,511	25,172	33,859	31,072
Virginia	15,128	15,184	16,628	17,414	20,938	16,758	18,447	18,039	26,456	24,021	21,762	19,055	18,589	16,026	17,097
Washington	29,743	20,234	26,795	23,232	21,772	17,052	15,523	19,338	18,341	19,635	20,203	22,041	20,167	22,743	19,541
Wisconsin	15,268	14,092	16,620	21,673	19,573	11,013	10,555	16,701	12,578	12,058	13,711	23,839	24,488	22,475	26,266
Other															
Totals	496,442	439,523	469,002	492,003	481,769	374,666	385,826	394,985	421,289	458,572	480,131	520,984	554,430	519,685	498,420
Employment, mine and mill	---	---	---	---	---	---	---	---	---	---	---	---	---	---	---

Table A3: Mine Production Of Recoverable Zinc In The United States, By State, 1906-1994.
(Metric tons)

State	1968	1969	1970	1971	1972	1973	1974	1975	1976	1977	1978	1979	1980	1981
Alaska	---	---	---	---	---	---	---	---	---	---	---	---	---	---
Arizona	4,936	8,200	8,725	7,041	9,173	7,645	8,799	7,852	8,619	3,973	W	W	W	138
Arkansas	---	---	---	---	---	---	---	---	---	---	---	---	---	---
California	3,198	3,018	3,188	2,724	1,090	18	7	187	154	2	W	W	W	W
Colorado	45,593	48,729	51,432	55,502	57,879	52,924	44,896	43,692	45,923	36,530	22,208	9,910	13,823	W
Idaho	51,935	50,712	37,242	40,894	35,060	41,828	35,806	37,127	42,262	28,121	32,353	29,660	27,722	W
Illinois	16,494	12,487	15,238	11,527	10,322	4,763	3,723	W	W	W	W	W	W	W
Iowa	---	---	---	---	---	---	---	---	---	---	---	---	---	---
Kansas	2,732	1,724	1,076	---	1,615	248	---	37	54	---	---	---	---	---
Kentucky	4,176	4,525	3,800	4,779	5,280	17,817	9,457	7,546	7,085	6,594	52	W	W	W
Maine	---	6,930	8,268	5,307	56,176	74,707	83,449	67,918	75,777	74,107	59,038	61,682	62,886	52,904
Missouri	11,159	37,284	46,013	43,740	11	66	123	100	58	72	79	104	71	25
Montana	3,427	5,573	1,295	327	---	---	3,089	4,986	1,305	1,517	1,371	W	4	W
Nevada	1,909	854	115	64	---	---	---	---	---	---	---	---	---	---
New Hampshire	---	---	---	---	---	---	---	---	---	---	---	---	---	---
New Jersey	23,286	22,749	26,021	27,195	34,560	29,962	29,799	28,218	30,633	30,358	28,915	31,118	28,859	16,198
New Mexico	16,952	22,052	15,060	12,663	11,553	11,183	12,505	9,993	W	W	W	W	W	W
New York	60,050	53,277	53,140	57,534	55,111	73,895	84,438	69,501	66,833	64,264	26,463	12,133	33,629	36,889
North Carolina	---	---	---	---	---	---	---	---	---	---	---	---	---	---
Oklahoma	6,279	2,489	---	---	---	---	---	---	---	---	---	---	---	---
Oregon	---	---	---	---	---	---	---	---	---	---	---	---	---	---
Pennsylvania	27,562	29,969	26,811	24,891	16,641	17,107	18,405	19,133	20,212	20,706	19,099	21,447	22,556	24,732
South Dakota	---	---	1	---	---	---	---	---	---	---	---	---	---	---
Tennessee	112,526	112,974	107,284	108,223	92,281	58,216	77,719	75,562	74,854	83,185	87,906	85,119	128,722	117,684
Texas	---	---	---	---	---	---	---	---	---	---	---	---	---	---
Utah	30,076	31,662	31,468	23,316	19,825	15,241	11,448	17,817	20,394	16,111	3,505	W	W	1,576
Virginia	17,470	16,968	16,386	15,267	15,231	15,135	15,599	13,745	10,198	12,040	10,974	11,406	12,038	9,731
Washington	12,595	8,834	10,846	5,245	5,881	5,786	6,268	W	W	5,055	W	W	W	---
Wisconsin	23,325	20,775	18,719	9,657	6,235	7,867	7,926	W	W	W	W	W	W	---
Other	---	---	2,432	3	---	---	21	30,340	35,182	26,394	10,703	4,762	3,763	52,541
Totals	475,680	501,785	484,560	455,899	433,924	434,408	453,477	433,754	439,543	409,029	302,666	267,341	334,073	312,418
Employment, mine and mill	---	---	8700	8200	6800	6700	6700	6700	6700	6600	5700	4600	5500	4200

150

Table A3: Mine Production Of Recoverable Zinc In The United States, By State, 1906-1994.
(Metric tons)

State	1982	1983	1984	1985	1986	1987	1988	1989	1990	1991	1992	1993	1994	TOTALS
Alaska	---	---	---	---	---	---	---	---	---	---	---	---	w	51
Arizona	---	---	---	---	---	---	---	w	w	w	w	w	---	998,254
Arkansas	---	---	---	---	---	---	---	---	---	---	---	---	---	24,430
California	---	---	---	---	---	---	---	---	---	---	---	---	---	161,043
Colorado	w	w	w	w	w	w	w	w	w	w	w	w	w	2,377,476
Idaho	w	w	w	w	w	w	w	w	w	w	w	w	w	2,880,379
Illinois	w	w	w	w	w	w	w	w	w	w	w	w	w	573,896
Iowa	---	---	---	---	---	---	---	---	---	---	---	---	---	628
Kansas	---	---	---	---	---	---	---	---	---	---	---	---	---	2,242,013
Kentucky	w	w	w	w	---	10	w	---	w	w	w	w	---	67,980
Maine	---	---	---	---	---	---	---	---	---	---	---	---	---	74,284
Missouri	63,680	57,044	45,458	49,340	37,919	34,596	41,322	50,790	48,864	42,506	44,031	40,200	43,000	3,267,954
Montana	w	---	---	---	---	w	18,935	w	w	w	20,588	w	---	2,620,161
Nevada	---	---	---	---	---	---	---	---	7,889	w	---	---	---	472,554
New Hampshire	---	---	---	---	---	---	---	---	---	---	---	---	---	128
New Jersey	16,800	16,475	w	w	w	w	---	---	---	---	---	---	---	4,161,251
New Mexico	---	---	---	---	---	---	---	---	---	---	---	---	---	1,345,969
New York	52,237	56,748	w	w	w	w	w	w	w	w	w	w	w	2,391,868
North Carolina	---	---	---	---	---	---	---	---	---	---	---	---	---	150
Oklahoma	---	---	---	---	---	---	---	---	---	---	---	---	---	4,730,428
Oregon	---	---	---	---	---	---	---	---	---	751	---	---	123	1,033
Pennsylvania	24,762	16,792	---	---	---	---	---	---	---	---	---	---	---	566,653
South Dakota	---	---	---	---	---	---	---	---	---	---	---	---	---	241
Tennessee	121,306	109,958	116,526	104,471	102,118	115,699	119,954	w	w	w	w	w	w	4,277,248
Texas	---	---	---	---	---	---	---	---	---	---	---	---	---	751
Utah	---	---	w	---	---	---	---	---	---	---	---	---	---	1,805,460
Virginia	---	---	---	---	---	---	---	---	---	---	---	---	---	689,216
Washington	---	---	---	---	---	---	---	---	---	w	w	w	---	583,789
Wisconsin	---	---	---	---	---	---	---	---	---	---	---	---	---	1,145,597
Other	24,375	18,277	90,784	72,734	62,595	66,316	64,103	225,093	458,602	474,547	458,811	448,112	526,900	3,157,390
Totals	303,160	275,294	252,768	226,545	202,632	216,611	244,314	275,883	515,355	517,804	523,430	488,312	570,023	40,618,275
Employment, mine and mill	3200	2900	2800	2600										

1/ Virginia is included in Tennessee total.
w Withheld to avoid disclosing company proprietary data; included in "Other".
e Estimated

Table A4: Estimated Oxidized Zinc Ore Production In The United States, By State, 1850-1994.

State	Ore Mined	Recoverable Zinc Content
Arizona 1/	48,000	10,000
Arkansas 1/	54,000	22,000
California	356,000	25,000
Colorado	816,000	136,000
Idaho	13,000	2,000
Kentucky	23,000	7,000
Maryland	600	200
Missouri, Central	9,000	3,000
Missouri, Southeastern	91,000	27,000
Montana	254,000	36,000
Nevada	490,000	59,000
New Jersey	91,000	27,000
New Mexico	220,000	73,000
Pennsylvania	684,000	112,000
Tennessee	363,000	109,000
Texas 1/	200	100
Tri-State (MO,KS,OK)	454,000	172,000
Utah	103,000	27,000
Virginia	241,000	84,000
Washington	10,000	· 600
Wisconsin (incl. IA, Northern IL)	159,000	42,000
Total:	4,480,000	974,000
Primary or Hypogene Ore:		
New Jersey	27,000,000	5,334,000
Grand Total:	32,000,000	6,308,000

Sources: Heyl and Bozion (1962); U.S. Bureau of Mines
1/ Production to 1934 only

Table A5: Lead Slag Fuming Plants in the United States, 1927-1995

State	Company	Plant Location	Date Opened	Date Closed
California				
	American Smelting and	Selby	1953	1970
Idaho				
	Bunker Hill & Sullivan Mining	Bradley/Kellogg	1943	
	The Bunker Hill			1981
Missouri				
	St Joseph Lead	Herculaneum	1943	1965
Montana				
	Anaconda Copper	East Helena	1927	
	ASARCO			1982
Texas				
	American Smelting &	El Paso	1948	
	ASARCO			1982
Utah				
	International Smelting Refining	Tooele	1942	1971

Table A6: Recoverable Zinc Extracted at Slag Fuming Plants in the United States, 1927-1994
(metric tons)

Year	Slag Processed	Recoverable Zinc Extracted
1927-1945	NA	NA
1946	441,000	37,700
1947	533,000	50,800
1948	463,000	48,400
1949	557,000	59,700
1950	592,000	57,600
1951	550,000	67,700
1952	568,000	66,500
1953	595,000	71,800
1954	661,000	73,100
1955	683,000	77,700
1956	725,000	88,100
1957	787,000	95,400
1958	717,000	88,900
1959	559,000	66,500
1960	564,000	67,400
1961	622,000	72,100
1962	625,000	76,100
1963	640,000	78,900
1964	663,000	84,700
1965	685,000	79,000
1966	743,000	79,000
1967	449,000	56,400
1968	556,000	66,100
1969	749,000	85,900
1970	634,000	79,500
1971	584,000	73,800
1972	NA	w
1973	NA	w
1974	NA	w
1975	NA	w
1976	NA	w
1977	NA	w
1978	NA	w
1979	NA	w
1980	NA	w
1981	NA	27,400
1982	NA	12,300
1983-1995	---	---
1/ Total		2,700,000

Source: Bureau of Mines
w Withheld. NA Not available.
1/ Total includes estimate for withheld and NA.

Table A7: Primary Zinc Smelter Production, By State, 1859-1994.
(Metric tons)

Year	AR	CO	ID	IL 1/	KS	MO	MT	OK	PA 2/	TN	TX 3/	OTHER	TOTAL
1859-1973e>	--	--	--	--	--	--	--	--	--	--	--	45,336	45,336
1874-1979e>	--	--	--	--	--	--	--	--	--	--	--	92,913	92,913
1880	--	--	--	9,892	2,224	5,145	--	--	--	--	--	3,822	21,083
1881	--	--	--	NA	NA	NA	--	--	--	--	--	27,450	27,450
1882	--	--	--	16,512	6,682	2,268	--	--	--	--	--	5,169	30,631
1883	--	--	--	15,233	8,174	5,198	--	--	--	--	--	4,844	33,449
1884	--	--	--	15,961	7,130	4,745	--	--	--	--	--	7,131	34,967
1885	--	--	--	17,624	7,713	4,243	--	--	--	--	--	7,332	36,912
1886	--	--	--	19,121	8,103	5,325	--	--	--	--	--	6,134	38,683
1887	--	--	--	20,211	10,845	7,856	--	--	--	--	--	6,755	45,667
1888	--	--	--	20,362	6,464	12,215	--	--	--	--	--	8,674	47,715
1889	--	--	--	21,645	12,390	10,049	--	--	--	--	--	9,312	53,396
1890	--	--	--	23,807	13,788	11,909	--	--	--	--	--	8,268	57,772
1891	--	--	--	25,139	20,636	14,744	--	--	--	--	--	11,941	72,460
1892	--	--	--	28,470	22,421	15,120	--	--	--	--	--	13,150	79,161
1893	--	--	--	23,220	20,697	12,462	--	--	--	--	--	11,507	67,886
1894	--	--	--	26,283	23,213	10,879	--	--	--	--	--	7,961	68,336
1895	--	--	--	32,416	23,383	13,606	--	--	--	--	--	11,958	81,363
1896	--	--	--	32,816	18,832	12,701	--	--	--	--	--	9,586	73,935
1897	--	--	--	34,361	30,296	16,443	--	--	--	--	--	9,601	90,701
1898	--	--	--	42,731	36,407	17,720	--	--	--	--	--	7,830	104,688
1899	--	--	--	45,466	47,193	16,426	--	--	--	--	--	7,988	117,073
1900	--	--	--	35,153	56,369	13,373	--	--	--	--	--	7,492	112,387
1901	--	--	--	40,729	67,349	11,869	--	--	--	--	--	7,805	127,752
1902	--	--	--	42,725	78,530	10,058	--	--	--	--	--	11,050	142,363
1903	--	796	--	43,236	80,184	9,066	--	--	--	--	--	11,159	144,441
1904	--	4,419	--	43,309	97,112	11,022	--	--	--	--	--	13,511	169,373
1905	--	5,987	--	42,280	103,679	10,745	--	--	--	--	--	22,238	184,929
1906	--	5,679	--	43,490	117,539	10,049	--	--	--	--	--	19,297	203,909
1907	--	4,811	--	50,853	121,661	10,643	--	4,568	7,855	--	--	23,913	226,669
1908	--	2,793	--	45,581	90,082	9,254	--	13,484	10,220	--	--	23,515	190,894
1909	--	5,487	--	61,374	93,711	6,384	--	26,111	6,185	--	--	25,146	232,022
1910	--	5,908	--	66,259	95,887	5,054	--	31,534	13,809	--	--	24,627	244,200
1911	--	6,710	--	75,414	89,279	3,734	--	42,016	14,931	--	--	28,865	259,931
1912	--	8,204	--	80,192	91,720	6,427	--	69,785	16,560	--	--	34,471	307,359

Table A7: Primary Zinc Smelter Production, By State, 1859-1994.
(Metric tons)

Year	AR	CO	ID	IL 1/	KS	MO	MT	OK	PA 2/	TN	TX 3/	OTHER	TOTAL
1913	---	7,760	---	96,758	67,228	4,613	---	75,490	24,434	---	---	38,219	314,502
1914	---	7,333	---	116,071	40,379	3,221	---	82,887	25,487	---	---	44,903	320,281
1915	---	7,953	---	145,107	92,009	---	229	99,072	40,068	---	---	59,642	444,080
1916	6,607	8,008	---	164,593	128,173	---	9,280	152,594	69,440	---	---	66,812	605,507
1917	23,278	---	---	156,479	68,990	---	NA	185,423	78,921	---	---	94,335	607,426
1918	24,270	---	---	128,679	26,444	---	32,070	126,159	70,164	---	---	62,071	469,857
1919	28,519	---	---	107,463	39,864	---	22,004	110,244	61,254	---	---	52,746	422,094
1920	28,559	---	---	98,934	37,235	---	45,375	100,244	67,344	---	---	42,679	420,370
1921	13,860	---	---	44,330	15,035	---	10,486	37,416	33,002	---	---	27,763	181,892
1922	12,362	---	---	69,187	21,924	---	50,230	67,619	51,999	---	---	48,074	321,395
1923	27,098	---	---	84,585	29,035	---	64,490	108,630	74,392	---	---	74,828	463,058
1924	25,806	---	---	84,704	25,982	---	70,300	107,490	75,310	---	---	79,729	469,321
1925	24,626	---	---	99,493	23,374	---	71,671	126,013	90,627	---	---	83,965	519,769
1926	29,676	---	---	100,136	30,546	---	101,238	123,885	91,207	---	---	84,336	561,024
1927	23,874	---	---	93,230	30,068	---	102,175	109,589	96,251	---	---	82,334	537,521
1928	18,602	---	1,759	94,134	30,774	---	143,536	96,867	98,704	---	---	62,477	546,853
1929	16,259	---	15,043	101,990	34,287	---	125,209	101,317	98,127	---	---	75,163	567,395
1930	12,625	---	8,626	93,740	12,412	---	102,428	72,341	92,457	---	---	57,189	451,818
1931	3,050	---	7,680	69,209	4,227	---	57,234	24,425	59,371	---	---	39,698	264,894
1932	580	---	5,402	61,335	---	---	15,649	24,699	50,381	---	---	---	158,046
1933	8,282	---	6,973	54,558	---	---	73,145	47,174	56,774	---	31,765	---	278,671
1934	10,712	---	9,013	50,596	---	---	60,529	55,983	91,379	---	51,631	---	329,843
1935	9,205	---	11,293	61,348	---	---	96,187	53,172	108,365	---	42,274	---	381,844
1936	16,334	---	19,253	73,840	---	---	96,118	57,119	136,463	---	47,527	---	446,454
1937	23,404	---	20,712	66,361	---	---	85,892	87,229	159,007	---	62,609	---	505,214
1938	18,576	---	14,183	61,840	---	---	70,432	61,892	126,912	---	51,079	---	404,914
1939	18,046	---	16,717	72,103	---	---	98,547	76,703	141,156	---	36,885	---	460,157
1940	32,202	---	33,999	92,369	---	---	135,681	87,715	159,077	---	71,557	---	612,600
1941	39,957	---	35,639	110,605	---	---	160,033	96,057	201,836	---	101,597	---	745,724
1942	37,783	---	36,211	159,170	---	---	175,528	92,916	209,888	---	97,597	---	809,093
1943	32,390	---	37,312	201,105	---	---	215,534	65,356	197,819	---	105,333	---	854,849
1944	28,440	---	33,168	140,942	---	---	203,564	97,399	187,166	---	99,918	---	790,597
1945	26,663	---	30,037	113,311	---	---	162,614	98,266	182,080	---	82,627	---	693,598
1946	16,982	---	31,599	94,349	---	---	169,337	94,461	162,215	---	91,725	---	660,668
1947	15,565	---	37,921	102,686	---	---	179,126	116,481	175,562	---	100,669	---	728,010
1948	14,139	---	38,160	84,576	---	---	188,438	125,050	155,379	---	108,906	---	714,648
1949	15,527	---	37,969	78,765	---	---	196,476	143,018	142,355	---	125,047	---	739,157

156

Table A7: Primary Zinc Smelter Production, By State, 1859-1994.
(Metric tons)

Year	AR	CO	ID	IL 1/	KS	MO	MT	OK	PA 2/	TN	TX 3/	OTHER	TOTAL
1950	18,768	--	48,917	98,249	--	--	196,046	131,648	147,453	--	124,099	--	765,180
1951	19,755	--	49,413	98,469	--	--	188,919	146,281	171,619	--	125,136	--	799,592
1952	19,635	--	49,296	104,627	--	--	195,027	146,276	175,822	--	129,846	--	820,529
1953	18,488	--	49,022	117,847	--	--	201,716	122,396	174,433	--	147,176	--	831,078
1954	7,780	--	43,004	83,699	--	--	139,728	139,567	163,934	--	150,236	--	727,948
1955	19,487	--	51,369	93,266	--	--	188,119	146,021	198,192	--	177,621	--	874,075
1956	25,085	--	52,434	92,375	--	--	194,823	150,750	180,501	--	196,349	--	892,317
1957	20,938	--	62,442	97,336	--	--	179,655	143,002	224,833	--	166,093	--	894,299
1958	16,286	--	50,307	75,155	--	--	135,099	110,802	169,864	--	151,222	--	708,735
1959	14,483	--	55,512	93,175	--	--	78,580	137,957	197,193	--	147,638	--	724,538
1960	1,380	--	23,994	80,096	--	--	120,012	146,868	176,460	--	176,499	--	725,309
1961	11,196	--	67,799	71,499	--	--	100,900	149,068	194,417	--	173,320	--	768,199
1962	13,105	--	69,632	89,861	--	--	117,157	133,705	212,316	--	161,998	--	797,774
1963	10,109	--	73,751	98,857	--	--	107,129	129,462	255,512	--	164,920	--	839,740
1964	--	--	83,244	104,866	--	--	113,701	136,401	238,472	--	189,408	--	866,092
1965	--	--	82,554	103,538	--	--	130,568	139,876	252,987	--	192,584	--	902,107
1966	--	--	82,538	87,824	--	--	158,595	149,832	264,356	--	186,779	--	929,924
1967	--	--	83,583	104,924	--	--	101,454	148,620	246,021	--	167,090	--	851,692
1968	--	--	93,391	108,551	--	--	129,663	156,194	274,772	--	163,566	--	926,137
1969	--	--	95,889	118,971	--	--	157,881	130,249	259,604	--	181,329	--	943,923
1970	--	--	86,760	100,548	--	--	134,896	113,227	201,482	--	159,424	--	796,337
1971	--	--	85,286	42,083	--	--	104,762	115,129	207,429	--	140,607	--	695,296
1972	--	--	92,300	--	--	--	63,280	103,566	191,289	--	123,977	--	574,412
1973	--	--	89,195	24,146	--	--	--	70,596	227,478	--	117,908	--	529,323
1974	--	--	83,752	50,373	--	--	--	39,179	218,533	--	111,821	--	503,658
1975	--	--	83,733	50,201	--	--	--	31,816	138,146	--	93,497	--	397,393
1976	--	--	91,348	61,876	--	--	--	20,323	198,518	--	80,489	--	452,554
1977	--	--	54,954	58,399	--	--	--	40,096	194,596	--	60,319	--	408,364
1978	--	--	76,175	55,277	--	--	--	42,414	175,473	1,661	55,698	--	406,698
1979	--	--	73,848	58,315	--	--	--	43,599	176,161	69,499	51,109	--	472,531
1980	--	--	69,990	62,608	--	--	--	W	W	73,646	42,600	91,612	340,456
1981	--	--	W	67,680	--	--	--	W	W	W	46,900	231,983	346,563
1982	--	--	--	W	--	--	--	W	W	W	37,400	190,776	228,176
1983	--	--	--	W	--	--	--	W	W	W	--	235,694	235,694
1984	--	--	--	W	--	--	--	W	W	W	28,500	224,632	253,132
1985	--	--	--	W	--	--	--	W	W	W	11,400	261,209	272,609
1986	--	--	--	W	--	--	--	W	W	W	--	253,369	253,369

Table A7: Primary Zinc Smelter Production, By State, 1859-1994.
(Metric tons)

Year	AR	CO	ID	IL 1/	KS	MO	MT	OK	PA 2/	TN	TX 3/	OTHER	TOTAL
1987	---	---	---	W	---	---	---	W	W	W	---	261,345	261,345
1988	---	---	---	W	---	---	---	W	W	W	---	241,294	241,294
1989	---	---	---	W	---	---	---	W	W	W	---	260,305	260,305
1990	---	---	---	W	---	---	---	W	W	W	---	262,704	262,704
1991	---	---	---	W	---	---	---	W	W	W	---	253,267	253,267
1992	---	---	---	W	---	---	---	W	W	W	---	271,867	271,867
1993	---	---	---	W	---	---	---	W	W	W	---	240,282	240,282
1994	---	---	---	W	---	---	---	W	W	W	---	216,600	216,600
Totals	900,553	81,848	2,644,101	7,325,037	2,267,679	324,566	6,628,495	6,988,613	10,133,712	144,806	5,743,304	5,251,453	48,434,167

Source: U.S. Bureau of Mines
e/ Estimate   NA Not Available
1/ Includes Indiana 1892-1903 and Missouri 1957,1958, and 1959.
2/ Includes West Virginia 1957-1971.
3/ Includes West Virginia 1933-1956 and Missouri 1950-1953, 1955 and 1956.

Table A8: Estimated Primary Zinc Metal Production By State In Various Time Periods, 1860-1994.
(Thousand metric tons)

Period	AR	CO	ID	IN	IL	KS	MO	MT	NJ	OK	PA	TN	TX	UT	VA	WV	WI	Totals
1860-1906	---	26	---	30	823	940	290	---	30	6	180	8	---	---	27	15	39	2414
1907-1920	111	77	---	25	1394	1109	60	129	3	1120	513	---	---	8	10	488	20	5067
1921-1940	345	2	170	30	1529	258	24	1631	---	1527	1891	---	380	---	---	705	---	8492
1941-1960	410	---	853	---	2114	---	16	3468	---	2449	3515	---	2240	---	---	430	---	15495
1961-1994	34	---	1690	---	2412	---	---	1420	---	2508	4408	1295	2743	---	---	370	---	16880
Totals	900	105	2713	85	8272	2307	390	6648	33	7610	10507	1303	5363	8	37	2008	59	48348

Table A9: Spelter Companies Operating in the United States, 1860-1906

State	Company 1/	Plant Location	Date Open	Date Closed
Arkansas				
	American Zinc Co.	White River		
Colorado				
	American Smelting and Refining Co	Denver	1902	1907
	Denver Zinc Co	Denver	1888	
	United States Smelting Co	*Canon City	<1905	
	United States Zinc Co	*Pueblo	<1904	
	United Zinc and Chemical Co	Denver	<1905	
	New Jersey Zinc Co	*Canon City	<1902	
Illinois				
	Collinsville and Lumaghi Zinc Works	Collinsville	<1898	
	Collinsville Zinc Co	*Collinsville	<1898	
	Empire Zinc Co	North Chicago	<1898	
	Excelsior Concentrating and Smelting Wk	Collinsville		
	Glendale Zinc Co	Collinsville		
	Illinois Zinc Co	*Peru	1870	
	Matthiesson & Hegeler Zinc Co	*La Salle	1860	
	Mineral Point Zinc Co	Depue	1906	
	Mineral Point Zinc Co	Waukegan	<1896	
	New Jersey Zinc Co	*Waukegan	<1896	
	New Jersey Zinc Co	*Depue	1906	
	Sandoral Zinc Co	*Sandoral	<1906	
	Swansea Vale Zinc Co	Sandoral	1898	
Indiana				
	Columbia Zinc Co	Marion	<1895	
	Humphrey Spelter Co	Upland	1897	
	Messrs. La Tourelle & Co	Marion	1897	
	Vulcan Spelter Co	Upland	1897	
Kansas				
	Altoona Zinc Co	Altoona	1904	
	American Sheet Steel Co	Girard		
	American Steel and Wire Co	Cherryvale	1898	
	American Zinc, Lead Smelting Co	*Dearing	1906	
	American Zinc, Lead Smelting Co	*Caney		
	Bruce Mining and Smelting Co	Bruce	<1900	
	Caney Zinc Co	Caney	1904	
	Chanute Zinc Co	*Chanute	1904	
	Cherokee Zinc Co	Weir	<1895	
	Cherokee Zinc Co	Pittsburg	<1895	
	Cherokee Zinc Co	Cherokee		
	Cherokee-Lanyon Spelter Co	Cherokee		
	Cherokee Lanyon Spelter Co	Pittsburg		
	Cherokee-Lanyon Spelter Co	Girard		
	Cherokee-Lanyon Spelter Co	Scammonville		1898
	Cherokee-Lanyon Spelter Co	Iola	1899	

Table A9: Spelter Companies Operating in the United States, 1860-1906 (cont.)

State	Company 1/	Plant Location	Date Open	Date Closed
Kansas				
	Cockerill A,B	Gas	1901	
	Cockerill Zinc Co	*Bruce		
	Cockerill Zinc Co	*Gas City	1903	
	Cockerill Zinc Co	*Altoona	1904	
	Cockerill Zinc Co	*La Harpe		
	Cockerill Zinc Co	*Pittsburg		
	Edgar Zinc Co	*Cherryvale	1898	
	Glendale Zinc Co	Cherryvale	1898	
	Gramby Manufacturing and Smelting Co	Pittsburg		
	Gramby Manufacturing and Smelting Co	*Neodesha		
	Girard Zinc Co	Girard	1889	1903
	Gross J.H. Co	Weir City	<1900	
	Kansas Zinc Mining and Smelting Co	Girard	<1898	1903
	La Harpe Smelting Co	La Harpe	1903	
	Lanyon Zinc Co	*Iola	1896	
	Lanyon Zinc Co	*La Harpe		
	Messrs W & J Lanyon	Iola	1898	
	Messrs W & J Lanyon	Pittsburg	<1883	
	Midland Coal and Smelting Co	Midland	1898	
	New Jersey Zinc Co	*Iola	1899	
	New Jersey Zinc Co	*Gas		
	Nicholson G.E.	Iola	1899	
	Ozark Oxide Co	*Coffeyville	<1905	
	Pittsburg and St. Louis Co	Pittsburg	<1891	
	Pittsburg Zinc Co	*Pittsburg	<1905	
	Prime Western Spelter Co	*Iola	1898	
	Prime Western Spelter Co	*Gas	<1901	
	R. Lanyon & Co	Pittsburg	<1883	
	Robert Lanyon's Sons Spelter Co	Iola	1897	
	Robert Lanyon's Sons Spelter Co	La Harpe	1898	
	S. H. Lanyon & Brothers	Pittsburg	<1883	
	S. H. Lanyon & Brothers	Scammonville	1888	
	Scammonville Zinc Works	Scammonville	1888	
	Standard Acid Co	Iola	1901	
	United Zinc and Chemical Co	*Argentine		
	United Zinc and Chemical Co	*Iola	1901	
	Weir City Company	Weir City	<1891	
	Weir City Company	Pittsburg	<1891	
Missouri				
	American Steel and Wire Co	Carondelet	<1883	
	Cherokee-Lanyon Spelter Co	Rich Hill		
	Cockerill Zinc Co	*Rich Hill		
	Cockerill Zinc Co	*Nevada	1897	
	Glendale Zinc Co	Carondelet	1869	
	Edgar Zinc Co	*Carondelet	<1883	

Table A9: Spelter Companies Operating in the United States, 1860-1906 (cont.)

State	Company 1/	Plant Location	Date Open	Date Closed
Missouri				
	Granby Mining and Smelting Co	*St Louis		
	Joplin Zinc Co	Joplin	<1884	
	Missouri Zinc Co	Carondelet		1883
	Nevada Spelter Co	Nevada	<1896	
	Nevada Spelter Co	Nevada	1897	
	Nicholson G.E.	Nevada	<1901	
	New Jersey Zinc Co	Nevada		
	Ozark Oxide Co	*Joplin	<1905	
	R. Lanyon & Co	Nevada	1887	
	Southwest Lead and Zinc Co	Rich Hill	<1883	
	West Joplin Lead and Zinc Co	Joplin	<1883	
Montana				
	Heinze-Coran Zinc Co	Basin	<1906	
New Jersey				
	Bergen Port Zinc Co	Bergen Port	1886	
	New Jersey Zinc and Iron Co	Newark	1884	
	New Jersey Zinc Co	Newark	1865	
	New Jersey Zinc Co	*Jersey City	<1880	
	New Jersey Zinc Co	*Franklin	<1870	
	Passaic Zinc Co	Jersey City	<1888	
Oklahoma				
	Lanyon-Starr Smelting Co	*Bartlesville	1906	
Pennsylvania				
	Friedensville Zinc Co	Friedensville	1888	
	Florence Zinc Co	Freemansburg	<1896	
	Florence Zinc Co	Florence		
	Lehigh Zinc and Iron Co	Bethlehem	1860	1910
	New Jersey Zinc Co	*South Bethlehem	<1871	1910
	New Jersey Zinc Co	*Freemansburg		
	New Jersey Zinc Co	*Palmerton	1898	
Tennessee				
	East Tennessee Valley Zinc Works	Clinton	1880	
	Edes, Mixter and Heald Zinc Co	Clinton		1894
Virginia				
	Bertha Zinc Co	*Pulaski	1879	
	Wythe Lead and Zinc Co	Pulaski		
Wisconsin				
	Mineral Point Zinc Co	Mineral Point	<1895	
	New Jersey Zinc Co	*Mineral Point	<1895	
West Virginia				
	Graselli Chemical Co	*Clarksburg	1903	

* Plants known to be available at the end of 1906
< Before the year listed
1/ Some smelters in the same location may have merely underwent a company name change or were taken over by a new company. In a number of cases it was not clear from the records examined what had occurred.

Table A10: Spelter Companies Operating in the United States, 1907-1920

State	Company	Plant Location	Date Opened	Date Closed
Arkansas				
	Arkansas Zinc & Smelting Corp	*Van Buren	1916	
	Athletic Mining and Smelting Co	*Fort Smith	1917	
	Fort Smith Spelter Co	*Fort Smith	1916	
Colorado				
	United States Smelting Co	Canon City		
	United States Zinc Co	*Pueblo		
	New Jersey Zinc Co	1/ Canon City	1909	
Illinois				
	American Zinc Co of IL	*Hillsboro	1912	
	American Zinc Co of IL	*East St Louis		
	Collinsville Zinc Co	Collinsville		
	Eagle Pitcher Lead Co	*Hillsboro		
	Granby Mining & Smelting Co	East St Louis	1915	
	Hegeler Brothers	Danville	1908	
	Hegeler Zinc Co	*Danville	1908	
	Illinois Zinc Co	*Peru		
	Matthiessen & Hegeler Zinc Co	*La Salle	1860	
	Mineral Point Zinc Co	*Depue		
	Missouri Zinc Co	*Beckmeyer	1913	
	National Zinc Co	Springfield		
	R.H. Lanyon Zinc & Acid Co	Hillsboro	1913	
	Sandoral Zinc Co	*Sandoral		
	United Zinc and Chemical Co	Springfield	1908	
	Wenona Zinc Works	Wenona		1910
Indiana				
	Grasselli Chemical Co	*Terre Haute	1917	
Kansas				
	Altoona Zinc Smelting Co	Altoona		
	American Metal Co	Iola		
	American Metal Co	La Harpe		
	American Spelter Co	Pittsburg	1915	1918
	American Zinc, Lead Smelting Co	Caney		
	American Zinc, Lead Smelting Co	Dearing		1918
	American Zinc, Lead Smelting Co	Neodesha		
	Beer Sondheimer & Co	Altoona		1918
	Chanute Spelter Co	Chanute		
	Chanute Zinc Co	Chanute		
	Cherokee Smelting Co	Bruce		
	Cockerill Zinc Co	Altoona		

Table A10: Spelter Companies Operating in the United States, 1907-1920 (cont.)

State	Company	Plant Location	Date Opened	Date Closed
Kansas				
	Cockerill Zinc Co	Bruce		
	Cockerill Zinc Co	Gas City		
	Cockerill Zinc Co	La Harpe		
	Cockerill Zinc Co	Pittsburg		
	Edgar Zinc Co	*Cherryvale		
	Granby Mining & Smelting Co	Neodesha		1918
	Iola Zinc Co	Concreto		1917
	J.B. Kirk Gas & Smelting Co	Iola		
	Joplin Ore & Spelter Co	Pittsburg		1918
	Kansas Zinc Co	Altoona		
	Kansas Zinc Co	Bruce		1918
	Kansas Zinc Co	Gas		1913
	Kansas Zinc Co	La Harpe		
	Kansas Zinc Co	Pittsburg		
	La Harpe Spelter Co	La Harpe		
	Lanyon Smelting Co	Iola		1911
	Lanyon Smelting Co	La Harpe		1911
	Lanyon Smelting Co	Pittsburg		1918
	L. Vogelstein & Co	Gas City		
	L. Vogelstein & Co	La Harpe		
	Owen Zinc Co	Caney		
	Ozark Oxide Co	Coffeyville		
	Pittsburg Zinc Co	Pittsburg		1918
	Prime Western Spelter Co	*Iola		
	Prime Western Spelter Co	Gas		1918
	United States Smelting Co	Altoona		1917
	United States Smelting Co	Iola	1915	1918
	United States Smelting Co	La Harpe	1915	1917
	United Zinc and Chemical Co	Argentine		1910
	United Zinc and Chemical Co	Iola		1910
	Weir Smelting Co	*Caney	1917	
	Weir Smelting Co	Weir City	1916	
Missouri				
	Cockerill Zinc Co	Nevada		
	Cockerill Zinc Co	Rich Hill		1910
	Edgar Zinc Co	Carondelet		1918
	Edgar Zinc Co	St Louis		

Table A10: Spelter Companies Operating in the United States, 1907-1920 (cont.)

State	Company	Plant Location	Date Opened	Date Closed
Kansas				
	Granby Mining and Smelting Co	St Louis		
	Kansas Zinc Co	Nevada		
	Kansas Zinc Co	Rich Hill		1910
	Missouri Zinc Smelting Co	Rich Hill	1916	
	Nevada Zinc Co	Nevada		1914
	Ozark Oxide Co	Joplin		1908
Montana				
	Anaconda Copper Mining Co	*Great Falls	1916	
New Jersey				
	New Jersey Zinc Co	Jersey City		
	New Jersey Zinc Co	Franklin		
Oklahoma				
	Bartlesville Zinc Co	*Bartlesville	1907	
	Bartlesville Zinc Co (Lanyon-Starr)	*Bartlesville		
	Bartlesville Zinc Co	*Blackwell	1916	
	Bartlesville Zinc Co	Collinsville	1911	1918
	Eagle-Pitcher Lead Co	*Henryetta	1916	
	Henryetta Spelter Co	Henryetta	1916	
	J.B. Kirk Gas and Smelting Co	Checotah	1916	
	Kusa Spelter Co	Kusa	1916	
	La Harpe Spelter Co	Kusa	1916	
	Lanyon-Starr Smelting Co	Bartlesville		
	National Zinc Co	*Bartlesville	1908	
	Oklahoma Spelter Co	Kusa	1916	
	Quinton Spelter Co	*Quinton	1917	
	Tulsa Fuel & Manufacturing Co	*Collinsville	1911	
	Tulsa Spelter Co	Sand Springs	1914	1920
	United States Smelting Co	Checotah	1916	
	United States Zinc Co	*Kusa		
	United States Zinc Co	*Henryetta		
	United States Zinc Co	*Sand Springs		
	Victory Metal Co	*Henryetta	1916	
	Western Spelter Co	Henryetta	1916	
Pennsylvania				
	American Steel & Wire Co	*Donora	1916	
	American Zinc & Chemical Co	*Langeloth	1915	
	New Jersey Zinc Co	South Bethlehem		1910
	New Jersey Zinc Co	Freemansburg		

Table A10: Spelter Companies Operating in the United States, 1907-1920 (cont.)

State	Company	Plant Location	Date Opened	Date Closed
Pennsylvania				
	New Jersey Zinc Co	*Palmerton		
Utah				
	Judge Mining & Smelting Co	*Park City	1917	
Virginia				
	Bertha Zinc Co	Pulaski		1911
Wisconsin				
	New Jersey Zinc Co	Mineral Point		
West Virginia				
	Clarksburg Zinc Co	Clarksburg		
	Grasselli Chemical Co	*Clarksburg		
	Grasselli Chemical Co	*Meadowbrook	1911	
	United Zinc Smelting Co	*Moundsville	1917	

* Plants known to be available at the end of 1920.
1/ Continued but not as a spelter producer.

Table A11: Spelter Companies Operating in the United States, 1921-1940

State	Company	Plant Location	Date Opened	Date Closed
Arkansas				
	Arkansas Zinc & Smelting Corp	Van Buren		
	Athletic Mining and Smelting Co	*Fort Smith		
	Falcon Zinc Co	*Van Buren		
	Fort Smith Spelter Co	Fort Smith		1924
	Van Buren Zinc Co	Van Buren		
Colorado				
	United States Zinc Co	Pueblo		1922
Idaho				
	Sullivan Mining Co	*Kellogg	1928	
Illinois				
	American Zinc Co of IL	Hillsboro		1924
	American Zinc Co of IL	*East St. Louis		
	American Zinc Co of IL	*East St. Louis		
	Eagle Pitcher Lead Co	Hillsboro		
	Evans-Wallower Zinc Co	East St Louis	1929	
	Hegeler Zinc co	*Danville		
	Illinois Zinc co	Peru		1940
	Matthiessen & Hegeler Zinc Co	*La Salle		
	Mineral Point Zinc Co	*Depue		1/ 1929
	Missouri Zinc Co	Beckmeyer		2/ 1930
	Sandoval Zinc Co	Sandoval		2/ 1931
Indiana				
	Grasselli Chemical Co	Terre Haute		
Kansas				
	American Steel & wire Co	*Cherryvale		
	Edgar Zinc Co	Cherryvale		
	Prime Western Spelter Co	Iola		
	Weir Smelting Co	Caney		1930
Montana				
	Anaconda Copper Co	*Anaconda	1928	
	Anaconda Copper Co	*Great Falls		
Oklahoma				
	Bartlesville Zinc Co	Bartlesville		1929
	Lanyon-Starr Smelting Co	Bartlesville		1923
	Bartlesville Zinc Co	Blackwell		
	Blackwell Zinc Co	*Blackwell		
	Eagle Pitcher Lead Co	Henryetta		
	Eagle Pitcher Mining and Smelting	*Henryetta		
	National Zinc Co	*Bartlesville		
	Wellman Corp	Kusa		
	Nicholson Corp	Kusa		1936
	Quinton Spelter Co	*Quinton		
	Tulsa Fuel & Manufacturing Co	Collinsville		1927
	United States Zinc Co	Kusa		1928
	United States Zinc Co	Henryetta		1928
	Victory Metal Co	Henryetta		1922

Table A11: Spelter Companies Operating in the United States, 1921-1940 (cont.)

State	Company	Plant Location	Date Opened	Date Closed
Pennsylvania				
	American Steel & Wire Co	*Denora		
	American Zinc & Chemical Co	*Langeloth		
	New Jersey Zinc Co	*Palmerton		
	St. Joseph Lead Co	*Josephtown	1931	
Texas				
	American Smelting & Refining Co	*Amarililo	1923	
	Illinois Zinc Co	*Dumas	1936	
West Virginia				
	Grassellli Chemical Co	Clarksburg		1928
	Grassellli Chemical Co	*Meadowbrook		
	United Zinc Smelting Corp	*Moundsville		

* Plants available at the end of 1940
1/ Horizontal closed; vertical started up.
2/ Became secondary zinc plant.

Table A12: Spelter Companies Operating in the United States, 1941-1960

State	Company	Plant Location	Date Opened	Date Closed
Arkansas				
	Arkansas Smelting Co	Van Buren		1947
	Athletic Mining and Smelting Co	*Fort Smith		
Idaho				
	Bunker Hill Co	*Kellogg		
	Sullivan Mining Co	Kellogg		
Illinois				
	American Zinc Co of IL	*East St Louis	1/1941	
	American Zinc Co of IL	Fairmont City		1958
	Hegeler Zinc Co	Danville		1947
	Matthiessen & Hegeler Zinc Co	*La Salle		
	New Jersey Zinc Co	*Depue		
Kansas				
	American Steel & Wire Co	Cherryvale		<1945
	Eagle Pitcher Co	Galena		1954
Missouri				
	St Joseph Lead Co	Herculaneum	1955	
Montana				
	Anaconda Copper Mining Co	*Anaconda		
	Anaconda Copper Mining Co	*Great Falls		
Oklahoma				
	Blackwell Zinc Co	*Blackwell		
	Eagle Pitcher Mining and Smelting	Henryetta		
	Eagle Pitcher Co	*Henryetta		
	National Zinc Co	*Bartlesville		
	Quinton Spelter Co	Quinton		
Pennsylvania				
	American Steel & Wire Co	Denora		1957
	American Zinc & Chemical Co	Langeloth		1948
	New Jersey Zinc Co of PA	*Palmerton		
	St Joseph Lead Co	*Josephtown		
Texas				
	American Smelting and Refining Co	*Amarillo		
	American Smelting and Refining Co	*Corpus Christi	1941	
	American Zinc Co of IL	*Dumas		
West Virginia				
	E.I. du Pont Nemours & Co	Meadowbrook		
	Grasselli Chemical Co	Meadowbrook		
	Matthiessen & Hegeler Zinc Co	*Meadowbrook		
	United Zinc Smelting Co	Moundsville		1948

* Plants known to be available at the end of 1960.
1/ Reconditioned Evans-Wallower electrolytic plant closed in 1931.

The U.S. Zinc Industry

Table A13: Spelter Companies Operating in the United States, 1961-1994

State	Company	Plant Location	Date Opened	Date Closed
Arkansas				
	Athletic Mining and Smelting Co	Fort Smith		1963
Idaho				
	Bunker Hill Co	Kellogg		1981
Illinois				
	Amax Zinc Co, Inc	Sauget		
	American Zinc Co	Sauget		
	American Zinc Co of IL	East St Louis		1971
	Big River Zinc Co	Sauget		
	Big River Zinc Co (Korea Zinc Co)	*Sauget		
	Matthiessen & Hegeler Zinc Co	La Salle		1971
	New Jersey Zinc Co	Depue		1971
Missouri				
	St Joseph Lead Co	Herculaneum 1/		1965
Montana				
	Anaconda Copper Mining Co	Anaconda		1973
	Anaconda Copper Mining Co	Great Falls		1972
Oklahoma				
	Blackwell Zinc Co	Blackwell		
	Blackwell Zinc Co	Blackwell		1973
	Continental Resources & Dev. Inc	Bartlesville		
	Eagle Pitcher Co	Henryetta		
	Eagle Pitcher Industries Inc	Henryetta		1968
	National Zinc Co	Bartlesville		1973
	National Zinc Co	Bartlesville	1976	
	National Zinc Co, Inc	Bartlesville		
	Pibro Resources Corp	Bartlesville		
	St Joe Resources Co	Bartlesville		
	Zinc Corp of America	*Bartlesville		
Pennsylvania				
	New Jersey Zinc Co	Palmerton		1981
	St Joseph Lead Co	Josephtown		
	St Joe Resources Co	Monaca		
	Zinc Corp of America	*Monaca		
Tennessee				
	Jersey Miniere Zinc Co	Clarksville	1978	
	Savage Zinc Co.	* Clarksville		
Texas				
	American Smelting and Refining Co	Amarillo		1975
	American Smelting and Refining Co	Corpus Christi		
	ASARCO Inc.	Corpus Christi		1985
	American Zinc Co of IL	Dumas		1971
West Virginia				
	Matthiessen & Hegeler Zinc Co	Meadowbrook		1971

* Plants available at the end of 1994.
1/ Electrothermic slag-fuming plant producing PW grade slab zinc. Test production beginning in 1943; commercial production in 1955,1956, 1957, 1958 and 1960; it was little used thereafter.

Table A14: Vertical Retort Companies Operating in the United States, 1927-1995

State	Company	Plant Location	Date Opened	Date Closed
Illinois				
	Mineral Point Zinc Co	Depue	1/ 1929	
	New Jersey Zinc Co			1971
Missouri				
	St. Joseph Lead Co 2/	Herculaneum	1943	1966
Pennsylvania				
	New Jersey Zinc Co of PA	Palmerton	1/ 1929	
	New Jersey Zinc Co			1980
	St Joseph Lead Co	Josephtown/ Monaca	1936	
	St Joe Zinc Co			
	St Joe Minerals Co			
	St Joe Resources Co			
	Zinc Corp of America			Operating
West Virginia				
	E. I. du Pont De Nemours & Co	Meadowbrook	1/ 1932	1971
	Matthiessen & Hegeler Zinc Co			

1/ Startup of vertical retorts at operating horizontal retort plant
2/ Intermittent operation of electrothermic lead slag-fuming unit.

Table A15: Electrolytic Companies Operating in the United States, 1915-1994

State	Company	Plant Location	Date Opened	Date Closed
California				
	Mammoth Copper Mining Co	Kennett	1917	1918
Idaho				
	Sullivan Mining Co	Kellogg	1928	
	The Bunker Hill Co	Kellogg		1981
Illinois				
	AMAX Zinc Co Inc	Sauget 2/		
	American Zinc Co	Sauget		
	American Zinc Co of IL	East St Louis		
	American Zinc, Lead & Smelter Co	Monsanto		
	Big River Zinc Co	Sauget		
	Big River Zinc Co (Korea Zinc Co)	Sauget		
	Evans-Wallower Zinc Co	East St Louis	1929 1/	
Iowa				
	River Smelting & Refining Co.	Keokuk	1916	1918
Maryland				
	Electrolytic Zinc Co	Baltimore	1916	1918
Montana				
	Anaconda Copper Mining Co	Anaconda	1915	1916
	Anaconda Copper Mining Co	Anaconda	1928	1969
	Anaconda Copper Mining Co	Great Falls	1916	1973
Oklahoma				
	National Zinc Co	Bartlesville	1976	
	St Joe Resources	Bartlesville		
	Zinc Corp. of America	Bartlesville		
Tennessee				
	Jersey Miniere Zinc Co	Clarksville	1978	
	Savage Zinc, Inc	Clarksville		
Texas				
	American Smelter & Refining Co	Corpus Christi	1942	
	ASARCO Incorporated	Corpus Christi		1985
Utah				
	Judge Mining & Smelting Co	Park City	1917	1920

1/ Plant closed in 1930. In 1940, it was sold to The American Zinc Co of IL.
2/ Same plant but location has been referred to as East St Louis. Monsanto, and Sauget.

Table A16: Secondary Zinc Recovery and Form of Recovery, 1907-1994
(Thousand metric tons)

Year	Slab Zinc 1/	Copper-base Alloy	Other 2/	Total
1907	6	NA	6	12
1908	6	NA	6	12
1909	8	7	7	22
1910	12	16	9	37
1911	13	22	9	44
1912	24	26	12	62
1913	24	23	12	59
1914	19	22	15	56
1915	27	31	20	78
1916	26	56	17	99
1917	15	73	17	105
1918	9	75	25	109
1919	18	57	25	100
1920	19	59	29	107
1921	16	36	11	63
1922	30	55	24	109
1923	36	59	23	125
1924	32	57	11	122
1925	36	48	38	122
1926	37	57	37	131
1927	39	57	37	133
1928	44	60	40	144
1929	43	60	41	144
1930	32	38	33	103
1931	20	27	34	81
1932	13	20	30	63
1933	27	29	28	94
1934	18	26	35	79
1935	26	27	39	92
1936	38	39	45	122
1937	47	47	44	138
1938	29	35	34	98
1939	46	90	32	168
1940	44	102	53	199
1941	54	130	63	247
1942	48	192	50	290
1943	44	230	52	326
1944	44	202	60	306
1945	45	206	65	316

Table A16: Secondary Zinc Recovery and Form of Recovery, 1907-1994 (con.t)
(Thousand metric tons)

Year	Slab Zinc 1/	Copper-base Alloy	Other 2/	Total
1946	40	148	74	262
1947	54	133	86	273
1948	56	145	84	285
1949	50	95	63	208
1950	61	146	79	286
1951	44	150	84	278
1952	50	160	66	276
1953	48	146	70	264
1954	62	120	59	241
1955	60	138	72	270
1956	65	111	71	247
1957	65	96	73	234
1958	42	90	72	204
1959	52	109	85	246
1960	62	97	78	237
1961	50	93	79	212
1962	53	107	74	234
1963	54	113	72	239
1964	64	125	77	266
1965	74	143	98	315
1966	75	156	88	319
1967	66	133	85	284
1968	71	148	98	317
1969	62	178	96	336
1970	68	147	89	304
1971	72	156	88	326
1972	67	175	104	346
1973	75	163	108	346
1974	71	135	99	305
1975	53	131	72	256
1976	62	181	93	336
1977	46	179	103	328
1978	35	198	113	336
1979	52	210	105	367
1980	29	172	100	301

Table A16: Secondary Zinc Recovery and Form of Recovery, 1907-1994 (con.t)
(Thousand metric tons)

Year	Slab Zinc 1/	Copper-base Alloy	Other 2/	Total
1981	50	171	117	338
1982	74	111	97	282
1983	69	172	105	346
1984	78	201	117	396
1985	73	181	92	346
1986	63	180	98	341
1987	83	153	116	352
1988	88	159	90	337
1989	98	149	100	347
1990	96	148	95	341
1991	122	NA	NA	354
1992	128	NA	NA	w
1993	141	NA	NA	w
1994	139	NA	NA	w

Source: U.S. Bureau of Mines

w Withheld. NA Not Available

1/ Excludes remelt

2/ Zinc dust, oxide, chemicals, light metal alloys, and miscellaneous.

Table A17: U.S. Apparent Consumption of Slab Zinc, 1900-1994
(Thousand metric tons)

Year	Galvanizing	Brass Products	Zinc-base Alloys	Rolled Zinc	Other Uses	Total
1900	49	21	NA	9	10	89
1901	63	32	NA	18	15	128
1902	71	32	NA	22	14	139
1903	71	36	NA	23	10	140
1904	82	41	NA	27	14	164
1905	91	47	NA	31	12	181
1906	112	52	NA	33	8	205
1907	135	36	NA	27	6	204
1908	108	30	NA	24	11	173
1909	149	44	NA	30	15	238
1910	147	49	NA	27	22	245
1911	150	54	NA	41	22	267
1912	177	82	NA	40	33	332
1913	159	74	NA	32	27	292
1914	154	82	NA	27	27	290
1915	172	127	NA	36	22	357
1916	181	159	NA	43	25	408
1917	172	154	NA	52	25	403
1918	168	150	NA	51	25	394
1919	148	141	NA	50	26	365
1920	174	131	NA	48	32	385
1921	125	68	NA	28	17	238
1922	186	132	NA	49	33	400
1923	213	159	NA	51	44	467
1924	217	141	NA	55	51	464
1925	257	150	NA	65	54	526
1926	263	163	12	78	48	564
1927	254	145	16	67	46	528
1928	264	158	27	67	53	569
1929	263	168	33	62	50	576
1930	197	109	20	47	37	410
1931	152	89	18	45	31	335
1932	99	60	15	36	24	234
1933	134	85	24	37	37	317
1934	138	89	29	37	34	327
1935	177	112	50	51	38	428
1936	220	150	65	50	44	529
1937	232	153	80	53	35	553
1938	180	93	44	42	24	383
1939	249	159	76	56	27	567

Table A17: U.S. Apparent Consumption of Slab Zinc, 1900-1994 (cont.)
(Thousand metric tons)

Year	Galvanizing	Brass Products	Zinc-base Alloys	Rolled Zinc	Other Uses	Total
1940	300	148	133	55	29	665
1941	318	207	138	65	23	751
1942	225	290	73	60	13	661
1943	230	380	69	44	18	741
1944	283	345	77	69	26	800
1945	303	234	119	89	24	769
1946	290	135	192	84	25	726
1947	328	102	195	64	25	714
1948	288	240	130	70	24	752
1949	309	184	159	75	25	752
1950	401	126	263	62	25	877
1951	363	130	269	58	27	847
1952	343	141	215	47	29	775
1953	369	162	279	50	35	895
1954	366	98	264	43	31	802
1955	409	133	391	47	36	1016
1956	398	112	327	43	34	914
1957	333	102	341	37	35	848
1958	346	92	287	37	26	788
1959	328	117	353	39	30	867
1960	337	90	307	35	27	796
1961	347	117	310	37	34	845
1962	352	118	384	38	43	935
1963	381	116	425	38	42	1002
1964	414	123	476	40	43	1096
1965	438	115	579	42	55	1229
1966	462	168	550	48	64	1292
1967	429	119	485	41	60	1134
1968	452	146	511	44	71	1224
1969	448	163	523	44	79	1257
1970	430	116	421	37	73	1077
1971	431	137	468	35	67	1138
1972	470	174	526	41	75	1286
1973	512	179	554	37	83	1365
1974	475	165	399	36	94	1169
1975	342	105	303	25	65	840
1976	393	151	387	27	71	1029
1977	396	128	367	27	80	998
1978	454	141	354	25	76	1050
1979	453	141	314	22	70	1000

Table A17: U.S. Apparent Consumption of Slab Zinc, 1900-1994 (cont.)
(Thousand metric tons)

Year	Galvanizing	Brass Products	Zinc-base Alloys	Rolled Zinc	Other Uses	Total
1980	379	99	254	21	58	811
1981	411	113	243	23	50	840
1982	342	81	198	37	51	709
1983	373	108	213	56	56	806
1984	375	126	233	57	58	849
1985	362	78	218	47	59	764
1986	529	135	231	34	74	1003
1987	550	140	229	43	97	1059
1988	549	146	249	50	95	1089
1989	552	130	220	52	106	1060
1990	506	142	205	48	91	992
1991	465	133	200	42	92	932
1992	505	140	200	48	99	992
1993	532	140	222	1/	141	1035
1994	608	161	250	1/	169	1188

1/ Included in Other.

GLOSSARY

ACCELERATOR A substance added to increase the rate of a chemical reaction. Said of zinc oxide in hastening the curing process of rubber.

ACID PLANT BLOW-DOWN Waters that have been used in an acid plant and that have accumulated contaminants to such an extent that they must be removed from the system.

ACTIVATION A process whereby the surface of a mineral particle is modified so as to make it react more readily or more strongly with a collector.

ACTIVATOR [1] In flotation, a chemical added to the pulp to increase the floatability of a mineral in a froth or to refloat a depressed mineral. In differential flotation of lead-zinc ores and other ores containing sphalerite, sphalerite is depressed so it will not float when the lead, for example, is floated. The sphalerite in the flotation tails is then typically reactivated by adding copper sulfate, and then again becomes floatable. [2] In rubber compounding, a substance that acts on an accelerator to enhance its reaction time. Zinc oxide was found to accelerate the curing of rubber in 1849, but the principal breakthrough in rubber accelerators came in 1906 with the development of organic accelerators. Despite this development, rubber compounders found that zinc oxide was an essential component in rubber in that it activated the organic accelerators improving the curing rate. Zinc oxide is used as an accelerator and as an activator by the rubber industry.

ADIT A horizontal or nearly horizontal passage driven from the surface for the working or dewatering of a mine.

ADOBE SHOT A dobe shot. A stick of dynamite is laid on the rock to be broken and covered with mud to add to the force of the shot. A mudcap shot.

ADVANCE ROYALTY Payment whereby the lessee must pay the owner of an operating interest in a mineral deposit royalties on a specified amount of ore extracted annually whether or not the minerals are extracted.

ADVERSE To oppose granting a patent to a mining claim.

AEROFROTH FROTHERS Trademark for a group of surface-active agents used primarily as foaming agents or frothers in flotation processes.

AGITATION RATIO In older type gravity concentrators, such as tables and vanners, the ratio between the average diameter of a mineral particle and the diameter of a gangue particle that travels at equal speed.

AICH'S METAL A yellow malleable alloy of copper, zinc, and iron invented in 1860.

AIR DRILL A small diamond drill driven by either a rotary or reciprocating-piston air-powered motor, used principally in underground workings. As used by miners, a percussive or rotary-type rock drill driven by compressed air.

AIR DUCT Tubing which conducts air, usually from an auxiliary fan, to or from a point as required in a mine, often into a dead-end heading for ventilation purposes. A fan line.

AIR JIG [1] Machine in which the ore feed is stratified by means of a pulsating currents of air and from which the stratified products are separately removed. [2] Apparatus for separating ores without water, by intermittent puffs of air.

AIR KNIFE A device to remove or reduce excess zinc from galvanized strip as it emerges from the bath by jets of air or super-heated steam.

AIR LEG A cylinder operated by compressed air for keeping a rock drill pressed into the hole being drilled.

AIR SHOT A shot prepared by charging in such a way that an air space is purposely left in contact with the explosive for the purpose of lessening the shattering effect.

AIR SHRINKAGE The decrease in volume which a clay undergoes in drying.

AIR-LEG SUPPORT An appliance to eliminate much of the labor when drilling with hand-held drilling machines. It consists of a steel cylinder and an air-operated piston, the rod of which extends through the top end of the cylinder and supports the drilling machine. The air leg and the drill machine can be operated by one man.

AIR-REDUCTION PROCESS See *Roasting and Reaction Process*.

AJAX High-strength, high-density gelatinous permitting explosive having good water resistance.

ALCHEMY The immature chemistry of the Middle Ages, characterized by the pursuit of the transmutation of base metals into gold, and the search for the alkahest and the panacea.

ALGAL LIMESTONE Rock composed largely of the remains of calcium-secreting algae.

ALGAL REEF An organic reef composed largely of algal remains and in which algae were the principal lime-secreting organisms. Reefs form deposition sites for some important Mississippi-Valley-type zinc deposits. Also see *Stromatilite*.

ALIDADE A surveying instrument; level; clinometer; also used with a plane table in mapping.

ALKAHEST In alchemy, an imaginary liquid, reputed to be a universal solvent, capable of resolving all bodies into their constituent elements.

ALKALINE CELL A dry primary battery that employs an alkaline electrolyte and zinc powder as the anode rather than the metal zinc can as in the zinc-carbon battery. The first successful alkaline cell was the mercury cell widely used during World War II for walkie-talkies, communication equipment, etc., but it was excessively expensive. In 1949, Rayovac introduced the first commercially successful alkaline cell; it used manganese rather than mercury, as the cathode.

ALL-FLOTATION Concentration of ores using only the flotation process.

ALLOMORPH In mineralogy, a pseudomorph formed without change in composition, as calcite after aragonite.

ALLOTROPHY; ALLOTROPISM The capacity of an element or compound to exist in two or more forms that are distinguished by differences in properties.. If not reversible, the phenomenon is termed polymorphism.

ALLOY PLATING Codeposition of two or more metallic elements.

ALLOY POWDER Powder of which each particle is composed of the same alloy of two or more metals.

ALLOYING ELEMENT Element added to a metal to effect changes in properties and which remain within the metal.

ALLOY Substance having metallic properties and being composed of two or more elements of which at least one is an elemental metal.

ALPHA The first letter of the Greek alphabet, commonly used with beta and gamma, and other letters, to represent phases of metals, minerals, alloys, etc. For example, the zinc-iron phases resulting from hot-dip galvanizing are designated by Greek letters, gamma, delta, etc.

ALUMINUM BRASS Brass to which aluminum is added to increase its corrosion resistance. Aluminum content can range from 1 to 6 percent.

AMALGAM ARC An arc in a vacuum tube having electrodes of mercury amalgamated with zinc, cadmium, and other metals. The spectra of such arcs contain the bright lines of the metals in the electrodes.

AMBIENT The environment surrounding a body but undisturbed or unaffected by it.

AMERICAN-BELGIAN FURNACE A direct-fired Belgian employed in the United States, conforming essentially to the Liege design, but presenting minor differences because of local adaptations.

AMMONIUM NITRATE GEL-IGNITE Explosives similar to the straight gelatins except that the main constituent is ammonium nitrate instead of sodium nitrate. Nitroglycerin varies from 25 to 35 percent and ammonium nitrate from about 30 to 60 percent. These explosives are all-purpose and are widely used in metal mines, nongassy coal mines, quarries and construction work.

AMORTIZATION The repayment of a debt, principal and interest, in equal monthly, annual, etc. payments as a means of paying off the debt over some specific time.

AN-FO Ammonium nitrate-fuel oil blasting agents.

ANGLE OF BITE In rolling metals where all the force is transmitted through the rolls, the maximum attainable angle between the roll radius at the first contact and the line of roll centers. If the operating angle is less, it is called the contact angle or rolling angle.

ANGLE OF NIP In roll, jaw, or gyratory crushing, the entrance angle formed by the tangents of the two points of contact between the working surfaces and the assumed spherical particle. The angle included between two approaching faces at or below which the particle is seized. Approximately 23 degrees for most minerals.

ANGLESITE Mineral composed of lead sulfate. Common associate of calamine minerals, but tends to form above them in the oxidation zone.

ANGLEUR FURNACE Furnace for the distillation of zinc.

ANION A negatively charged ion; it flows to the anode in electrolysis.

ANISOTROPY The characteristic of exhibiting different values of a property in different directions with respect to a fixed reference in a material.

ANNEALING Heating to and holding at a suitable temperature and then cooling at a suitable rate [a metal] to produce a desired microstructure or for obtaining desired mechanical, physical or other properties. Generally, when applied to nonferrous alloys the term annealing implies a heat treatment designed to soften a cold worked structure by recrystallization or subsequent grain growth or to soften an age hardened alloy by causing a nearly complete precipitation of the second phase in relatively coarse form.

ANODE The electrode where electrons leave (current enters) an operating system such as a battery or an electrolytic cell. In a battery or electrolytic cell, it is the electrode where oxidation occurs. Contrast with *cathode*.

ANODE BED A group of anodes connected to a common header cable for cathodic protection.

ANODE CORROSION. The dissolution of a metal acting as an anode.

ANODE EFFICIENCY Current efficiency at the anode. See *current efficiency*.

ANODE MUD Deposit of insoluble residue formed from the dissolution of the anode in commercial electrolysis. Sometimes called anode slime.

APEX In a vein, the top or highest part of all veins, lodes, and ledges within the boundary lines of a claim extended downward vertically. At its apex a vein must have a dip and a strike.

APPLE Code word for a group of contractual terms and conditions that will prevail, unless otherwise changed, in a specific contract dealing with the purchase or sale of scrap metal of specified grades that are also known by code words, such as scrub, seal, or seam that denote types of zinc scrap of a specified source and quality. The codes, developed by the Institute of Scrap Recycling Industries, Inc., became effective in 1991 and are internationally accepted.

ARBITRAGE A form of hedged investment meant to capture slight differences in the prices of two related securities; for example, buying gold in London and immediately or soon after selling it at a higher price in New York.

ARC FURNACE Furnace in which material is heated either directly by an electrical arc between an electrode and the work or indirectly by an arc between two electrodes adjacent to the material.

ARC MELTING Melting metal in an arc furnace.

ARCH [1] One of the fire chambers of certain kinds of furnaces and ovens, from the arched roof. [2] The roof of a reverberatory furnace.

ARGALL FURNACE A reverberatory roasting furnace of which the hearth has a reciprocating movement whereby the ore is caused to move forward by the action of rabbles extending across the hearth.

ARGALL TUBLAR FURNACE A tubular roasting furnace consisting of four brick-lined steel tubes 30 feet long nested inside two steel tires which revolve upon steel-faced carrying rolls.

ARSENIC PURIFICATION PROCESS A method to extract deleterious cobalt and, sometimes, nickel in purifying the leach solution in electrolytic zinc refining by addition of zinc dust in the presence of soluble copper and arsenic. The arsenic is often prepared by dissolving arsenic trioxide in a caustic solution.

ARSINE Arseniureted hydrogen. A colorless, inflammable, extremely poisonous gas having a garlic odor. It may arise as a problem in the electrolytic zinc process being formed when hydrogen is liberated in a solution containing arsenic.

ASCENSION THEORY The theory of infiltration by rising solutions from below. When heated ore-bearing solutions rise, the solutions deposit their minerals

The U.S. Zinc Industry

at diminished temperature and pressure. The theory that the minerals filling fissure veins were introduced in solution from below.

ASH METAL A low-grade brass made from metal skimmings and ash from brass foundries.

ASH'S FURNACE A furnace for refining spelter.

ASSAY To test ores and minerals by chemical or blowpipe examination; to determine the proportions of metals in ores. An examination of an ore, mineral, or alloy to determine only certain components as opposed to an analysis of everything it contains.

ASTM Abbreviation for American Society for Testing and Materials.

ATHLETE'S FOOT Infectious foot disease caused by a fungus; common in mining change houses.

ATOMIC NUMBER The number of protons in the nucleus of an atom.

ATOMIC PER CENT The number of atoms of an element in a total of 100 representative atoms of a substance.

ATOMIC WEIGHT The weight of an atom of a chemical element as compared with that of an atom of hydrogen or oxygen.

ATOMIZATION The dispersion of a molten metal into particles by a rapidly moving stream of gas or liquid. Zinc powder is made by atomizing a molten stream of zinc metal with a gas followed by cooling in a large chamber.

AUGER MACHINE Machine for the manufacture of zinc-distillation retorts. Similar to machines used for manufacture of drain pipes.

AURICHALCITE Mineral, a basic carbonate of zinc and copper.

AUSMELT LANCE The pipe that delivers air and process fuel beneath the surface of a liquid slag bath when extracting zinc by the *Sirosmelt process.*

AUTOGENOUS [1] In the dense-media process, fluid media partly composed of a mineral specie, such as galena, selected from the ore being treated. [2] Sized lumps of ore to used as a grinding media.

AUTOGENOUS GRINDING The grinding of an ore by tumbling it in a revolving cylinder with no bars or balls. Self grinding.

AUTOGENOUS ROASTING Roasting in which the heat generated by oxidation of the sulfides is sufficient to propagate the reaction.

AXMAN; AXEMAN One who clears the ground and stakes the ground for the rodman.

AZA The American Zinc Association, Inc.

AZTEC STONE A name for greenish smithsonite.

B

BACK The roof or upper part of any underground working.

BACK WORK Any kind of operation in a mine not immediately concerned with production or transportation.

BACK OF ORE The ore between two levels which has to be worked from the lower level.

BACK STOPING Mining a stope by taking successive slices of ore beginning at the bottom. Comparable to *overhand* and *shrinkage stoping.*

BACKFILL Process and/or the material used to fill a mine opening.

BACKSIGHT In surveying, an observation made for verification from one station to the one behind. Opposite of foresight.

BACKWARDATION A backwardation occurs when the forward price is less than the nearby price. There is no limit to the value of a backwardation. Compare with *Contango*.

BAD GROUND Rock formations in which mine openings cannot be safely maintained unless heavily timbered or supported in some manner.

BAELEN PROCESS A zinc calcine sintering process whereby dead-roasted ore is mixed with 6% coke and fired on a sintering machine. This process and the Rigg process were introduced into the United States in the early 1920's; both were successful.

BAFFLE BOARD Board fitted across a compartment in an ore washer to retain the heavy ore and allow the light material to flow away.

BAGHOUSE Chamber containing bags for filtering solids out of gases. In the production of zinc oxide, baghouses are the principal means for collecting the dust. Zinc-oxide-laden gases that have been cooled and if necessary, cleaned by cycloning, enter the baghouse under low pressure and fall, by weight, into large bags where the oxide clings to the bag walls. Bags are generally made of cotton come in many sizes; some of the largest are 20 inches in diameter and 45 feet long. In early practice they were shaken by hand to make the oxide fall into hoppers for treatment or packaging, now they are automatically shaken at set intervals. A baghouse may have more than 1,000 bags.

BAKE To dry, harden, or vitrify by exposure to heat, as in a furnace or kiln; as, to bake pottery or bricks.

BALL CLAY A plastic, white-burning clay used as a bond in chinaware.

BALL MILL A rotating cylindrical mill containing pebbles or steel balls in which ores and a variety of other materials are finely ground.

BALLON A metal prolong fixed to a zinc condenser.

BALLOT Damp, moldable, clay-grog-etc. A preparation used to machine manufacture zinc retorts.

BAND WONDER In concentration on a shaking table, the movement of a segregated band of mineral so that it no longer discharges from the table deck at the desired point and is not collected correctly.

BANDED ORE Ore composed of bands as layers that may be composed of the same mineral differing in color, texture or proportions, or different minerals. Synonym for banded texture.

BANDED TEXTURE See *Banded Ore*.

BANDED VEIN A vein made up of layers of different minerals parallel with the walls. Also called ribbon vein.

BANK OF CELLS A row of flotation cells.

BAR [1] A length of timber placed horizontally for supporting the roof. [2] Heavy steel rod with either pointed or flattened ends used as a pry by miners to dislodge loose rock in underground workings. [3] Early term used at Joplin for solid bodies of blende before zinc was commercial ore in the district. Miners excavated or gophered around bars to save on the expense of blasting them out.

BARNEY Small car or truck, attached to a rope and used to push cars up a slope or incline plane. Also called bullfrog, donkey, ground hog, larry, ram, mule, and truck.

BARREL PLATING Method of zinc plating a multitude of parts loosely placed in a nonconductive container (the barrel) which has a cathode to maintain intermittent contact with the parts during plating. A filled barrel is first submerged and rotated in a cleaning bath, followed by similar treatment in pickling and plating solutions. In the plating bath, the tumbled parts are gradually electroplated as they make contact with the cathode. The process was developed because of the cost and impracticably of electroplating small parts, which may involve thousands of units.

BARREL Piece of small pipe inserted in the end of a cartridge to carry the squib of the powder.

BARRING DOWN [1] Removing loose rocks in the roof of a mine by means of a bar. [2] Loosening ore in a bin by means of a bar, so it will flow through the chute.

BARRING SCRAP Prying adhering scrap metal from runners, ladles, or skimmers.

BARTLETT PROCESS Used in the early 1900's to make zinc-lead pigment by roasting low-grade calamine ores in Colorado. The product consisted of zinc oxide and lead sulfate and was used for mixed paints.

BARTLETT TABLE A three-shelf table driven by an eccentric that gives it a vanning motion. Ore and water are fed on the upper shelf giving two products, heads and tailings. The former are treated on the second shelf and the latter pass on to the lower shelf for retreatment.

BASE BULLION Crude lead containing recoverable silver, with or without gold.

BASE CHARGE The charge or tariff made by a smelter for roasting and smelting the concentrate, refining, and transport of metals or products. There may be premiums for easy ore and penalties for impure ore. Also see *Smelter Schedule*.

BASE METAL [1] The metal present in the largest proportion in an alloy; brass, for example is a copper-base alloy. [2] Metal to be brazed, cut or welded. [3] Any metal (iron, lead, zinc, etc.) which is altered by exposure to air, etc., in contrast with noble or precious metals.

BASE ROCK [1] Used by some drillers for the solid rock immediately underlying the soil and weathered rock. [2] As used by drillers in the Midwestern United States, the igneous or basement rocks underlying the sedimentary rocks.

BASEL CONVENTION International agreement, implemented in May 1992, which places restrictions on the movements of hazardous waste and recyclable material between one country and another unless the material is transported, treated, and/or disposed of by the importing country in a manner that is consistent with the protection of the environment and human health. As of the end of 1995, the United States and most other members of the Organization of Economic Cooperation and Development (OECD) had not become signatories to the Convention, but the OECD members agreed to abide by a set of controls governing the transboundary movement of recyclable waste until Basel was approved.

BASIC BOTTOM OR LINING The inner bottom and lining of a melting furnace consisting of materials like crushed burnt dolomite, magnesite, magnesite brick or basic slag that give a basic reaction at the operating temperature.

BASIC LINING A lining for furnaces, converters, etc., formed of nonsiliceous material, usually limestone, dolomite, lime, magnesia, or iron oxide.

BASIS METAL The original metal to which one or more coatings are applied.

BASKET BARREL Core barrel fitted with a basket core lifter.

BASSETING [1] Outcropping. [2] The cropping out or the appearance of rock on the surface.

BATCH GALVANIZING Hot-dip galvanizing after fabrication, as opposed to in line or continuous hot dip galvanizing.

BATCH PROCESS Process in which the feed is introduced as discrete charges, each of which is processed to completion separately.

BATCH TREATMENT Treatment of a parcel of material in isolation, as distinct from the treatment of a continuous stream of ore. See *Batch Process*.

BATH METAL Any one of several varieties of brass.

BAUME' GRAVITY The specific gravity of a liquid relative to pure water, based on a scale developed by the French chemist, Antoine Baume' (1728-1804).

BAUME' SCALE Device for determining the specific gravity of liquids; formerly widely used to indicate the concentration of soluble zinc compounds, such as zinc chloride, in commercial marketing. The weight percentage of the compound in solution is now the commercial norm.

BEAN SHOT Small zinc metal globules formed when molten zinc is poured into hot water. See *feathered-shot metal*.

BEARER BAR One of the bars which support the gratebars of a furnace.

BEAT DISEASES Disorders of the hands, knees, and elbows caused by repeatedly applied or continuous pressure, sudden strain, or jarring, such as when using a pick or pneumatic tool.

BEDDED DEPOSIT A mineral deposit, like coal, of tabular form that lies horizontally and is commonly parallel to the stratification of the enclosing rocks. Few zinc deposits are of this type. The Kupferschiefer copper-lead-zinc deposits in Europe are of this type and some geologists have interpreted the Sterling Hill and Mine Hill zinc-iron-manganese deposits at Franklin, New Jersey as originally being stratiform deposits of sedimentary origin that were later transformed and deformed by metamorphism.

BEDE Miner's pickax.

BELGIAN PROCESS Process formerly employed in the smelting of zinc. Roasted zinc ore, mixed with a reducing material, such as coal or coke, is placed in retorts which consist of cylindrical pipes of refractory material closed at one end, of a length and diameter convenient for charging and cleaning them. A number of these retorts are placed slightly inclined in a properly constructed furnace. The open ends of the retorts are fitted with a condenser which in turn is covered with a sheet-iron hood (a prolong) to which are connected short conical pipes for discharging the molten zinc.

BELGIAN ZINC-FURNACE Furnace in which zinc is reduced and distilled from calcined ores in tubular retorts. These furnaces may be classified as direct-fired and gas-fired, but there is no sharp division between these systems which merge into one another by difficult definable gradations. Each class of furnace may be divided into recuperative and nonrecuperative, but heat recuperation in connection with direct firing is rare.

BELL METAL Hard bronze, containing sometimes small proportions of iron, zinc or lead, but ordinarily consisting of 78 parts copper to 22 parts tin.

BELLS Signals for lowering and hoisting the bucket, cage, or skip in a shaft are usually given by bells, the number of which indicate the load, the place for stopping, etc.

BENCH MARK A permanently fixed point of known position and elevation used as a reference in surveying.

BENEFICIATION Concentration or other preparation of ore for smelting by drying, flotation, magnetic separation, etc.

BERGENPOINT SPELTER Metal produced by the Bergenpoint Zinc Co. between 1875? and 1886 at Bergenpoint, New Jersey from the high quality, largely impurity-free zinc ores near Friedensville, PA. The metal gained great fame in Europe and the United States for its great purity and superior quality. The Friedensville Zinc Co. opened a small smelter at one of the area's mines in 1888 and produced Bergenpoint spelter for a few years, closing down in the early 1890's.

BERTHA PURE SPELTER An early high purity (99.98% zinc) spelter brand made from calamine ores mined at Austinville and Bertha, Va. by the Bertha Mineral Co. and smelted at Pulaski, VA. from 1879 to about 1911. The high purity gained a wide reputation for its purity, similar to that attained by Bergenpoint spelter. Other grades sold by the company were Old Dominion and Southern, which contained from 0.2-0.4% and 0.8-1.0% lead, respectively.

BETA BRASS Copper-zinc alloys containing from 46 to 49 percent zinc, which consist of the intermediate constituent known as beta, when at room temperature.

BETHANIZING A steel wire plating process in which insoluble anodes and an electrolyte prepared by dissolving zinc ore, calcine or dross in sulfuric acid are employed. The solution is carefully purified and the wire, the cathode, is passed successively and continuously through cleaning, pickling, and plating baths. The process involves rapid movement of the wire, high acidity of the bath and very high current densities.

BETTS PROCESS For the electrolytic refining of lead in which the electrolyte contains lead fluosilicate and fluosilicic acid.

BEVILL AMENDMENT (EXCLUSION) Amendment added to the Resource, Conservation and Recovery Act in October 1980 that excluded solid waste from the extraction, beneficiation and processing of ores and minerals from the stringent regulations of the Act until their hazardous waste status could be determined. In 1991, the Environmental Protection Agency found that slag from primary zinc processing had low risk and was exempt from stringent waste management regulation.

BICHROMATE CELL Zinc-carbon cell having, an acid bichromate solution as an electrolyte. The cell was also provided with the means of raising the zinc, or both the zinc and carbon electrodes from the fluid when not in use. E.M.F. is about 2 volts.

BIMBO A somewhat controversial term to describe a deal involving both existing and outside managers; a buy-in/management buy-out (or Bimbo).

BINARY ALLOY Containing two component elements.

BIOHERM Moundlike or circumscribed mass built exclusively or mainly of sedentary organisms, such as corals, stromatoporoids, algae, etc., and enclosed in normal rock of different lithological character.

BIT Any device that may be attached to, or is, an integral part of a drill string and is used as a cutting tool to bore into or penetrate rock by using power applied to the bit percussively or by rotation.

BLACK ASH Term used for crude barium sulfide made by roasting barite with petroleum coke under reducing conditions. Black ash is used to manufacture lithopone.

BLACK LIGHT Electromagnetic radiation not visible to the human eye. The portion of the spectrum generally used in fluorescent inspection falls in the ultraviolet region between 3300 and 4000 A, with the peak at 3650 A. Also known as *ultraviolet light*.

BLACK ORE Used at the Franklin and Sterling Mines in New Jersey for ore containing black willemite, which microscopically is a fine mixture of willemite and franklinite that could not be separated from each other. It was treated separately from the other (sometimes referred to as brown ore) ore as it was unacceptable in either the willemite or the franklinite concentrates. Zinc was extracted from the black willemite by treating it in a furnace.

BLACK SAND; BLACK ROCK Term used in the Tri-State zinc district for dark colored sphalerite ore that had been strongly cemented by jasperoid (silica). In the early years of mining, this material was produced but not desired. Crushing was difficult and satisfactory separation was impossible. Discerning ore buyers tended to avoid buying black sand containing concentrates. Ores of this type were locally referred to as chatty.

BLACK SOLDER A solder composed of copper, zinc, and a little tin.

BLACKJACK A dark variety of zinc-blende or sulfide of zinc.

BLACKPOWDER An explosive mixture of potassium nitrate, powdered charcoal, and sulfur.

BLAKE JAW CRUSHER The original crusher of the jaw type.

BLANC FIXE; PERMANENT WHITE Precipitated barium sulfate; white powder used in the paint industry and as a filler in textiles and rubber.

BLANK ISLANDS In East Tennessee zinc district, noncommercial rock of sundry sizes and shapes found in zinc ore shoots. Also known as limestone islands.

BLANKET DEPOSIT A flat deposit of ore of which the length and breadth are relatively great as compared with the thickness. The term is more or less synonymous with flat sheets, bedded veins or flat masses. Such deposits are frequently intercalated between rocks of different lithological origin and character, and may have been deposited in a regular sedimentary series or subsequently introduced between the beds or impregnating them.

BLANKET VEIN A horizontal vein or deposit. Sheet ground. A sheet deposit.

BLANKING Cutting desired shapes out of metal to be used for forming or other manufacturing operations, such as blanks for penny production or for forming dry cell battery cans.

BLAST DRAFT Produced by a blower, as by blowing in air beneath a fire, or drawing out the gases from above it. A forced draft.

BLAST FURNACE A shaft furnace in which solid fuel is burned with an air blast to smelt ore in a continuous operation. Attempts to produce zinc metal by this method were unsuccessful until 1952 when the Imperial Smelting Process was developed.

BLAST ROASTING A generic term given by A. S. Dwight to a process of forcing air through finely divided sulfides with the object of roasting and agglomerating in a single operation. A number of processes fall into this category; they can be updraft or downdraft operations. Updraft processes tended to be intermittent whereas downdraft operations typically are continuous.

BLENDE Zinc blende, sphalerite, or sulfide of zinc. German miners used the term *blenden*, to dazzle, because it occurred as bright, shinning crystals; whereas others defined the term as to deceive, alluding to the fact that it looked like galena but yielded no lead.

BLENDS Used for leaded zinc oxide in which the separate components, zinc oxide and basic lead sulfate, are physically blended. See *Cofumed*.

BLINDED Said of filters in mineral processing when clogged.

BLOCK CAVING A method of caving in which a thick block of ore is partly cut off from surrounding blocks by a series of raises; it is then undercut by removing a slice of ore or a series of slices separated by small pillars underneath the block. The isolated, unsupported block of ore breaks and caves under its own weight. The broken ore is drawn off from below.

BLOCK ORE Local term in Wisconsin for large cubical crystals of galena.

BLOCKED-OUT ORE The amount, content, and mineability of ore which has been proven by development work or by drilling developed ore.

BLOW HOLES Voids or pores which may occur due to entrapped air or shrinkage during solidification of heavy sections in die casting.

BLOWING-UP FURNACE A furnace used for sintering and the volatilization of lead and zinc.

BLUE POWDER; BLUE DUST [1] Finely divided and partly oxidized metallic zinc formed by the condensation of zinc vapor into micron-size droplets. [2] That portion of vaporized zinc which does not condense as a liquid, but passes directly to the solid state in finely divided bluish powder.

BODIES SEVEN In alchemy, the metals corresponding to the planets, being gold, silver, iron, quicksilver, lead, tin and copper, answering respectively to the sun, the moon, Mars, Mercury, Saturn, Jupiter and Venus.

BOETIUS FURNACE An early gas-fired Belgian furnace with Boetius regenerators.

BOILING POINT [1] Temperature at which a liquid begins to boil or to be converted into vapor by bubbles forming in the mass. [2] Temperature at which the vapor pressure inside the bubbles of heated liquid equals the vapor pressure of the liquid.

BONAMITE Jeweler's trade name for an apple-green smithsonite, resembling chrysoprase in color, from Kelly, New Mexico.

BONANZA In miner's phrase, good luck, or a body of rich ore.

BONDING LAYER Used in hot-dip galvanizing for the first layer formed on steel in the process.

BONE Miner's term for masses of smithsonite ore.

BOOK FASHION A method of arranging core in a core box. Core representing the shallowest depth is placed in the first groove starting at the left with progressively deeper core placed in order of depth in the grooves toward the right side of the box.

BORGNET FURNACE An early Belgian zinc distillation furnace with a single combustion chamber.

BOTTLE RETORT A tilted, drinking-glass-shaped retort used mainly for distilling zinc from metallic zinc scrap materials to make zinc dust. The retort is open-mouthed but is sealed by grout after the charge is placed in the retort. Zinc vapors pass through the grouted mouth via a duct into a condensing chamber. The larger retorts of this type are approximately 3 feet in diameter and 4 feet tall and can hold about 2.5 tons of charge. Also called a *zinc retort*.

BOTTOMED [1] Completed borehole, or the point at which drilling operations in a borehole are discontinued. [2] Said of shafts on reaching completion.

BOTTOMING; BOTTOMING OUT Thinning out or ending of an ore body with depth.

BOULDER (BOWLDER) ORE In the Tri-State area, ore in which blue and white chert boulders or nodules are cemented by jasperoid or calcite and sulfides.

BOX METAL A brass, bronze or anti-friction alloy used for the journal boxes of axles or shafting.

BRADFORD PREFERENTIAL SEPARATION PROCESS A flotation process for the separation of mixed sulfides, in which certain mineral salts, such as thiosulfates, are added to the water used in the cell. The addition causes the zinc sulfide to be wetted while the lead and pyrite float.

BRAKE BLOCK Device for checking by friction the speed of a rope, as in a hoist.

BRAKE INCLINE Incline in which the full trucks descend by gravity and pull up the empty ones.

BRANDS Registered metal and metal compound products of producers that by their presence on a label or imprinted in the metal indicate the class, quality, or some standard specification of the metal or compound.

BRANGLE Local term in Upper Mississippi Valley zinc/lead district for a saddle-form core of limestone containing disseminated sulfides in a pitches and flats structure.

BRASS FOUNDER'S AGUE Mild malady found in brass workers caused by freshly formed zinc oxide in the manufacture of brass and other alloys. Under conditions of continuous exposure, the human system can develop a high degree of resistance. When not freshly formed, the dust tends to be practically harmless. Similar terms used are zinc chills and oxide shakes.

BRASS ORE Early name for aurichalcite. A basic carbonate of zinc and copper.

BRASS SPECIAL Specification grade of zinc composed of 99% zinc and a maximum of 1.0% of lead, iron and cadmium and used mainly in the production of brass. As a rolled product it was used as dry cell cans and weather stripping. Its use was gradually replaced by HG metal and it was dropped by ASTM as a specification metal in 1967.

BRASS An alloy consisting mainly of copper (over 50%) and zinc, to which smaller amounts of other elements may be added.

BRAZE To solder with hard solder which usually is copper and zinc, half and half.

BRAZIER An artificer who works in brass.

BREAKING JOINTS Unscrewing drill rods, casing, etc., where they are joined by threaded couplings.

BREAKING-IN SHOT In blasting a solid face, the first shot or group of shots of a round to be fired, essentially to provide space into which material from subsequent shots may be thrown.

BREAKTHROUGH Opening made, either accidentally or deliberately, between two underground workings.

BREAKTHROUGH-TYPE DEPOSIT Used in Tennessee zinc districts for a zinc deposit in which mineralized collapse breccias extend through the limestone beds into overlying dolomitic strata.

BRECCIA Fragmental rock, the components of which are angular. Also see *Tectonic and Collapse Breccias.*

BRECCIATED VEIN Fissure filled with fragments of rock and in the interstices, deposited mineral matter, not uncommonly lead and zinc sulfides.

BRECCIATED Rock composed of angular fragments held together in a matrix.

BRIDGE FINANCING Short-term funding provided when a company is about to raise a new round of equity, or is about to go public.

BRIGHT ROPE Rope of any construction, whose wires have not been galvanized, tinned or otherwise coated.

BRIGHTNER Agent or combination of agents added to an electrolytic bath to produce a fine-grained lustrous deposit.

BRISTOL GLAZE Raw glaze containing zinc oxide, often used on terra cotta.

BRISTOL METAL Alloy of copper and zinc in the proportion of about 16 to 6.

BRONZE Copper-rich, copper-tin alloy with or without small portions of other elements such as zinc and phosphorus. By extension, certain copper-base alloys containing considerably less tin than other alloying elements, such as manganese bronze (copper-zinc plus manganese, tin and iron) and leaded tin bronze (copper-lead plus and sometimes zinc). Also trade designations for certain specific copper-base alloys that are actually brasses, such as architectural bronze (57% copper, 40% zinc, 3% lead) and commercial bronze (90% copper, 10% zinc).

BROWN HORSESHOE FURNACE Furnace of the annular turret type for roasting sulfide ores.

BROWN ORE Term used at the Sterling Mine, NJ for typical franklinite-willemite-calcite ore that could be separated into its component parts, as opposed to *black ore.*

BRUCKNER FURNACE Horizontal revolving, cylindrical furnace for roasting pulverized sulfide ores.

BRUNOING Local for Arkansas and Missouri. Pulling fine ore down from the working place, especially with the hands. From its similarity to the action of a bear.

BUCKING ORE Hand process for crushing ore.

BUCKSTAY Upright iron or steel brace resting upon or built into a furnace wall to support the brickwork.

BUDDLE [1] Inclined vat, a stationary platform, or a round shaking table which has a motion initiating that of hand panning, onto which is fed a mixture of ground ore and flowing water to affect a crude separation of the ore components into light and heavy fractions. [2] To separate ore or stamp work by means of a buddle.

BUG LIGHT Slang for a miner's electric cap lamp.

BULL JIGS Jigs used at Mississippi Valley type deposits for preconcentration of zinc and zinc-lead ores. The jigs separated coarse fractions (plus one-quarter to minus three-fourths inch) of ore-containing fragments from very low grade and barren ore fragments (gangue). The gangue fragments went directly to waste so as to reduce expensive crushing and grinding of barren and low grade material. Bull jigs were gradually replaced in mills by heavy-media separation (sink-float processes) beginning in the late 1930's.

BULLING SHOVEL Triangular, sharp-pointed shovel used in ore dressing. Also called a vanning shovel.

BULLING The firing of explosive charges in the cracks of loosened rock. The clay stemming is forced around the charge by a bulling bar.

BUMPER Person who pushes loaded cans or cars into the station for the hooker and takes the empties away.

BUMPING AND JERKING TABLES Machines using mechanical agitation to bring the light and heavy ore grains into their respective layers on a washing surface. They use a bumping or jerking action to convey the heavy grains to one side or the other of the machine, while the current of water conveys the light grains to another side or to the end. They may be side-bump, end-bump, side-jerk or end-jerk relative to the flow of water.

BUMPING BAR Heavy bar used to break the loam from the mouth of each retort so the condenser can be fitted on the next cycle.

BUNKER HILL SCREEN A rotating screen shaped like a funnel. Material is delivered inside the funnel, the undersized passing through the screen while the oversize is discharged through the funnel neck.

BUNKY In metal mines in Illinois and Wisconsin, a partner.

BURLEIGH; BURLEY Miner's term for any heavy two-man drill. The Burleigh, invented by an American of the same name, was the first successful self-rotating rock drill.

BURN IN To deliberately run a bit with reduced amount of coolant until core is jammed inside the bit. See also dry block.

BURNING HOUSE The furnace in which sulfide ores are calcined to sublime the sulfur; a kiln.

BURNT BRASS Synonym for copper sulfate.

BURNT ORE Roasted ore.

BURST Explosive breaking of rock in a mine due to pressure. In metal mines, also known as rock burst.

BUS BAR Heavy metal conductor for high-amperage electricity, used in electrolytic refining.

BUSINESS PLAN Document created by managers to justify their application for finance.

BUSTER SHOT Same as *breaking-in shot*.

BUTCHERING [1] Term used at Joplin, MO for the sacrificing of either the smelting furnace or the zinc content in the ore for the best financial return. In the early years the furnaces were roughly constructed of inferior materials that would not long sustain the heat required to extract the zinc, such that it was accepted opinion that there was no economy in "butchering" the furnace for the sake of a small additional percentage of metal. It was preferred to increase the production of the furnace and save on labor and fuel costs by increasing the ore charge; in other words to "butcher" the ore and save the furnace. [2] Excessive deterioration of retorts in horizontal retort furnaces owing to continuation of the firing rate when there is no longer sufficient charge within to absorb much of the heat endothermically. Good practice was to reduce the firing rate near the end of the run.

BUTTON METAL An alloy, one-fifth copper and four-fifths zinc, used for brass buttons.

BUTTON Globule of metal remaining in an assaying crucible or cupel after fusion has been completed.

BUYING ON MARGIN Purchase of shares in which the purchaser supplies cash or collateral for a certain margin or percentage of the cost, and the broker lends or undertakes to borrow the balance, charging his client interest.

BYPRODUCT Secondary or additional product extracted in mining or processing the primary metal or metals. In zinc production, for example, the trace quantities of germanium and indium in the concentrate that are recovered are byproducts. Also see *Co-product*.

C

CADGER Little pocket oil can for miners.

CADMIA Impure zinc oxide that forms on the walls of furnaces in the smelting of ores containing zinc.

CADMIUM OCHRE The mineral greenockite.

CAGE Mining term for elevator.

CAKE [1] Solidified drill sludge. [2] Solid residue left in a filter press after the liquid has been drawn off. [3] Mass formed when ore sinters. [4] Residues resulting from the purification processes in preparing the pregnant zinc solution prior to electrowinning, i.e., copper cake, cobalt cake, etc.

CALAAEM Early name for Far East spelter.

CALAMINE BRASS Name given to brass made by the cementation process.

CALAMINE LOTION Liquid medicinal zinc oxide compound used mainly to treat skin rashes and itches.

CALAMINE STONE English term for the carbonate of zinc. More properly, smithsonite.

CALAMINE [1] Commercial, mining and metallurgical term comprising the oxidized ores of zinc, primarily smithsonite, hemimorphite and hydrozincite. [2] Synonymous with hemimorphite, the hydrous silicate of zinc. This was not always the case. American mineralogists used the term calamine to mean hemimorphite, whereas British mineralogists used it for smithsonite, the zinc carbonate. The disparate usages continued for many years until agreement as to what calamine would mean in a mineralogical sense was finally decided. Generally the term should be avoided for hemimorphite and used strictly in the sense of definition [1] above. [3] Type of galvanized iron. Also spelled kalamin.

CALCINATION Heating ores, concentrates, precipitates or residues to decompose carbonates, hydrates or other compounds.

CALCINATION Reduction of ore or other material to calx or friable condition by the action of fire.

CALCINER; CALCINING FURNACE Furnace or kiln used for roasting ores in order to drive off certain impurities.

CALCINE Product of roasting if in loose form like a powder. See *Sinter*.

CALCIUM CARBIDE Compound that decomposes in water, generating acetylene that is burned in miner's lamps for light.

CALLOW PROCESS Flotation process embodying the usual principles but in which agitation is secured by air forced into the pulp through a canvas-cover at the bottom of the cell.

CALMIA Ancient term for oxidized ores of zinc from which the term calamine was probably derived.

CALX The friable residue left when a metal, ore, or mineral has been subjected to calcination. Metallic calxes are called oxides.

CALYX A long cylindrical vessel of the same diameter as the core-barrel, which guides the drill bit, and receives the debris resulting from the action of the cutter. Its action is not unlike that of the diamond drill and necessitates the use of a powerful water flush. The cutter, which takes the place of the diamond crown, has a number of long teeth which produce a chipping action when rotated by hollow flushing rods in the presence of a constant flow of water.

CAMPAIGN Period during which a furnace is continually in operation.

CAN Used in the Tri-state zinc and lead district for a bucket used in hoisting. A can ranges from 1200 to 1400 pounds capacity. In the early days of mining at Joplin, cans were commonly oil cans and whiskey barrels.

CAN HOISTING SYSTEM Method of hoisting in shallow lead/zinc mines in some areas of the United States. Instead of the conventional enginehouse, operation is controlled at the top of the shaft. An onsetter below hooks the ore filledcan on, and then signals by a lamp attached to his wrist to the hoistman sitting above. Can is hoisted, swinging free. At the surface a tail rope is snapped to underside, a deflection plate is swung into place, and the can lowered. It capsizes and discharges its load to a surge bin, and again hoisted, freed of the tail rope, and wound down the shaft, where it is replaced by a full can.

CANADIAN SPELTER BOUNTY In August 1915, the Canadian Government established a bounty on spelter made in Canada from Canadian ores to help solve the problem of insuring at a reasonable prices, a Canadian supply of zinc for use in making brass for quick-firing cartridge cases and shells. It was at this time that the price of high-purity spelter reached a historically high price of 40 cents per pound or more than U.S. $5 in 1987 constant dollars. The bounty, up to 2 cents per pound based on a sliding scale adjusted to the zinc price, was to insure producers against too great a fall in the zinc price. The government must have been very optmistic about the war's end , in that the bounty would continue from the end of the war to July, 31, 1917.

CANNONS Thick, open-ended tubes, about the same size as Belgian retorts, placed below the lowest row of retorts in some Belgian furnaces to protect that row from the radiant heat of the fire.

CAP LIGHT, WET CELL Rechargeable cap light comprised of a belt mounted, wet-cell battery connected by insulated wire to an electric light mounted on the miner's hard hat. Two bulbs or two filaments in one bulb assure the wearer a constant source of light.

CAP [1] Piece of plank or timber placed on top of a prop, stull, or post. [2] Flat piece of wood inserted between the top of a prop and the back. [3] A detonator or blasting cap.

CAPITAL EXPENDITURE Amount of money required for the purchase and/or development and/or purchase of equipment and plant to operate a mine, and for working capital.

CAPITAL PROJECT Development scheme which is not financed by the revenue of a mine.

CAPPEAU FURNACE A modification of the Ropp furnace for calcining sulfide ore.

CAPPED FUSE A length of safety fuse with a cap or detonator crimped on before it is taken to the place of use.

CAPTIVE MARKET In mining, an independent operation that is dependent on a nearby mill or smelter for its viability.

CARBIDE BIT Steel bit which contains inserts of tungsten carbide. Such bits are commonly used in underground zinc mining.

CARBIDE INSERTS Shaped pieces of a hard metal, sometimes inset with diamonds, formed by the pressure molding and sintering of a mixture of powdered tungsten carbide and other binder metals, such as iron, cobalt, or nickel. Inset into slots, grooves, or slots in bits, reaming shells or core barrels, the hard metal pieces become cutting points or wear-resistant surfaces. Also called carbide slugs.

CARBIDE LAMP Lamp charged with calcium carbide and water (and sometimes spit) and burns the acetylene generated.

CARBIDE Commercial term for calcium carbide used in miner's lamps.

CARBOLOY-SET Diamonds inset in pressure-molded and sintered matrix metal composed of a cobalt-bonded, powdered tungsten carbide mixed with varying amounts of other powdered metals, such as iron, copper and zinc, for the purpose of making diamond drill bits and cutting tools.

CARBON MONOXIDE A colorless, odorless, very toxic gas that burns to carbon dioxide with a blue flame. Carbon monoxide and carbon are the principal reducing agents in reducing zinc oxide to metal. Also known as *White damp*.

CARBON SPAR Name given to several mineral carbonates, as carbonate of zinc, magnesium, etc.

CARBON STEEL Steel containing carbon up to about 2% and only residual quantities of other elements except those added for deoxidization. Also termed plain carbon steel, ordinary steel and straight carbon steel.

CARBONATE ROCK General term for limestone, magnesian limestone, dolomite, or magnesite.

CARBONET See *Briquette*.

CARE AND MAINTENANCE The continued maintenance of a mine or plant that is temporarily closed so that it can be reopened fairly quickly when conditions improve.

CARINTHIAN FURNACE An early zinc-distillation furnace with small vertical retorts closed at the top and ending in tubes leading to a single condensation chamber.

CARTRIDGE Cylindrical waterproof, paper shell, filled with high explosive and closed at both ends. Used in blasting.

CASTAWAYS Sterile veinstone.

CASTELLANOS POWDER Blasting powder containing nitroglycerin and either nitrobenzene or a picrate, mixed with other materials.

CASTILLITE Impure variety of bornite, containing zinc, lead and silver sulfides.

CASTING WHEEL Large turntable with molds mounted on the outer edge, primarily used for casting ingots of base metal.

CASTING [1] Object at or near finished shape obtained by solidification of a substance in a mold. [2] Pouring molten metal into a mold to produce an object of desired shape.

CATALYST Substance capable of changing the rate of a reaction without itself undergoing any net change.

CATARACTING Motion of crushing bodies in a ball mill in which some fall freely after breaking away from the top of the crop load and fall with impact to the toe of the load.

CATCH PIT Reservoir for saving tailings from reduction works. A catch basin.

CATCHALL Tool for extracting broken implements or junk from boreholes; a fishing tool.

CATHODE Electrode where electrons enter [current leaves] an operating system such as a battery, an electrolytic cell, an x-ray tube or a vacuum tube. In the battery, it is positive; in the others, it is negative. In a battery or electrolytic cell, it is the electrode where reduction occurs. Contrast with *anode*.

CATHODIC PROTECTION Partial or complete protection of a metal from corrosion by making it a cathode, using either a galvanic or impressed current. The former is accomplished by affixing an active sacrificial anode, such as zinc metal to the item to be protected whereas the latter is accomplished by driving current through electric cabling into less active anodes installed on the item.

CATION Positively charged ion; it flows to the cathode in electrolysis.

CATIONIC COLLECTORS In flotation, amines and related organic compounds capable of producing positively charged hydrocarbon-bearing ions, hence the name, for the purpose of floating miscellaneous minerals, including silicates.

CATRAKE Hydraulic brake or controller of a Cornish pumping engine, first introduced by Boulton and Watt.

CAZIN Brass containing 82.6% copper and 17.4% zinc. Used to braze steel.

CELESTE BLUE Ceramic color made by softening the normal cobalt blue by the addition of zinc oxide.

CELL FEED Material supplied to the cell in the electrolytic production of metals.

CELL [1] Compartment in a flotation machine. [2] Battery unit consisting of two electrodes separately contacting an electrolyte so there is a potential difference between them.

CEMENTATION [1] Process by which brass is made by mixing calamine, furnace zinc oxides, etc., with molten copper. The process is generally thought to be the first intentional metallurgical use of zinc despite the fact that the metal was unknown to the makers of the brass. The cementation process was used to make virtually all brass produced through the mid-19th century, even though processes using zinc metal were known as early as 1781 in England. [2] Process for the precipitation of a more noble metal by displacement with a more active element. Zinc dust, for example, is used to precipitate (cement out) copper, cobalt, nickel, etc. in the purification of the zinc sulfate electrolyte prior to electrowinning. The resulting deposits are known as *cakes*.

CENTER CUT Boreholes, drilled to include a wedge-shaped piece of rock, and which are fired first to make room for the rock of subsequent blasts in a heading, tunnel, drift or other working place.

CENTER SHOT A shot in the center of the face of a room or entry.

CENTRIFUGAL CASTING Made by pouring metal into a mold that is rotated or revolved.

CENTRIFUGAL PUMP Form of pump which displaces fluid by whirling it around and outwardly by vanes rotating rapidly in a closed case.

CERMAK-SPIREK FURNACE An automatic reverberatory furnace of rectangular form divided into two sections by a longitudinal wall. Used for roasting zinc and quicksilver ores.

CHAIN Surveyor's steel tape measure or a measuring instrument that consist of 100 links joined together by rings

CHAIRS Movable supports for the cage arranged to hold it at the landing when desired. Also called catches, dogs, and keeps.

CHALCOPHANITE Hydrous manganese-zinc oxide. See also *Hydrofranklinite*.

CHAMOTTE [1] Burned clay used in zinc smelters. [2] The refractory portion of a mixture used in the manufacture of fire-brick, composed of calcined clay or of reground bricks. In the manufacture of zinc retorts, burnt clay (chamotte) can range from 10 to 55% of the raw retort mix; in cases where low amounts of chamotte were used, ground-up old and broken retorts were substituted.

CHANGE HOUSE Special building or housed area at mines or other works where laborers may wash and change their clothes. Also called dry house, dry room, changing house or moorhouse.

CHANNEL SAMPLE Material from a level groove cut across an ore exposure in order to obtain a cross-sectional sample of the exposed ore.

CHAPMAN FURNACE SHIELD Track-mounted heat shield patented in 1900 used for the protection of workers in extraction of zinc metal from horizontal retorts. First used at a smelter in Pittsburg, KS.

CHAR To reduce to charcoal or carbon by exposure to heat.

CHARCOAL Amorphous carbon prepared from vegetal and animal substances; charcoal is made by charring wood in a kiln or retort from which air is excluded.

CHARGE [1] Liquid and solid material fed into a furnace for its operation. [2] Weights of various liquid and solid material put into a furnace during one feeding cycle. [3] Explosive loaded into a bore hole for blasting.

CHARGING [1] The loading of a borehole with explosives. [2] Feeding raw or partially treated ore, fluxes, fuel, etc., into an apparatus, such as a furnace for treatment or conversion.

CHARK To burn to charcoal; to char; or to coke, as coal.

CHASER See *Chilean mill*.

CHATS; CHATS PILE Loosely used

term in Missouri for tailings and waste rock from the concentration of lead and zinc ores. Also see *Chatty Ore*.

CHATTY ORE At Joplin, used for finely divided sphalerite in siliceous gangue or chert. Because of the requirement for fine grinding resulting in slimes and the ample supply of coarse material for jigs and tables, chatty ores were generally not processed at Tri-State mines until the introduction of flotation processes after World War I. As a result, zinc recoveries were poor at most Tri-state mines until the early 1920's.

CHECK TANK Storage tank in which purified zinc electrolyte is held until its purity is verified.

CHECKING Temporary reduction of the temperature in a furnace.

CHEMICAL EXTRACTION A term synonymous with hydrometallurgy.

CHENHALL FURNACE Gas-fired furnace for the distillation of zinc from zinc-lead ores.

CHERT Compact, siliceous rock formed of chalcedonic or opaline silica, one or both, and of organic or precipitated origin. Commonly occurs in limestones. Common constituent in the Tri-state ores. Also called flint and hornstone.

CHICKEN LADDER Notched log or pole used as a ladder.

CHILEAN MILLS Mill having rollers running in a circular enclosure having an iron base. Such mills, also known as edge runners, pan mills or chasers were used to further mix and densify the ingredients (roasted zinc concentrate, coal, and binders) before they were briquetted for feed in the vertical retort process.

CHIMNEY [1] Used in Virginia for limestone pinnacles bounding zinc ore deposits. [2] Any ore body of pipe-like shape, as a vertical or highly inclined, elongated ore shoot or similar-shaped replacement orebody. The term was used for large funnel-shaped, replacement zinc-copper-silver ore bodies at the Eagle Mine in Colorado. [3] Used in Iowa for large sulfide ore shoots or veins that occur at principal crevice or joint intersections; the ore tends to be continuous but blossoms out in certain formations.

CHINESE SILVER Alloy used as an imitation of silver containing 58% copper, 17.5% zinc, 11.5% nickel, 11% cobalt and 2% silver.

CHLORINE DEZINCING Process developed to extract the small amount of zinc remaining in refined lead after it has been desilverized. The process consists of pumping gaseous chlorine into molten lead, resulting in the formation of molten zinc chloride which rises in the bath and can be extracted. The zinc chloride is retreated to eliminate metallics and combined lead and packaged as a commercial product.

CHOCK Square pillar for supporting the roof, constructed of prop lumber laid up in cross-layers, in log-cabin style, the center filled with waste. Commonly called crib in Arkansas.

CHOKE DAMP Mine atmosphere that causes choking or suffocation, due to insufficient oxygen. As applied to air that causes choking, does not mean any single gas or combination of gases.

CHOKER Chain or cable so fastened that it tightens on its load as it is pulled. For example, such chokers are used to pull timbers and equipment up raises in the timber way or chute.

CHURN DRILL Boring tool employed when solid cores are not required, stratigraphic thicknesses need not be measured accurately, and only vertical holes are desired. A churn drill consist of a cutting bit that is suspended from a rope or cable that is alternately raised and lowered by hand or by a power-driven mechanism, chopping a hole in the rock. The cuttings are removed at intervals and can be examined and assayed.

CHUTE LOADER; CHUTE TENDER; CHUTE MAN In metal mining, one who opens gates of finger raises and ore chutes of raises and stopes to convey rock through grizzlies or to load cars.

CHUTE Channel or shaft underground, or an inclined trough above ground, through which ore falls or is shot by gravity from a higher to lower level. Also spelled shoot.

C.I.F. In commercial transportation, it stands for cost, insurance and freight. Intended to cover the cost of certain goods at point of destination. Usually applied to maritime freight.

CIRCLE In the central United States, a nearly circular lead and zinc deposit developed in the clayey chert breccias in old sinkholes in Paleozoic limestones and dolomites.

CIRCULATION FLUID Pumped through and to the end of the drill string and back to the surface in the process of drilling a borehole.

CIRE-PERDUE PROCESS Used in bronze casting;. See *Lost-wax Process*.

CLAD METAL Composite metal containing two or three layers that have been bonded together. The bonding may have been accomplished by co-rolling, welding, casting, heavy chemical deposition or heavy electroplating.

CLAIM Portion of mining ground held under the Federal and local laws by a claimant or association by virtue of one location and record. Also known as location and mining claim.

CLAMSHELL Twin-jawed bucket without teeth, usually hung from the boom of a crane, that is dropped in the open position onto the material to be excavated. It is then closed, encompassing the material between the two closed jaws. Clamshells are used to remove rock during shaft sinking and such operations as removing concentrate from barges and ships.

CLARIFICATION In leaching, removal of the last traces of solid matter from the pregnant solution, for example the zinc solution before electrowinning.

CLARIFIER Centrifuge, settling tank, or other device for separating solid matter from a liquid.

CLARK CELL Form of cell used as a standard of electromotive force. The elements are mercury and zinc, and the excitant and depolarizer is mercurious sulfate. The EMF at 15 degrees C. is 1.4322 volts. Also called a *zinc standard cell*.

CLASSIFICATION Separation of a mixture of particles into fractions based on particle size.

CLASSIFIER [1] Machine or device for separating constituents in ore according to relative size and density. In mineral dressing, the classifier is a device that takes the ball-mill discharge and separates it into a finished product which is ground as fine as desired, and oversized material for regrinding. [2] Machine for grading the feed to concentrators so that each individual concentrator will receive its proper feed. Classifiers are used to separate sand from slime, water from sand, and water from slime.

CLASSIFY Separation into fragments of different dimensions into classes on different size limits as affected by screens and classifiers.

CLAY Fine-grained, natural, earthy material composed primarily of hydrous aluminum silicates. It may be a mixture of clay minerals with varying amounts of nonclay material, or it may be predominantly one clay mineral. The physical and chemical properties (plasticity, natural and firing color, drying and firing characteristics, etc.) relate to the types of clay minerals and other materials present. A clay's suitability for commercial use is a function of its properties and not on its origin or occurrence. Compare with *Clay Mineral*.

CLAY MATERIAL Natural substance or soft rock which, when finely ground and mixed with water, forms a pasty moldable mass that preserves its shape when air dried; the particles soften and coalesce upon being highly heated and form a stony mass upon cooling. Clays differ greatly mineralogically and chemically and consequently in their physical properties. Most of them contain many impurities but ordinarily their base is hydrous aluminum silicate. Contrast with *Clay Mineral*.

CLAY MINERAL General term for several groups of crystalline, micron-sized, hydrous, aluminum-silicate minerals. The most common clay minerals belong to the kaolinite, montmorillonite, attapulgite, and illite groups. Compare with *Clay*.

CLEANER CELLS; RECLEANER CELLS Secondary flotation cells for the retreatment of the concentrate from the primary cells. Cleaner cells typically produce the final concentrate.

CLEANER CONCENTRATE High-grade concentrate derived from retreating rougher concentrate.

CLEANING [1] Retreatment of the rough concentrate to improve its quality. [2] Removal of grease and other foreign material from a surface.

CLIMB Tendency of an inclined diamond-drill hole to follow an upward-curving, increasingly flat course; also the tendency of a diamond or other rotary-type bits to drill a hole in an updip direction in alternating hard and soft rock having bedding planes that cross the borehole at an angle other than at right angles.

CLINKER Product of the fusion of the earthy impurities of coal during combustion. A slag/ash product.

CLOSURE [1] Relative inward movement of the walls and/or the floor and back of a stope or drift owing to pressure release causing a flowage of rock, etc., to fill the space; not uncommon in fault zones. May also be said of drill holes. [2] A surveying term. Also see *Rock Burst*.

CLOTTING Sintering and semi-fusion of ores during roasting.

COAL ASH Noncombustible matter in coal.

COARSE JIGS Used to handle the larger sizes and heavier grades of ore or metal.

COB To break ore into small pieces preliminary to hand sorting, esp. to break off waste or low grade material from ore lumps with hand hammers.

CODE OF FEDERAL REGULATIONS (CFR) Codification by subject of all regulations of the United States. Consists of 50 titles arranged by subject matter and an index volume. Every year each title is brought up-to-date with all Federal regulations in force at the time of publication.

COFUMED [1] Practice in which two or more components are fumed in the same furnace operation so as to yield a uniform fume product, such as in the manufacture of leaded zinc oxide. Compare with *blends*. [2] In lead slag fuming, lead and zinc are said to be cofumed.

COINAGE BRONZE Copper-based alloy containing 2 to 4 percent tin and 1 to 2 percent zinc, used for copper coins.

COINING [1] Closed-die squeezing operation, usually preformed cold, in which all surfaces of the work are confined or restrained, resulting in a well-defined imprint of the die upon the work. [2] Restriking operation used to sharpen or change an existing radius or profile.

COKE BREEZE Fine screenings of crushed coke, normally less than 3/4-inch in size.

COKE Bituminous coal from which the volatile constituents have been driven off by heat, so that the fused carbon and the ash are fused together. Largely artificial, but natural coke is also known.

COKEY HERDER Foreman of a shovel gang. Local to Joplin, Missouri.

COKING COAL Most important of the bituminous coals, which burns with a long yellow flame, giving off more or less smoke, and creates an intense heat when properly attended. Usually quite soft and does not bear handling well. In the fire it swells, fuses, and finally runs together in large masses, which are rendered more or less porous by the evolution of the contained gaseous hydrocarbons.

COLD CHAMBER MACHINE Die-casting machine where the metal chamber and plunger are not heated.

COLD NOSE In the western United States, term used for a mining expert who underrates the value of mining properties.

COLD ROLLING Rolling of strip zinc where the temperature of the strip is maintained under 300 degrees F. Strip rolled between 350 and 400 degrees F. is considered hot rolling.

COLD-SHORT Condition of brittleness existing in some metals at temperatures below the recrystallization temperature.

COLD WORKING Deforming metal plastically at a temperature lower than the recrystallization temperature.

COLLAPSE Complete cave-in of walls of a borehole or mine workings.

COLLAPSE BRECCIA Rock composed of angular rock fragments owing to the collapse of the roof of a solution opening, most commonly in carbonate rocks. The mass is often mineralized and may be of large size, often extending several hundred feet upward. A number of important zinc deposits in Tennessee and Virginia occur in collapse breccias, where some are known as *cave-collapse breccias* and *crackle* and *founder* breccias.

COLLAR [1] Applied to the timber or concrete around the mouth or top of a shaft. [2] The beginning point of a shaft or drill hole.

COLLARED Starter hole drilled sufficiently deep to confine the drill bit and prevent slippage of the bit from normal position.

COLLECTING AGENT Reagent added to the pulp to bring about adherence between solid particles and air bubbles. Also referred to as a collector.

COLORADO SILVER; COLORADO METAL Misleading name for a German silver containing 57% copper, 25% nickel, and 18% zinc.

COMBINATION DIE Die having two

or more different cavities for different castings.

COMBINATION SHOT Blast made by dynamite and permissibles or permissible explosives and blasting powder in the same hole. It is bad practice and prohibited by law.

COMBINED OVERHAND AND UNDERHAND STOPING Working of an ore block simultaneously from the bottom to its top and from the top to the bottom. Modifications are distinguished by the support used, as open stopes, stull-supported stopes, or pillar-supported stopes.

COMING BACK Said of cadmium that returns to the neutral solution after it had been precipitated by zinc dust treatment. Attributed to the fact that finely divided particles of cadmium oxidizes rapidly if air agitation is used, thereby making them soluble in neutral zinc sulfate solution. Term not used in cases where cadmium redissolves, as when excess copper is in the neutral solution.

COMING OUT Process of withdrawing or hoisting the drill string or tools from a borehole.

COMMERCIAL ORE Mineralized material profitable at current prices.

COMMINUTE To reduce to minute particles, or to a fine powder; to pulverize; triturate.

COMPACT Object produced by the compression of metal or nonmetal powder generally while confined in a die. Synonymous with *Briquette*.

COMPANY MAN Person who works for the company by the hour or by the day, such as track layers, timbermen, drivers, and cagers, as distinguished from miners who work under contract, as by the ton, yard, etc.

COMPANY STORE Sells groceries and general merchandise, owned and run by an industrial company. Common in early mining and lumber camps.

COMPLEX ORE The term has no precise meaning. Generally stands for an ore containing two or more metallic ore minerals that are difficult to extract or separate.

COMPOSITION METAL Yellow alloy of copper, zinc, etc., used for sheathing vessels.

COMSTOCKITE Zinc-magnesia chalcanthite from the Comstock Lode of Nevada.

CONCENTRATE [1] To separate metal and non-metal ore minerals from gangue or associated rock. [2] The product of concentration.

CONCENTRATION TABLE Table on which a stream of finely crushed ore and water flows downward and the heavier metallic minerals lag behind and flow off into a separate compartment.

CONCENTRATION Process for enrichment of valuable minerals in an ore by separation and removal of non-valuable constituents or gangue minerals.

CONCENTRATOR An apparatus or entire plant containing various concentrating devices where ore is separated into values (concentrate) and rejects (tailings). Concentration is carried out in appliances, such as flotation cells, jigs, tables, electromagnets, etc. Also called a mill or ore dressing plant.

CONCRETE Term used by sinter-plant operators for a hard, dense material that forms in the sinter operation and appears to be sinter but is actually unsintered material cemented together by iron and/or zinc sulfate.

CONCUSSION TABLE Inclined table, agitated by a series of shocks, and operating at the same time as a buddle.

CONDENSER MAKER In zinc smelting and refining, one who operates an automatic machine in which fire clay condensers, used in smelting zinc ores, are made.

CONDENSER SETTER One who sets up condensers in which zinc vapor is collected and condensed.

CONDENSER Vessel or chamber in which volatile products of roasting or smelting (e. g. mercury or zinc vapors) are reduced to solid form by cooling, or in which the fumes of furnaces, containing mechanically suspended as well as volatile metallic matters, are arrested.

CONDITIONING A stage of froth-flotation process in which the surfaces of the mineral species in the pulp are treated with appropriate chemicals to influence their reaction when the pulp is aerated.

CONE CRUSHER; GYRATORY CRUSHER A machine for reducing the size of rock by means of a truncated cone revolving on its vertical axis within an outer chamber, the annular space between the outer chamber and the cone being tapered.

CONE A three-sided pyramid made of unfired ceramic material whose composition is such that when heated at a controlled rate, it will deform and fuse at a known temperature. Also known as pyrometric cones. Cones are used to test fusion temperatures of clays, etc., and to indicate temperatures in furnaces.

CONGOS Describes diamonds used in diamond drill bits, named after a variety of diamonds found the Republic of the Congo (Zaire). The term applies to diamonds of similar character as those from Congo.

CONGRESS Name given to a huge 15-inch brass mortar captured by an American privateer from a British ship in 1775, and subsequently used by General Washington in the siege of Boston.

CONING AND QUARTERING Old method for sampling bulk ore. The sample is thoroughly mixed by hand shoveling and formed into a coned pile that is flattened out and divided into quarters. Two opposite quarters are removed and the procedure repeated until the desired sample weight is reached.

CONSTITUTIONAL WATER Water bound in and necessary for the crystal structure of a hydrated mineral or compound.

CONTACT PROCESS For the manufacture of sulfuric acid, based on the catalytic action of finely divided platinum. Conducted by passing the well-dried and purified burner gases through the contact apparatus at a temperature of 350 degrees C and absorbing the sulfur trioxide formed by the direct union of sulfur dioxide and oxygen, in water.

CONTACTORS Devices that electrically contact the work in the electrogalvanizing process to complete the circuit. In electrogalvanizing, such devises have been in the form of rollers or as numerous finger contacts that touch the work as it passes through the process. In rack plating, the rack and connecting hanger wires serve that purpose and in barrel plating, the cathodic contact is in the barrel and only intermittently contacts the work.

CONTANGO Price situation in which the forward price is greater than the nearby price. The value is limited to the cost of financing, warehousing, and insurance. Compare with *Backwardation*.

CONTENT That which is contained. Often used in mining, as ore-content, mineral-content, zinc-content, etc.

CONTIGUOUS CLAIMS Mining claims which have a side or end line in common.

CONTINUOUS THICKENER OR CLARIFIER [1] Large cylindrical tank with a conical base. Rakes rotate on a shaft and move settled sludge (tailings, precipitates, etc.) toward the central discharge to be drained and discharged. [2] Settling tank or other device for separating suspended solids from a liquid.

CONTRACT DRILLER One who contracts with a company to carry out a specific type and quantity of drilling for a specific amount of money. Bonuses might be included if the work is completed ahead of schedule or in the case of core drilling, if core recovery is above a certain percentage.

CONTRACT MINER Miner, generally in development or production work, paid on a contract basis by the foot advance in a heading or raise, tonnage or volume of ore broken or moved, etc.

CONTROL ASSAY An assay made by an umpire to settle differences between the buyer and seller of ore, concentrate, scrap, or metal.

CONTROLS In sampling ore, concentrate, or scrap, controls are samples that are analyzed separately by or for the buyer and the shipper for each important component. If the control assays are close enough for agreement as defined in the smelter or contract schedule, the average values are used for settlement, if not, a reserve sample is analyzed as an empire.

CONVENTIONAL POT GALVANIZING Applied to the method for galvanizing sheet material in cut lengths. See also zinc-bath process and lead-zinc process.

COOK-NORTEMAN PROCESS An in-line, continuous process for sheet galvanizing, except that the coils are annealed in a separate operation before the coils are put on the continuous galvanizing line.

COOLING CELL A cell containing coils of lead pipe through which zinc electrolyte (neutral solution) flows and is cooled by externally circulated water.

COOLING TOWER A tower-like structure for the cooling of electrolyte at an electrolytic zinc plant. The solution is pumped to the top of the tower and is cooled by allowing it to flow down in thin streams.

COONTAIL ORE Banded ore consisting mainly of fluorite and sphalerite in alternate light and dark layers found in the Cave-in-Rock district of southern Illinois.

COPPER SULFATE In flotation, a reagent used to activate sphalerite so it will float. *The benefits of copper sulfate in zinc flotation were realized independently in Australia and the United States at about the same time when sulfuric acid began to be used in flotation processes. The U. S. discovery was realized when laboratory flotation tests using sulfuric acid on Tennessee and Mississippi valley sphalerite ores worked very well, whereas similar tests in the field did not. It was found that copper sulfate was generated owing to the corrosive effects of the acid on the brass and bronze equipment used in the laboratory flotation cells.*

CORDITE An explosive of nitroglycerin and a dope, used chiefly as a propellant.

CORE BARREL A hollow cylinder attached to a specially designed bit and which is used to obtain and to preserve a continuous section of core of rock penetrated in drilling.

CORE BIT A hollow, cylindrical boring bit for cutting a core in rock drilling.

CORE BOX A parallel-grooved box specially made to hold core for study and storage.

CORE SHACK or SHED or SHANTY A roofed structure in which filled core boxes are stored.

CORE SLUDGE Slurry produced from grinding and cutting during drilling. Collected sludge from long-hole and diamond-core drilling can provide varying information, such as changes in rock types, faulting, hidden veins, etc., and in core drilling may provide the only sample for a section of rock or ore.

CORE SPLITTER Tool employing a chisel to split core longitudinally in half so that a sample can be physically and chemically tested while at the same time preserving half for the record and perhaps test work in the future.

CORE A cylindrical sample of rock obtained in core drilling.

CORNISH PUMP; ENGINE A single-acting engine in which the power for pumping operations was transmitted through the action of a cumbersome beam. These pumps were introduced early in the 19th century and held the field for practically 100 years.

CORROSION EMBRITTLEMENT Severe loss of ductility in a metal resulting from corrosive attack, usually intergranular and often not visually apparent.

CORROSION Deterioration of a metal by chemical or electrochemical reaction with its environment.

COST DEPLETION An allowed deduction from Federal income taxes for a wasting asset. It is computed by dividing the amount of money expended to acquire a mineral property by the mineral units (tons of ore) in the mineral deposit at the end of the taxable year, giving the cost depletion per unit of reserve. This value times the number of mineral units sold during the year yields the depletion allowance. See also *Percentage Depletion*.

COSTEANING Tracing or proving an orebody by trenching or surface pits.

COTTON ROCK In Missouri, a local name for a soft, fine-grained siliceous magnesian limestone of the Lower Silurian.

COTTRELL PRECIPITATOR An electrostatic device whereby negatively charged dust or fume particles are attracted to a positively charged bar or wire electrode enclosed in a flue, the walls of which act as the other electrode. In the process the accumulated deposits are shaken from the electrodes and collected in a hopper below.

COVER HALF Die casting term for the stationary half of a die.

COYOTING See *Gophering*.

CREVICE CORROSION A type of concentration-cell corrosion; corrosion of a metal that is caused by the concentration of dissolved salts, metal ions, oxygen or other gases, and such, in crevices or pockets remote from the principal fluid stream, with a resultant building up of differential cells that ultimately cause deep pitting.

CREVICE DEPOSIT A term used in the Upper Mississippi Valley lead-zinc district for vein-like ore occurrences. Synonymous with *pitches and flats*.

CRIBBING Close timbering, as the lining of a shaft, or the construction of cribs of timbers, or of timber and rock to support the roof.

CRIB A structure composed of frames of timber laid horizontally upon one another, or of timbers built as in the walls of a log cabin.

CRITICAL MINERALS Minerals and metals essential to the national defense, the procurement of which in war, while difficult, is less serious than those of strategic minerals because they can be either domestically produced or obtained in more adequate quantities or have a lesser degree of essentiality, and for which some degree of conservation and distribution control is necessary. Compare with *Strategic Minerals*.

CROPPING OUT A natural exposure of bedrock at the surface. That part of a vein which appears at the surface is called the cropping or outcrop.

CROSS ROLLING A pack-rolling process whereby the zinc sheet pack is rolled at right angles to the direction of the first rolling the minimize the development of grain in the sheets. Reducing the grain improves bending without cracking because cracking tends to be more prevalent with the grain than across it.

CROSSCUT In general, any drift driven across between any two openings for any mining purpose.

CROWN PILLAR Pillars left between stopes for support in dipping veins. Because these pillars are generally ore, they are recovered by top slicing with square setting or by cut and fill mining depending to some extent the character of fill in the adjacent stopes.

CRUDE ORE Unconcentrated ore as it leaves the mine.

CRUDE In the natural state; not altered, refined or prepared for use by any process, as crude ore.

CRUSHER Machine for crushing rock or other materials. As a gyratory crusher, jaw crusher, stamp mill, etc.

CRYSTAL RECTIFIER Point contact between a metal and a crystal or two crystals, such as zincite and bornite. It has marked unidirectional conductivity.

CUNDERED HOLE A lifter hole or a hole drilled to throw the burden upward.

CUP A sheet metal part, the product of the first deep-drawing operation.

CURRENT DENSITY The amount of electric current per unit of cross-section area of the conductor, at any part of the circuit.

CURRENT EFFICIENCY [1] Proportion of current used in a given process to accomplish a desired result. In electroplating, the proportion used in depositing or dissolving metal. [2] Percentage derived by comparing actual weight of zinc plated to the theoretical weight predicted by Faraday's Law for the measured amount of electricity.

CUSTOM MARKET The market involving that portion of zinc concentrates actually sold on the open market.

CUSTOM MILL [1] A mill which buys ores for treatment or which treats ores for customers. [2] A plant receiving ore for treatment from more than one mine.

CUSTOM ORE Ore bought by a mill or smelter, or treated for customers.

CUSTOM SMELTER A smelter which buys ores or treats them for customers.

CUT HOLES The first round of holes fired in a tunnel or shaft. So placed to force out a cone-shaped core in the center of the heading and relieve the burden of the second round of shots.

CUT-AND-FILL STOPING A stoping method in which the ore is mined by successive flat or inclined slices, working upward from the level as in shrinkage stoping. However, after each slice is blasted down, all broken ore is removed, and the stope is filled with waste up to within a few feet of the back before the next slice is taken, with just enough room between the top of the fill and the back to provide working space.

CUTOFF GRADE [1] The lowest grade of mineralized rock that qualifies as ore and is included in an ore estimate. [2] Term used to define the assay grade below which an orebody cannot be profitably exploited.

CYCLONE CLASSIFIER A device for classification by centrifugal means of fine particles suspended in water, whereby the coarser grains collect at and are discharged at the apex of the vessel, while the finer particles are eliminated with the bulk of the water at the discharge orifice.

CYCLONE SEPARATOR A funnel-shaped device for removing material from an air stream by centrifugal force.

D **DAMPING DOWN** In pyrometallurgy, reduction of air supply to a furnace, to lower temperature and reduce working rate.

DANIELL CELL Primary cell with a constant electromotive force of about 1.1 volts, having as its electrodes copper in a copper sulfate solution and zinc in dilute sulfuric acid or zinc sulfate, the two solutions being separated by a porous partition.

DAVIS FURNACE A long, one-hearth reverberatory furnace, heated by lateral fireplaces for roasting sulfide ore.

DE BAVAY PROCESS A flotation process invented by Auguste J. F. De Bavay in 1904, in which a freely flowing pulp is brought to the surface of a vessel of water, where advantage is taken of the surface tension of the liquid, and the sulfide floated. A film of carbonate on the sulfide from weathering is detrimental, and is removed by soaking the ore in a weak solution of carbonate of ammonia, or by passing carbon dioxide through the pulverized wet ore, or by friction. In the original process no oil or acid was used. Later these were used.

DE SAULESITE A green amorphous hydrous silicate of nickel and zinc.

DEACTIVATION In froth flotation, treatment of one or more species of mineral particles to reduce their tendency to float; modification of action of an activating agent for similar purposes.

DEAD AIR Air in a mine when it contains carbonic acid (black damp), or when ventilation is sluggish.

DEAD ROAST [1] Roasting for complete elimination of sulfur. [2] Roasting carried to the furthest practical degree in the expulsion of sulfur.

DEAD-DIPPING Act or process of imparting a dead or dull surface to brass or other metal by dipping it in an acid.

DEADENING ROAST A slight roast of lead, copper, and iron sulfides to oxidize their surfaces but not that of zinc sulfide. This was an early preferential flotation system for treating mixed sulfide ores. When the ore is subjected to flotation, the zinc sulfide floats and is easily separated from the "deadened" sulfides. The term *dimming roast* is sometimes used for this process.

DEADMAN A buried log or the like, serving as an anchor as for a guy rope. Also a wooden block used to guard the mouth of a mine against runaway cars.

DEADS Barren rock which encloses the ore on every side. The wall rock.

DE-AIRING Removal of air from damp plastic clay to improve its properties before being extruded or fed to the pug or auger. De-aired cylindrical ballots of clay for zinc retort manufacture were generally prepared in hammering or vacuum-extrusion machines; other processes de-aired the clay in a vacuum chamber just prior to passing into the extrusion chamber (See *Vacuum Pug*).

DEEP DRAWING The forming of deeply recessed parts by means of plastic flow of the material.

DEFENSE MINERALS EXPLORATION ADMINISTRATION (DMEA) Government office established in 1951, by provision of the Defense Production Act to encourage mineral exploration of critical and strategic minerals by providing loans up to 50 percent of the total cost of approved exploration projects. Zinc projects were temporarily made ineligible for loans beginning May 15, 1953 owing to large stocks of available metal, but were restored to the eligible list in March 23, 1954 owing to a redefinition of Government zinc objectives. In 1958 the DMEA legally expired but its functions and obligations were transferred to the Office of Minerals Exploration (OME), a new group established in the Department of the Interior by the Congress also in 1958. Zinc was removed effective July 1, 1962, from the list of minerals eligible for exploration assistance under the program. Through the program, DMEA and OME had entered into 290 contracts for zinc and/or lead exploration, at an estimated costs of about $13.7 million. Exploration was judged to have been successful at 97 of the projects, with ore discoveries ranging from a few hundred tons to one of 35 million tons in eastern Tennessee.

DELAY DETONATOR A blasting cap so designed as to delay the firing by a few milliseconds of certain boreholes to increase the break and fragmentation of ore or rock.

DELEADING Elimination of lead from a material. In the slag fuming process, deleading of the initial zinc oxide product from the extraction of fuming lead smelter slags, occurs during the zinc oxide densification step, if this step is required for shipping or processing factors. Deleading is carried out in kilns at 1,200 degrees C., to yield a zinc product containing about 1% lead, from 5-10% lead in the product treated.

DELPRAT PROCESS Pioneering flotation method (1903). See *Potter-Delprat Process*.

DELTA-METAL A non-rusting, copper, zinc, and iron alloy resembling Aich's metal and sterro metal.

DEMONSTRATED RESOURCES Sum of measured and indicated.

DENSE-MEDIA SEPARATION See *Sink-Float Process*.

DENVER CELL Flotation cell of the subaeration type, widely used.

DEPLETION ALLOWANCE Portion of income derived from mining that is considered to be a return of capital not subject to income tax.

DEPLETION Act of emptying, reducing, or exhausting, as in the depletion of a natural resource; specifically said of ore reserves.

DEPRECIATION Reduction of assets of a working mine through rundown of ore reserves, and obsolescence, wear, and tear of equipment.

DEPRESSING AGENT In the froth flotation process, an agent that reacts with particle surfaces to render them less prone to stay in the froth, thus causing them to wet down as a tailings constituent.

DEPRESSION A period of severe economic decline.

DEPRESSORS Substances (usually inorganic) whose presence in the pulp prevents the anchoring of the collector molecules on the mineral surface, thus preventing flotation of the mineral.

DESCENSION THEORY Theory that the material in veins entered from above.

DESCLOIZITE A vanadate of lead and zinc found only in the oxidized zone of ore deposits. Locally abundant in some zinc deposits, especially in the southwestern U.S., where it has been mined only for its zinc content. Other deposits, along with associated vanadium minerals, were mined for lead and vanadium.

DESERT RAT In western United States, a prospector, especially one who works and lives in the desert, or one who has spent much time in arid regions. The name is derived from a small rodent common throughout much of the west.

DESILVERIZING KETTLE Circular kettle 3 to 4 feet deep used for the desilveration of base metal.

DESLIMING Removal of slimes, however accomplished, in any stage of processing ores.

DESULFURIZATION Removal of sulfur from sulfide ores.

DETONATOR A device for producing detonation in an explosive charge, and initiated by a safety fuse or by electricity.

DEVELOPED ORE Ore which is so completely exposed that its yield with respect to tonnage and tenor is essentially certain and which, in addition, is available to immediate withdrawal by mining.

DEVELOPMENT CAPITAL Later-stage finance for more established companies which are profitable or nearly so.

DEVELOPMENT WORK Undertaken to open up an ore body, as distinguished from the work of actual ore extraction. Sometimes development work is distinguished from exploratory work on one hand and stope development work on the other.

DEWATERING In mineral processing, removal of part of the liquid from the pulp. Generally carried out in classifiers, thickeners, and filters.

DEWAXING Removing the expendable wax pattern from an investment mold by heat or solvent.

DEZINCIFICATION Corrosion of some copper-zinc alloys involving loss of zinc and the formation of a spongy porous copper.

DIAMOND BIT Rotary drilling bit studded with bort-type diamonds.

DIAMOND DRILL A form of rotary rock drill in which the work is done by abrasion of diamonds set in a hollow drill bit. The diamond drill is typically used in prospecting and development work where core is desired. In coring, a circular section of rock works its way through the center of the drill bit as the bit penetrates the rock, and passes into a core barrel where the core is protected until extracted for examination and assaying.

DIATREME Vent occurring in a surface fissure in volcanic regions.

DIE CASTING [1] Casting made in a die. [2] Casting process where molten metal is forced under high pressure into the cavity of a metal mold.

DIFFERENTIAL FLOTATION Separation of a complex ore by flotation into two or more concentrates and gangue. Separate concentrates are made possible by the use of suitable depressors, activators, and frothing agents and by varying pulp conditions.

DILUTION Contamination of ore with barren rock in mining; often expressed as a percentage in the ore after mining, as in dilution of the ore is less than 10%.

DIMMING ROAST See *deadening roast*.

DIMORPHOUS Having the same chemical composition but crystallizing in two different crystal systems; e.g., zinc sulfide crystallizes in the isometric (sphalerite) and hexagonal (wurtzite) crystal systems.

DIP PILLARS Support pillars in stoping moderately dipping (20 to 40 degrees) orebodies, such as those employed in mining at Austinville, VA.

DIP Angle of a bed, stratum, vein, etc. from the horizontal.

DIRECT ARC FURNACE One in which heat is supplied by an arc is struck between an electrode and the material charged into the furnace.

DIRECT PROCESS Manufacturing zinc oxide directly from the ore or concentrate or from oxidized scrap. More commonly called the American Process. Compare with *Indirect* or *French Process*.

DIRT At Joplin, Missouri and elsewhere, the term applied to crude lead-zinc ore. The concentrate was called ore.

DISCOVERY [1] First finding of a mineral deposit in place upon a mining claim. A discovery is necessary before the location can be held by a valid title. The opening in which it is made is called discovery-shaft, discovery-tunnel, etc. [2] Finding of mineral in place as distinguished from float rock constitutes a discovery.

DISK FILTER; AMERICAN DISK FILTER Continuous dewatering filter in which the membrane (filtering cloth) is stretched on segments of a disk. These disks rotate through a tank of slurry. The vacuum inside the disk draws the liquid through the cloth to discharge; the solids forming a cake on the filter cloth are lifted clear of the slurry tank and separately discharged, by application of air pressure behind the filter cloth.

DISSEMINATED DEPOSIT an ore deposit in which the ore minerals occur dispersed as small particles or veinlets scattered in the country rock.

DISTILLATION COLUMN Columns

used in processes to purify zinc metal that are based on the differences in the boiling points of zinc and associated impurities. Zinc boils at 907 degrees C., cadmium at 778 degrees C., and lead at 1620 degrees C. Lead and other high-boiling-point metals are readily separated from zinc and cadmium by distilling the zinc and cadmium off; similarly, cadmium is distilled from zinc. Such columns are used to upgrade the quality of the metal produced and ,in a few cases, are used to upgrade metal prior to French-process, zinc oxide production.

DISTILLATION Volatilization, followed by condensation to the liquid state.

DIVIDEND Payment by a company to its shareholders, generally from the profits earned by the company.

DMEA Defense Minerals Exploration Administration.

DOG Any of various devices for holding, gripping, or fastening something.

DOGHOUSE [1] A term used in Joplin, Missouri for a washroom, changehouse, and dryhouse. [2] Also used in Joplin, Missouri, for a box or platform on which a can or bucket rests at the bottom of a shaft.

DONKEY A winch with drums which are controlled separately by clutches and brakes.

DOR FURNACE A regenerative zinc-distillation furnace with heat-recuperating chambers at ends of the furnace instead of beneath the combustion chamber.

DOR PRESS; DORR HYDRAULIC PRESS A hydraulic machine invented by E. Dor, a Belgian engineer, about 1877 for molding retorts. Because one of the key factors in successful zinc smelting was the manufacture of the densest and most refractory retort possible, the Dor press and later, similar type machines gained widespread use.

DORR RAKE CLASSIFIER Mechanical classifier consisting of an inclined settling tank and a rake-type conveying agitating mechanism. Feed introduced at the low end of the tank flows over a distributing apron toward the high end of the tank. The heavier particles of sand settle in the rake zone and are raked up the slope and out of the tank; finer sand and slimes are carried over the rear wall in suspension.

DORR THICKNER Large cylindrical vat with peripheral overflow and central bottom discharge. Ore pulp is fed in at top center, gravitates down and is moved to the discharge area by slowly circling plows while relatively clear liquid overflows.

DOUBLE CORE BARREL Core barrel with an inner tube to hold the core. The inner tube does not rotate during drilling, thereby giving a better core recovery.

DOUBLE DECOMPOSITION The name given to a chemical reaction in which two chemical compounds take part, both are decomposed and two new substances are formed by an exchange of radicals. An example of this type of reaction would be the double decomposition required to make zinc lithopone. See *Lithopone*.

DOUBLE JACKING Rock drilling by hand preformed by two men, one holding and rotating the steel bit and the other swinging the sledge hammer.

DOUBLE LEACHING In electrolytic zinc refining, calcine leaching is generally carried out by a single or double leach. The first leach extracts the easily soluble zinc in the calcine followed by precipitation of soluble impurities as iron, silica, aluminum, etc. A second leach is carried out on the residues and additional calcine to maximize the recovery of zinc and neutralize the solution.

DOUBLEHAND DRILLING Manual rock drilling with a long handled sledge hammer requiring both hands. A second man holds the drill and turns it between strokes. Two or even three strikers may work together.

DOUGLAS FURNACE A horizontal, revolving cylindrical roasting furnace having a central flue.

DOUSE; DOWSE [1] Search for deposits of ore, for lodes, or water, by aid of the dousing or divining rod. [2] Used by drillers as a synonym for devices, as divining rod, forked tree limb, or other nonscientific contraptions, supposedly useful in locating subsurface water, oil, or minerals.

DOWNDRAFT SINTERING Process to produce coarse sintered furnace feed from fine materials using a grate-type sintering machine in which air is pulled through the charge into underlying windboxes.

DOWNGRADE To classify a substance as lower quality than warranted. Also commonly used in scrap transactions to indicate scrap below specifications.

DOWNTIME Production time lost through mechanical breakdown, adjustments, maintenance, lack of power or of feed, etc.

DOWNWARD ENRICHMENT A term which is synonymous with secondary enrichment as the latter has applied to enrichment of ore bodies by the downward percolation of waters.

DOWSON PRODUCER A furnace used for the manufacture of producer gas.

DRAWABILITY A measure of the workability of a metal to the drawing process. Usually expressed to indicate a metal's ability to be deep-drawn.

DRAW To remove ore from the work places, as in draw ore from the stope.

DRAWBACK Money or duty returned upon the exportation of that on which it was levied.

DRAWING [1] Forming recessed parts by forcing the plastic flow of metal in dies. [2] Reducing the cross section of wire or tubing by pulling it through a die.

DRAWPOINT A place or chute where gravity fed ore from a higher level is loaded into hauling units.

DRESS To clean ore by breaking off fragments of the gangue from the valuable minerals. Compare with *Ore Dressing*.

DRESSING A MINE A method of fraud carried out by a seller of a mine, by systematically mining out all the low-grade ores and barren spots in a vein, leaving only the high-grade areas. This method is used on deposits of copper, lead and zinc where the values are in the form of sulfides distributed in a coarse and irregular manner in the vein. In a mine dressed for sale, there is a lack of straight lines which is in itself suggestive. The back of a drive or stope tends to have a billowy appearance which is the result of gouging out the lower-grade places.

DRESSING WORKS See *Concentrator*.

DRESSING Originally referred to the picking, sorting and washing of ores preparatory to reduction. The term now includes the more elaborate processes of milling and concentration of ores.

DRIFT Horizontal passage underground, which typically parallels the vein or is driven along and in the vein itself.

DRIFTER An excavator of mine drifts.

DRILL CARRIAGE [1] A movable platform, stage, or frame which incorporates several drills, used generally for heavy drilling work. [2] Mobile machine on which several drills of drifter type are mounted. A jumbo.

DRILL CORE A solid, cylindrical core of rock cut out by a diamond or shot drill.

DRILL STRING; DRILLING STRING An assemblage of drill rods, core barrel and bit which is connected to and rotated by a drilling machine.

DRILL TOWER A machine-mounted adjustable swing-jib or crane with a platform from which high places in stopes and mine workings can be drilled. A cherry picker.

DRIVE To excavate horizontally, or at an inclination, as in a drift, adit, or entry. Distinguished from sinking and raising.

DRIVING Extending excavations horizontally. Distinguished from sinking and raising.

DROP ZINC Zinc in the form of small globules.

DROSS The scum that forms on the surface of molten metals largely because of oxidation but also from fluxes, reaction products and rising impurities from the work or melt. Hot-dip galvanizer's dross occurs at the top and the bottom of the zinc melt. The bottom dross is a metallic iron and zinc alloy derived from the steel being treated and the molten zinc metal. The alloy forming the bottom dross has a greater specific gravity than zinc and liquidates to the bottom, and in doing so, entrains some unalloyed zinc, which becomes part of the dross. The dross formed on the top of a galvanizing bath is generally referred to as skimmings or ashes if formed by oxidation when no flux blanket is used and as sal skimmings when the surface scum consists of spent flux containing oxides, free zinc, and foreign material carried into the bath. A top dross is formed on the surface of the bath in the continuous sheet galvanizing process, mainly as a result of minor additions of aluminum to the process. This addition results in the formation of a zinc-alum-inum-iron alloy which has a specific gravity less than zinc, and therefore, it rises to the top mixing with surface and bath-sourced oxides. On skimming this material off the top of the bath, bath metal is also entrained.

DRY CELL Primary cell in which the electrolyte is held in a gel or some absorbent material so that it may be used in any position. Adopted in the 1880's to distinguish such cells from the wet cells then in use.

DRY CHARGE Metallurgy term for a charge that is not slagged in processing.

DRY GALVANIZING The process whereby a steel product that is fluxed in hot ammonium chloride and subsequently dried by hot air before being passed through a bath of molten zinc.

DRY MAN Person in charge of the building in which workmen change their clothes.

DRY-BONE ORE [1] Miner's term for an earthy, friable carbonate of zinc, smithsonite. Also often applied to the hydrated silicate, so-called calamine. Usually found associated in veins or beds in calcareous rocks accompanying zinc, iron and lead sulfides. [2] At Joplin, the term was sometimes used for lead carbonate; Smithsonite and calamine were called "silicate".

DUALIN A variety of dynamite consisting of 4 to 5 parts nitroglycerin, 3 parts sawdust, and 2 parts saltpeter.

DUCTILE Metal capable of being permanently drawn out or hammered thin.

DUCTILITY Ability of a material to deform plastically without fracturing, being measured by elongation or reduction of area in a tensile test, by height of cupping in an Erichsen test or by other means.

DUMB'D Choked or clogged, as a grate or sieve in which ore is dressed.

DUMP CART Cart or car having a body that can be tilted, or a bottom opening downward, for emptying.

DUMPER [1] A tilting car used on dumps. [2] One who operates or dumps a dump cart.

DUST CHAMBER Enclosed flue or chamber filled with deflectors, in which the products of combustion from an ore-roasting furnace are allowed to settle, the heavier and more valuable portion being left in the dust chamber and the volatile portions passing out through the chimney or other escape.

DUTCH WHITE Pigment consisting of one part of white lead and three parts of permanent white.

DWIGHT-LLOYD MACHINE Sintering machine in which feed moves continuously on articulated grates pulled along by chains in belt-conveyor fashion. Controlled combustion on these grates causes the minerals to sinter.

DWIGHT-LLOYD ROASTER A multi-hearthed circular roasting furnace, through which horizontal rabbles revolve and move the feed across each hearth, so that it falls peripherally to the one below and then works inward to central discharge to the next hearth below. Rising heat and air provide the roasting conditions.

DYNAMITE [1] Industrial explosive that is detonated by blasting caps, and one whose principal ingredient is nitroglycerin or specially sensitized ammonium nitrate. [2] To blow up or shatter with dynamite. [3] Originally, an explosive made of 75% nitroglycerin absorbed in 25% kieselguhr; but now generally any high explosive used for blasting purposes.

DYNAMITER One who uses, or is in favor of using, dynamite for unlawful purposes.

DYSLUITE A zinc-bearing gahnite from New Jersey.

 EARTHY CALAMINE Early name for hydrozincite.

EAF Electric arc furnace

EAF DUST Dust resulting from the

recycle of iron and steel scrap at steelmaking minimills or plants that produce marketable iron and steel products by remelting metal scrap. Typically in the process, about 1% to 2% of the furnace charge components are converted to dust and fume, which are collected as particulates by scrubbers or baghouse systems. Because scrap contains numerous components, including galvanized items, iron alloys, brass, solder, plastics, paint and oils, the resulting dust consist of volatilized elements and compounds and mechanically transported materials, including considerable iron oxide. The particles and fume are extremely fine (generally less than 1 micron in size), spherical, and of multi-element composition. The zinc content ranges from trace amounts to 40% but typically ranges between 5% and 20%. Because the dusts contain toxic elements, they are classed as a hazardous material; as such they must be treated before it can be dumped in a non-hazardous facility. The nature of the particles preclude component separation by physical means. The preferred treatments in the mid-1990's are thermal methods whereby zinc is the economic product extracted and the toxic elements are extracted and stored or are combined in environmentally safe slag compounds that can be dumped as non-hazardous landfill or commercially used.

EASTERN WETHERILL FURNACE See *Wetherill Furnace*.

ECONOMIC MINERAL Any mineral having commercial value.

ECONOMIC STABILIZATION PROGRAM Price-control program that became effective in August 1971, instituted mainly as a way to slow inflation. The price of zinc was froze at 17 cents per pound and a surtax was imposed on all basic zinc imports. The surcharge on imports was lifted in late December of that year, but price controls continued until they were rescinded at yearend 1973.

EDGE RUNNER See *Chilean mill*.

EFFICIENCY MINER Frequently applied to a boss miner, or a contract miner.

EFFLUENT Liquid, solid, or gaseous product, frequently waste, discharged or emerging from a process.

EJECTOR HALF Movable half of a die containing the ejector pins.

EJECTOR ROD A rod used to push out a formed part in a die.

EJECTOR Device which is mounted in such a way that it removes or assists in removing a formed part from a die.

ELECTRIC ARC FURNACE (EAF) See *Arc Furnace*.

ELECTRIC CALAMINE Zinc silicate, or calamine; so called on account of its strong pyroelectric properties.

ELECTRIC EXPLODER Former designation for electric blasting cap.

ELECTRIC SQUIB Small shell containing an explosive compound that is ignited by the electric current brought in through lead wires. Used for firing single small holes filled with black powder.

ELECTROCHEMICAL CORROSION Corrosion which occurs when current flows between cathodic and anodic areas on metal surfaces.

ELECTROCHEMICAL SERIES Same as *Electromotive Series*.

ELECTRODE Electrical conductor

for leading current into or out of a medium.

ELECTRODEPOSITION Deposition of a substance upon an electrode by passing electric current through an electrolyte. Electroplating, electroforming, electrorefining and electrowinning result from electrodeposition.

ELECTROGALVANIZING Process of electroplating zinc upon iron or steel. Sometimes restricted to continuous plating as opposed to batch plating.

ELECTROLYSIS Act or process of chemical decomposition by the action of an electric current.

ELECTROLYTE [1] Ionic conductor. [2] Liquid, most often a solution, that can conduct an electric current.

ELECTROLYTIC CORROSION Galvanic action caused by electrical contact of two different metals in the presence of an electrolyte, so that an electromotive force is set up.

ELECTROLYTIC PROCESS Employing an electric current, either for separating and depositing metals from solution, or as a source of heat in smelting and refining.

ELECTROLYTIC PROTECTION See preferred term, *Cathodic Protection*.

ELECTROLYTIC Pertaining to electrolysis or an electrolyte; deposited by electrolysis.

ELECTROMOTIVE SERIES A list of elements arranged according to their standard electrode potentials. In corrosion studies the analogous but more practical galvanic series of metals is generally used. The relative position of a given metal is not necessarily the same in the two series.

ELECTROPLATING Electrodepositing metal or alloy in an adherent form upon an object serving as a cathode. Sometimes applied only to batch plating as opposed to continuous plating (as with wire and strip) or electrogalvanizing.

ELECTROSTATIC SEPARATION Method of separating materials by dropping feed material between two electrodes, positive and negative, rotating in opposite directions. Nonrepelled particles drop in a vertical plane; susceptible charged particles are propelled in a forward position somewhat removed from the vertical plane.

ELECTROWINNING Recovery of a metal from a solution by means of electrochemical processes.

ELECTROZINCING See *Electrogalvanizing*.

ELECTRUM Alloy of copper, zinc and nickel (German silver). Amber and natural alloy of gold and silver are also called electrum.

ELUTRIATION Classification and separation of particles by size, shape or weight by a rising stream of gas or liquid.

EMBOSSING Raising a design in relief against a surface.

EMPIRICAL FORMULA Simplest formula of a compound which expresses its composition by weight.

EMPLOYEE BUY-OUT Deal involving not just the top management but also all, or a large number, of the junior employees of the organization. The difficulty of involving large numbers of employees without disclosing the deal prematurely has resulted in few of these deals being done. Some managers get around this by staging the buy-out and then involve other staff at a later date.

EMPLOYEE SHARE OWNERSHIP PLAN (ESOP) Trust established to acquire shares in a company for subsequent allocation to employees over a number of years.

END-BUMP TABLE Mechanically operated, sloping table by which light and heavy minerals are separated. The end motion imparted to the table tends to drive all minerals up the slope of the table, but a flow of water carries the light minerals down the slope faster than the mechanical motion carries them up, whereas the heavy minerals settle to the bottom and migrate to the top of the slope and are collected as concentrate.

ENDOTHERMIC Pertaining to a chemical reaction which occurs with absorption of heat.

ENGLISH ZINC FURNACE Zinc is reduced and distilled from calcined ores in sealed crucibles. Zinc vapor passes through an opening in the bottom of the crucible and is conducted by means of a tube into a condensing chamber beneath. The distillation takes place *per descensum* instead of *per ascensum* as in all later pryrometallurgical zinc production methods. This furnacing method was the first to be used in Europe for intentional zinc production. The first works were erected at Bristol, England about 1740; the English furnace became obsolete in the mid-1800's.

EPIGENETIC DEPOSIT A mineral deposit formed later than the enclosing rock; most zinc deposits are of this type.

ESCAPE CLAUSE Provision in a contract absolving a signatory from penalty in specific circumstances.

ESCAPEWAY Opening through which miners may leave the mine if the ordinary exit is obstructed.

ESCROW Deed, bond, money, or property delivered to a party of a contract to be put into trust and returned only upon the performance or fulfillment of some condition of the contract or to insure fulfillment by some other disposition.

ESPERANZA CLASSIFIER A classifier of the free-settling type in which the settled material is removed by dragging it up an inclined plane by means of a continuous belt of flat blades or paddles.

ESTORAQUE Mexican term for yellow zinc blende.

EUTECTIC [1] An Isothermal reversible reaction in which a liquid solution is converted into two or more intimately mixed solids on cooling, the number of solids formed being the same as the number of components in the system . [2] Alloy having the composition indicated by the eutectic point on an equilibrium diagram.

EUTECTOID Alloy structure of intermixed solid constituents formed by a eutectoid reaction.

EXHAUSTION In mining, the complete removal of ore reserves.

EXIT ROLLS In continuous hot-dip galvanizing, rolls that limit the thickness of the zinc coating on the sheet or strip. Exit rolls to some extent have been replaced by air or superheated steam jets or knives.

EXOTHERMIC Pertaining to a chemical reaction which occurs with the evolution of heat.

EXPLOITATION Extraction and utilization of ore.

EXPLORATION Work involved in looking for ore.

EXPLOSIVE OIL Nitroglycerin.

EXSOLUTION Separation of individual minerals or phases in solid solution when the temperature is lowered.

EXTENDER Mineral substance used to dilute and to reduce the cost of paint but which may also be useful in improving the paint's properties.

EXTRACTOR BOX See *Zinc-Box*.

EXTRALATERAL Situated and extending beyond the sides; specifically noting the right of a mine owner to the extension of a lode or vein from his claim but within the vertical planes through the end lines.

EXTRUSION The act or process of extruding; thrusting or pushing out; also a form produced by the process.

F

FABER DU FAUR FURNACE A cubical crucible furnace built into cast-iron framework, mounted on trunnions in order that the furnace may be turned over and the contents emptied. Used in the desilverization of zinc crusts.

FACE In any adit, tunnel or stope, the vertical surface exposed by excavation of the work in progress or last done.

FALDING FURNACE A mechanically raked muffle furnace having three hearths with combustion flues under the lowest hearth.

FALSE GALENA Sphalerite.

FAN LINE See *Air Duct*.

FARADAY'S LAW [1] The quantity of a substance liberated at the cathode or anode is proportional to the quantity of current passed. [2] Quantities of different substances liberated by the same quantity of current are proportional to their chemical equivalents.

FAT BACK In Arkansas, term used for secondary carbonate and/or silicate zinc ores, alluding to the similarity in appearance of fat from a pig's back.

FAT CLAYS Sedimentary clays that have lower melting points than lean or residual clays. Fat clays are generally more plastic than lean clays and in retort mixtures the two generally were blended to gain the advantages of each.

FAUSTITE Apple-green mineral, the zinc analogue of turquoise.

FAUVELLE A system of drilling, that was invented by Beart, an Englishman, and Fauvelle, a French engineer, in 1846, providing for the continuous removal of detritus from the well by means of a water flush or current of water. Water-flushing systems developed later were largely modifications of the Fauvelle system.

FEATHERED-SHOT METAL Granules of zinc metal formed when molten zinc is poured into cold water. Same as *Bean Shot*.

FEED Material, as ore, upon which a crusher or grinding mill operates. The material supplied to a furnace or other metallurgical process.

FERRARIS FURNACE An inclined reverberatory furnace for roasting sulfide ore.

FERRITE SPINELS Collective name for minerals of the magnetite series in the spinel group. During the roasting of zinc concentrates, zinc and iron com-

The U.S. Zinc Industry

bine with oxygen to form zinc-iron spinel, generally referred to as ferrite (sometimes called ferrate). Ferrite formation results in the loss of zinc in the typical roast-leach-electrowin process because ferrites are insoluble in the leach solution unless. it undergoes special treatment. See *jarosite, hematite,* and *goethite processes.*

FERRITE [1] A solid solution of one or more elements in body-centered cubic iron. [2] In the field of magnetics, substances having the general formula of magnetite. The trivalent metal is most often iron, and in the case of zinc ferrite, zinc is the divalent metal.

FETTLE To clean or smooth a metal after casting.

FIASCO An ignominious failure of any kind; a complete breakdown. Said of a mining venture which has resulted in failure.

FIDUCIAL POINT Triangulation point; a bench mark.

FIELD A large tract or area of many square miles containing valuable minerals.

FILL Tailings, waste, etc., used to fill underground space left after extraction of ore. See *Sand Fill.*

FIELD HOPPERS Bins at custom or central mills in the Tri-State district in which the ores from different mines were kept separate for the purposes of sampling and batch processing through the mill. Field hoppers were generally large enough to hold 350 to 500 tons of ore.

FILLED STOPES Stopes which have been filled with barren rock, low-grade ore, sand or tailings after the ore has been extracted.

FILTER CLOTH Fabric used as a medium for filtration; for example, nylon cloth, blanket cloth, finely woven wire mesh, etc.

FILTER PRESS A machine for removing a liquid from crushed ore, flotation concentrate, or other mineral pulp by forcing the liquid under pressure or vacuum through canvas or cloth, leaving a damp de-watered mass behind.

FIND Thing found or discovered; especially, a valuable discovery, as a find of minerals.

FINE ZINC Arbitrary term, no longer used, for zinc metal of higher zinc purity than common spelter grades. Generally applied to metal containing 99.8% or better zinc.

FINES [1] Very small particles produced in breaking ore. [2] Ore minerals in too fine or pulverulent condition to be recovered, treated or smelted in the same way as coarser ore.

FINISHING JIG The jig used to recover the smaller particles of ore in a concentrator or stampmill.

FIRE BRICK A refractory brick of fire clay or siliceous material used to line furnaces.

FIRE CLAY A clay comparatively free of iron and alkalies and not easily fusible, used for fire bricks.

FIRE SETTING Ancient method of tunneling through rock. A fire was built against the rock face, which was then quenched with water, thus cracking the rock.

FIRING CYCLE Used in American horizontal retort smelters to reference a complete cycle of production from charging to discharging the retorts. In American practice, the firing cycle was either a 24-hour or 48-hour cycle.

FIRING [1] In mining, the igniting of explosive charges. [2] Starting up a furnace or kiln.

FIRST CLASS Used in the Tri-State district for large sheet-ground deposits. Before 1900 only a few of these were mined because the ore grades were low and extraction difficult, but after that time, they constituted the bulk of the production in the older Tri-State mining districts because most of the second class ore had been depleted. See *Second Class*.

FIRST CONTACT ORE At Leadville, CO, term used for the silver-rich, lead-zinc carbonate orebodies that replace limestone at the contact with an overlying sheet of intrusive porphyry.

FIRST-CLASS ORE; SHIPPING ORE Ore of sufficient value that it can be shipped to a smelter directly from the mine without processing.

FIRSTS The best ore picked from a mine.

FISSURE VEIN A mineral mass, tabular in form as a whole, frequently irregular in detail, occupying or accompanying a fracture or fault in the enclosing rock; the mineral mass formed later than the country rock, gaining access through open spaces or through alteration of the adjoining rock.

FLAGGING A SQUIB Uncoiling the end of the paper which is impregnated with sulfur or some other combustible substance. Flagging the squib permits more time to elapse from the ignition of the unrolled paper and the firing of the charge.

FLAKE POWDER Flat or scale-like particles, relatively thin.

FLAME REACTOR A compact oxy/fuel-fired flash smelting furnace developed to recover metals from a variety of materials, including crude zinc oxide from EAF dusts. Natural gas is fired with oxygen-enrich air in a water cooled burner under fuel-rich conditions at temperatures exceeding 2500 degrees C. Metal oxides are reduced and volatilized and reoxidized, cooled, and captured.

FLASH ROAST Rapid oxidation of finely divided sulfide minerals in a roasting furnace in which they drop or pass through a heated oxidizing atmosphere. Also called *Suspension Roast*.

FLAT OF ORE A horizontal ore deposit occupying a bedding plane in the rock.

FLAT [1] In mine timbering, a horizontal crosspiece or cap used in roof support. [2] In the Wisconsin and Illinois zinc district, flat is used for the horizontal joints or bedding planes along which ore occurs.

FLATS AND PITCHES In the Upper Mississippi Valley zinc district (Wisconsin, Illinois, and Iowa), term is used to describe a form of ore occurrence in which a flat ore deposit, conformable with the strata, is modified by a series of step-like ore veins that are alternately transverse to and parallel with the strata. The largest mines were in this class of deposit.

FLATTING MILL [1] Rolling mill for producing sheet metal. [2] Mill in which grains of metal are flattened by steel rolls and reduced to metallic flakes or dust.

FLINT CLAY [1] Flintlike clay

which, when ground and wetted, develops no plasticity. [2] Hard refractory clay which is largely composed of well-crystallized kaolinite that breaks with a concoidal fracture like that of flint, hence its name.

FLOATABILITY In mineral concentration, the word is used in connection with response of a specific mineral to the flotation process.

FLOAT Surface pieces of ore or rock which have fallen from veins or strata, or have been separated from the parent vein or strata by weathering agencies. Not usually applied to stream placers.

FLOTATION MIDDLINGS Flotation products, containing particles consisting of ore minerals attached to gangue or poorly separated low grade concentrate, which may be re-treated.

FLOTATION PROCESS The process or processes by which the valuable minerals in a mass of finely ground ore can be caused to float on a liquid into which the finely ground ore is fed. Classified as *Film Flotation* and *Froth Flotation*.

FLOTATION Concentration of valuable minerals from ores by agitation of the ground material with water, oil, and flotation chemicals. The valuable minerals are generally wetted by the oil, lifted to the surface by clinging air bubbles in a froth and then floated off.

FLOWERS OF ZINC A old name for zinc oxide.

FLUE DUST [1] Particles passing into the flues of a smelter or metallurgical furnace. [2] Zinc-bearing flue dust generated from furnacing copper-zinc alloys, galvanized steel parts or zinc metal and recovered as a dust or sludge.

FLUE Passage for air, gas or smoke.

FLUID BED ROASTING Process in which finely divided solid sulfides are kept in suspension by a rising current of air. Reaction between mineral and air is maintained at a desired exothermic level by control of oxygen entry, by admission of cooling water, or by added fuel.

FLUKE Rod used for cleaning drill holes before they are charged with explosives.

FLUORESCENCE [1] Emission of characteristic electromagnetic radiation by a substance as a result of the absorption of electromagnetic or corpuscular radiation having a greater unit energy than that of the fluorescent radiation. It occurs only so long as the stimulus responsible for it is maintained. [2] Emission of light from within a substance while it is being exposed to direct radiation. A number of zinc minerals and compounds fluoresce when exposed to ultraviolet light; willemite from Franklin, NJ. is probably the most famous and best known zinc mineral having this property, and zinc sulfide is probably the most commercially known fluorescing zinc compound.

FLUX BLANKET In the hot-dip galvanizing process, the work passes through a blanket of flux foam, usually zinc ammonium chloride or sal ammoniac, on the surface of the zinc bath before entering the bath itself. Blanket acts as a cleaning and wetting agent to the work. Sometimes this form of galvanizing is known as wet galvanizing as opposed to dry galvanizing in which the work is dipped in the flux and dried before entering the bath. In the dry process, the flux protects the work from oxidation and acts as a wetting agent when it is galvanized.

FLUX[1] Substance used to promote fusion of metals and minerals. [2] In soldering and brazing, a substance which is applied to the parts to be joined and which on heating, aids in the ready flow of the solder and prevents the formation of oxides, on which the solder will not join.

FLY ASH Finely divided siliceous material formed during the combustion of coal, coke or other solid fuels.

FLY Gate or door in a hopper for diverting ore, rock or coal from one bin or conveyor to another.

FLYROCK Rock fragments which are thrown and scattered during blasting.

FOAM BLANKET Foam covering that is maintained on an electrolyte to prevent acid fumes and mists from escaping.

FOAMING AGENT Material that tends to stabilize a foam.

FOB; FOR Free on board; free on rail. See *Free on Board*.

FOIL Metal in sheet form less than 0.006 inch in thickness.

FONTAINEMOREAU BRONZE See *Reverse Brass*.

FOOTAGE Refers to length by running foot of work in drifting, raising, drilling, etc., preformed by a miner or group of miners for payment purposes. Generally refers to contract miners who are paid by the linear foot of drifting, etc.

FOOTWALL [1] Underside of a vein or lens in relation to the dip of the ore deposit. [2] Wall rock underlying the lode.

FORCING FAN A fan that blows or forces intake air into the mine workings.

FORCE MAJEURE A protective clause in a contract to guard against penalties in the event of certain unforeseen contingencies, such as acts of nature, fire, war, etc., which prevent fulfillment of the contract.

FORMABILITY Relative ease with which a metal can be shaped through plastic deformation.

FORMULA WEIGHT Sum of the atomic weights of the elements in a compound.

FORWARD PRICE Price for a specific time period in the future at which sellers are willing to sell their metal.

FOUNDER BRECCIA See *Collapse Breccia*.

FOUNDING The act or process of casting metals.

FOUNDRY A manufacturing establishment in which articles are cast from metal: as a brass foundry.

FOUR-PIECE SET Squared timber frame used in underground driving to give all around support to weak ground. Cap supported by two posts on a sill piece or sill.

FOUR-WAY TURN SHEET A metal plate placed at a right angle junction of hand-tramming tracks at a mine whereby the miner pushes the car onto the plate, physical rotates it to change directions, and resumes tramming but in a different direction.

FOWLERITE Zinc-bearing variety of rhodonite.

FRACTIONAL DISTILLATION Operation for separating a mixture of two or more liquids which have different boiling points. See *Distillation, Redistillation,* and *Distillation Column.*

FRANCISCI FURNACE A furnace for the extraction of zinc from roasted blende. It consisted of a series of superimposed muffles formed by the arches of magnesia brick and built into the walls of the furnace. The muffles fed into a common condensing chamber.

FRANKLINITE Iron-manganese-zinc oxide mineral. Rarely an ore but was a major zinc-ore mineral at Franklin, NJ where it was processed for zinc oxide and spiegeleisen. The mineral was named after its type locality, Franklin, and in honor of Benjamin Franklin after whom the town was named.

FREE ON BOARD Price of consignment when delivered with all prior charges paid onto a ship (railcar).

FREE-CUTTING BRASS Alpha-beta brass containing about 2 to 5 percent lead to improve the machining properties; used for engraving and screw machine work (CTD).

In the late 1980's, concerns over possible lead contamination in drinking water and the environment resulted in less lead being used in leaded brasses and in some cases, resulted in substitution of lead by bismuth to obtain similar properties.

FREEZE To solidify, as a molten charge in a furnace or molten metal in a mold.

FRENCH PROCESS; INDIRECT METHOD Manufacturing zinc oxide, whereby zinc metal or metallic zinc-containing scrap is the starting material. The metal is melted and the zinc distilled; the vapor is burned in air to produce the oxide. The purity of the oxide produced is controlled by the purity of the metal and/or the vapor oxidized.

FRONT RUNNING When a dealer buys or sells a metals position or a security ahead of a big order they believe will move the metal's or security's price.

FROTH FLOTATION A process in which the minerals floated gather in and on the surface of bubbles of air or gas driven into or generated in the liquid in some convenient manner.

FROTHER An oil which makes a foam or froth. An additive used in the flotation process.

FROTHING AGENT[1] A frother used in flotation. [2] A frothing substance used on the surface of electrolytic cell solutions to reduce acid misting above the solution. In zinc electrolyzing cells, a mixture of cresylic acid, sodium silicate, and gum arabic is often used as a frothing agent. Zinc soaps are used for misting control at some electrolytic copper refineries.

FRUE VANNER An ore-dressing apparatus consisting of a rubber belt traveling up a slight inclination. Material to be treated is washed by a constant flow of water while the entire belt is shaken from side to side.

FULMINATE Explosive mercury compound which is employed for caps or exploders to fire charges of gunpowder, dynamite, etc.

FUME Metals and metal compounds that volatilized in a furnace, condense at lower temperatures, and are carried by furnace gases into the flues. Sulfur trioxide and elemental sulfur, driven off from furnaces and condensed are also classified as fume. In general, all of the volatile constituents of the furnace charge are represented. The particles are very fine and do not settle easily.

FUMING PROCESSES Any number of industrially important processes that are based on the reduction of zinc oxide with carbon or with reducing gases to produce zinc vapor followed by oxidation of the vapor to produce zinc oxide fume.

FUMING SULFURIC ACID Strong acid made by dissolving sulfur trioxide in concentrated sulfuric acid.

FURNACE CADMIUM, OR CADMIA Impure zinc-cadmium oxide which accumulates in the chimneys of furnaces smelting zinciferous ores. *Cadmia* is a Greek name for calamine from which calamine may be a corruption.

FUSE CUTTER In metal mining, one who cuts blasting fuse to standard length; inserts the fuse into the open end of the detonators and attaches it by squeezing the open ends with crimpers.

FUSION Act or operation of melting or rendering liquid by heat.

FUTURES CONTRACT Agree-ment reached on an organized exchange calling for the delivery of a specified quantity of zinc at a specified time in the future for an agreed upon price. If a buyer's long position is not liquidated, the hedger is required to pay the full amount of the contract and take physical delivery at an approved warehouse.

G **GAHNITE** A naturally-occurring, green, zinc-bearing spinel.

GALENITE In some U. S. lead-zinc mining districts, an early term for galena.

GALFAN An alloy, composed of 95% zinc and 5% aluminum with a small content of mischmetal, used as a coating for sheet, wire, and tubing. In some aspects, a Galfan coating is superior to standard zinc galvanizing coatings.

GALLERY In mining, a level or drift.

GALMEI Synonym for calamine.

GALVALUME A 55% aluminum, 43.5% zinc, and 1.5% silicon alloy, patented by the Bethlehem Steel Corp., used for continuously coating steel strip. It is competitive with zinc galvanizing and other zinc-alloy coatings.

GALVANIC ANODE The electrode of zinc, or other metal, which, when placed in a common electrolyte and electrically connected to a more noble metal, causes an electric current to flow in such a direction so as to protect the less anodic metal.

GALVANIC CORROSION Associated with the current of a galvanic cell consisting of two dissimilar conductors in an electrolyte or two similar conductors in dissimilar electrolytes. Where the two dissimilar conductors are in contact, the resulting reaction is referred to as couple action.

GALVANIC SERIES Metals and alloys arranged according to their relative electrode potentials in a specific environment. Compare with electromotive series.

GALVANIZER'S DROSS Unsweated zinc dross from a hot dip galvanizing batch process and from the top and the bottom of continuous line galvanizing baths. This type of scrap is usually sold in slab or block form, assays between

85 and 96 percent zinc and is free of skimmings.

GALVANNEALING Hot-dip-ped sheet on emerging from the zinc bath is passed through a short oven before the coating has solidified, so that the zinc-iron alloy layer is allowed to continue to grow. The resultant coating consists entirely of zinc-iron alloys. Typical uses of galvannealed steel are road signs, automobile panels, switch boxes, stoves and refrigerators.

GANGUE Nonvaluable minerals in an ore.

GARNET BLENDE Synonym for sphalerite.

GASH VEIN Mineralized fissure that extends only a short distance vertically. Typically, it is confined to a single stratum of rock.

GATE Portion of the runner in a mold through which molten metal enters the mold cavity. Sometimes the generic term is applied to the entire network of connecting channels which conduct the metal into the mold cavity.

GEAT Hole in a mold through which the metal is poured in casting.

GEE To cause a draft animal to turn to the right, away from the driver: opposed to haw or turn left.

GELATIN DYNAMITE High explosive of varying composition consisting mainly of nitroglycerin.

GEOCHEMICAL PROSPECT-ING The search for concealed deposits of metallic ores by analyzing soils, surface waters, and/or organisms for abnormal concentrations of metals.

GEOLOGIC RESERVES An estimate of the total in-ground ore reserves in a deposit, based on detailed geologic studies, close-spaced drilling, sampling, underground exploration, metallurgical testing of the ore, etc., which provide data to estimate the size, shape, and characteristics (grade distribution, zoning, alteration type and intensity, mineralogy, etc.) of the orebody. The mineable or economic ore reserves are based on the evaluation of how the mining system will operate within the geologic reserve. Geologic reserves are sometimes referred to as in-situ reserves.

GEOPHYSICAL PROSPECTING Exploration by measuring the various physical properties of the rocks and interpreting the results in terms of geologic feature or the economic deposits sought. There are four main methods employed: gravity, magnetic, electrical, and seismic.

GERMAN Straw filled with gunpowder to act as a fuse in blasting operations.

GHURR; THURR OR THE MOTHER OF METALS Used by alchemists for the mineral substance which in time is supposed to ripen and become real ore.

GIANT POWDER A form of dynamite consisting of a mixture of nitroglycerin and kieselguhr or diatomite, which has a capacity to absorb about three times its weight in nitroglycerin. Widely used mining explosive that gain local fame in Idaho as a mill explosive in the 1890's when it was used to blow up the Frisco (1892) and the Bunker Hill (1899) mills, during labor disputes.

GILDING METAL High-copper red brass, with 90 to 97 percent copper and zinc as the remainder; used for jewelry and cartridge cap fabrication.

GOB See *Good Ordinary Brand.*

GOETHITE PROCESS Iron removal process for solutions obtained from the treatment of leach residues at electrolytic zinc plants. Zinc ferrite in the residue is dissolved by strong, hot sulfuric acid putting zinc and iron in solution; but to extract the zinc, the iron has to be removed. Although iron can be precipitated readily by neutralization of the solution, it is difficult to filter and wash. In the goethite process, ferric iron in the solution from the hot leach stage is reduced to ferrous iron by adding zinc sulfide concentrate. Unused solids from the process are separated from the solution now containing the iron as ferrous sulfate. When the solution is neutralized with calcine to a pH of about 2.5 in the presence of air or oxygen, the iron precipitates as goethite, the resulting low-iron zinc solution is added to the neutral leach.

GOOD DELIVERY Under metal exchange rulings, description of metal delivered at an agreed purity or of a defined quality.

GOOD ORDINARY BRAND; GOOD ORDINARIES Commercial spelter, often listed as *GOB*; the European brands of zinc metal corresponding to Prime Western grade in North America. Until the late 1980's, GOB was the metal on which virtually all zinc pricing outside the United States was based.

GOOD SINTER A term used in the early days of sintering indicating the sinter plant was operating at the proper temperature. The plant was making good sinter when the smoke emitted from the sinter plant was dense with a pronounced yellowish brown tinge, owing to the cadmium being driven out of the sinter feed. If the smoke was barely visible, or very light bluish, the sinter plant was doing poor work. *It was said that an initiated person could tell from a mile away whether or not the sintering plant was good sinter, by observing the color of the smoke from the stack.*

GOPHER HOLE Small irregular prospect hole in mining.

GOPHER; GOPHER-DRIFT Irregular prospecting drift following or seeking the ore without regard to maintenance of a regular grade or section.

GOPHERING Prospecting work confined to digging shallow pits or starting adits. Term is from similarity of this work to the crooked little holes dug in the soil by gophers.

GOSSAN The weathered or otherwise decomposed upper zone of a lode, characterized by an abundance of oxidized and hydrated alteration products such as limonite that is derived mainly from pyrite. Iron-hat is a synonym. Gossans are common features overlying sulfide ore bodies and when observed, they attract the interest of exploration geologist.

GOUGE [1] Layer of soft material along the wall or vein, favoring the miner, by enabling him after gouging it out with a pick, to attack the solid vein from the side. [2] Ground-up rock or clayey material in a fault zone. [3] To work a mine without a plan or to contract working a mine face by neglecting to keep the sides cut away.

GRAB SAMPLE Ore or metallurgical sample taken at random from a rock face, orebody, pile, car, etc., often to obtain a general example of the character, content, or some special feature of the sampled material.

GRAB An instrument for extricating broken boring tools from a borehole.

GRADE A (B,C,D,E) Terminology used mainly by the U.S. Government during and between the two World Wars for various grades of zinc metal. Although the War and Navy Departments sometimes had slightly different specifications, the grades corresponded to the grades established by the ASTM. During World War I Grade A corresponded to High Grade, B to Intermediate, C to both Brass Special and Selected, and D to Prime Western. During World War II, Grade A was applied to both High Grade and Special High Grade, B to Intermediate, C and D to both Brass Special and Selected, and E to Prime Western. Soon after the Wars, the domestic industry tended to rapidly revert back to using the ASTM terminology.

GRADE Percentage of valuable metals or minerals in an ore.

GRAIN REFINER Material added to a molten metal to attain finer grains in the structure.

GRANULATED METAL Small pellets produced by putting liquid metal through a screen or by dropping it onto a revolving disc, in both cases, chilling with water.

GRANULATION Production of coarse metal particles by pouring the molten metal through a screen into water or by agitating the molten metal violently during its solidification.

GRAPNEL Implement for removing the core left by an annular drill in a bore hole, or for recovering tools, fragments, etc., that have fallen into a drill hole.

GRASS-ROOTS MINING Inadequately financed operation, depending on hand-to-mouth existence. Mining on a shoestring.

GRATE BAR [1] A bar forming part of a fire grate. [2] One of the bars forming a coarse screen or grizzly.

GRAVEL JACK Miner's term in Joplin district for loose aggregates of sphalerite crystals found in rich ore pockets and seams.

GRAVEL POWDER Very coarse gunpowder.

GRAVEYARD SHIFT Night shift.

GRAVIMETRIC ANALYSIS Quantitative determination of the constituents of a compound by weight; contrasted with volumetric analysis.

GRAVITY CONCENTRATION Separation of various grains by any means based on differences in their specific gravity.

GRAVITY PLANE A tramline laid at such an angle that full skips running downhill will pull up the empties.

GRAVITY RAILROAD The cars descend by their own weight.

GREASE BOX A container that holds fat or grease to lubricate a bearing.

GREASER Person who oils or greases the mine cars.

GREEN CINNABAR Green pigment consisting of the fired oxides of cobalt and zinc.

GREEN COMPACT; GREEN BRIQUET Compressed briquette of material for coking, sintering, or smelting as made in the briquetting machine before other processing as drying and sintering.

GREEN SEAL See *Seals*.

GREENOCKITE Cadmium sulfide mineral, occurring largely as a minor constituent in zinc deposits. It is the principal mineral source for cadmium, which is recovered as a byproduct in smelting zinc ores.

GRINDERS' ASTHMA, ROT, OR PHTHISIS Disease of the lungs consequent upon inhaling the metallic dust produced in grinding metals.

GRIND To reduce to a powder by friction as in a mill.

GRIZZLY A set of parallel bars or grating used for the coarse screening of ores, rock and other material to limit the size of broken rock, etc. passing through.

GROG Ground-up pieces of burned clay or brick, added to the raw clay mixture for the purpose of decreasing the shrinkage and density of the burned ware. The amount of grog used in retort manufacture varied greatly from about 10% to more than 50% depending on the characteristics of the clay and old retorts, coke or other material if added to the raw mixture. The grog used in the silica retort accounted for from about 30% by weight of the dry raw mix; silica flour and plastic fire clay accounted for 20% and 50%, respectively.

GROSS RECOVERABLE VALUE The part of the metal recovered multiplied by its price. See also *Net Unit Value*.

GROUND HOG See *Barney*.

GROUNDMAN A person employed to work on the ground, as in digging or excavating.

GROUNDSILL Bed piece or foundation timber supporting a timber superstructure as a set of mine timbers. A ground plate.

GROUT [1] Thin mortar poured into the interstices between stones and brick. [2] Thin cement mixture forced into the crevices of a stratum or strata to prevent ground water from seeping or flowing into an excavation. Frequently employed in shaft sinking and bore-hole drilling.

GROUTING Filling in or finishing with grout. Also the grout thus filled in.

GRUB Slang for food.

GRUBSTAKE Supplies furnished to a prospector on promise of a share in his discoveries.

GUIDES [1] Timbers at the side of a shaft to steady and guide the cage. [2] Holes in a crossbeam through which the stems of the stamps in a stampmill rise and fall. [3] Pulley to lead a driving belt or rope in a new direction or to keep it from leaving its desired direction.

GUINEA ROCK Local term in East Tennessee zinc district for a sparsely mineralized, light grey, medium crystallized dolomite that occurs in recrystalline or replacement ores.

GUMBO [1] A sticky mud when wet. [2] In southwest Missouri, putty-like clay associated with lead and zinc deposits.

GUN METAL Am alloy of copper with tin or zinc, and sometimes a little iron. In some cases, known as Aich's metal. At one time, much used for cannons.

GUNTER'S CHAIN A chain commonly used in surveying, having 100 links, each 7.92 inches long.

GUT To rob, or extract, only the rich ore of a mine. See *High Grading*.

GUY Guide, rope, chain, or rod attached to anything to steady it.

GYRATORY BREAKER OR CRUSHER Rock crusher built on the principle of the old-fashioned coffee grinder. It consists of a vertical spindle, the foot of which is mounted on an eccentric bearing within a conical shell. The top carries a conical crushing head revolving eccentrically in a conical maw. There are three types of gyratory: those which have the greatest movement on the smallest lump; those that have equal movement for all lumps; and those that have the greatest movement on the largest lumps.

H

HAAS FURNACE A muffle furnace of the McDougall type.

HADE Angle of inclination of a vein measured from vertical; dip is measured from the horizontal.

HAIRPIN COOLERS Apparatus consisting of a series of large hairpin-shaped ducts for cooling gases and fume before they enter the baghouse.

HALL FURNACE A modification of the Wethey furnace for roasting sulfide ore.

HAMMER DRILL A jackhammer-type of drill.

HAND RABBLING Stirring roasting ore by hand using a rake or some similar tool.

HANDCOCK JIG A moving screen jig developed in the U.S. to treat lead-zinc ores of the Tri-state district. The sieve and ore are immersed in water and jigged with some forward throw. The heavy minerals settle down and are withdrawn through traverse slots.

HANDPICKING; SORTING Manual removal of a selected ore, waste, or scrap from an ore or scrap pile or picking belt.

HANG To have its charge choke up or arched in one part, while the part underneath falls away so as to leave a gap; said of a blast furnace.

HANGFIRE Said of a charge that explodes later than expected. A hangfire rarely occurs with electric firing but is not infrequent with blasting cap and fuse.

HANGING WALL Upper wall of an inclined vein or that which hangs over the miner at work.

HARD BRASS Brass which has not been annealed after drawing or rolling; used for springs, etc.

HARD BUCKFAT Term used in reference to the character of the clay matrix associated with the secondary zinc ores in Wythe County, VA. When hard it was called hard buckfat, and when soft, *soft buckfat*. In treating the ore the latter broke down in water when violently agitated, permitting separation of the calamine, but with substantial zinc losses. The hard buckfat did not break easily in water, but had a brittle character and lower specific gravity than the zinc minerals, resulting in easy separation by jigging.

HARD DRIVING Performance of zinc-retort distillation process at 1400 to 1450 degrees C. is hard driving, while performance at 1200 degrees C. is known as *slow driving*. Ordinary temperatures of distillation are 1250 to 1350 degrees C.

HARD FIRING Temporary increase in the temperature of a furnace.

HARD GALVANIZING Used in wire galvanizing, referring to the temper given the wire before it enters the zinc bath. The temper is controlled by varying the length of immersion or the temperature of the molten lead bath through which the wire is passed. The converse term is *Soft Galvanizing*.

HARD SPELTER [1] Formerly a common brand of metal, consisting of zinc contaminated with iron, resulting from the galvanizing process. [2] Unsweated bottom dross, containing a minimum of 92% zinc, from a continuous galvanizing bath.

HARD ZINC See *Hard Spelter*.

HARMONIZED TARIFF SCHEDULE OF THE UNITED STATES New tariff schedule that replaced the previous schedule, the Tariff Schedule of the United States (TSUS) on January 1, 1989. The Harmonized Tariff Schedule (HTS) was international in scope and was implemented to surmount the increasing difficulties arising from the use of different tariff classification systems by countries. The HTS provided a means for most countries to have a single modern structure for product classification, thereby easing duty determination, statistical accounting, and transportation documentation. Because of many, mostly minor, differences between the TSUS and the HTS, U.S. trade data before and after the change may not be comparable.

HARRIS PROCESS A process to remove and recover as zinc oxide the remaining eutectic zinc content, generally about 0.5%, from desilverized lead. Similar processes include *chlorine dezincing* and *vacuum distillation*.

HAULAGE [1] Act or labor of hauling or drawing. In mining, the drawing or conveying, in cars or otherwise, of the produce of the mine from where it is mined to where it is to be hoisted, treated, stored, or used. [2] The horizontal transport of broken ore along a level to an ore pocket near the shaft.

HAULAGEWAY Gangway, entry, or tunnel through which loaded or empty mine cars are hauled.

HAWLEYITE A cubic cadmium sulfide; dimorphous with greenockite.

HEAD GRADE Average percentage of a metal or metals in the ore feed to a concentrating mill.

HEAD VALUE Assay value of the heads or mill feed.

HEAD; HEADS [1] In mineral processing, the mill head or ore accepted by the mill for treatment. [2] In gravity separation of a feed, the heads are the concentrate.

HEADER Plank or timber, longer than a cap, supported by two props, one at each end.

HEADFRAME Structure erected over a shaft to carry the sheaves over which the cable runs for hoisting the cage or ore buckets.

HEADING-AND-BENCHING MINING Stoping method used in thicker ore where it is customary to take out a slice 7 or 8 feet high directly under the top of the ore and then to bench or stope down from the bottom of the heading. The heading is always kept a short distance in advance of the bench. Mill holes are sometimes employed to ease broken ore withdrawal.

HEADING The face that is being advanced by a mining operation.

HEAP ROASTING Burning the sulfur

out of ores piled in heaps, with a small amount of wood or other fuel. Heap roasting was an early method for roasting zinc ores and for the production of zinc sulfate from sphalerite. In the latter process, sphalerite was altered to zinc sulfate by the roast; the zinc sulfate was extracted by leaching and crystallization.

HEARTH Bottom portion of certain furnaces, such as the blast furnace, air furnace and other reverberatory furnaces, in which the molten metals, matte, etc., are collected or held.

HEAT OF COMBUSTION Heat evolved when a substance is completely burned in oxygen.

HEAVY GROUND Dangerous roof, back, or hanging wall, capable of falling in. Sometimes detected by a hollow sound when tapped with a bar.

HEAVY SPAR Synonym for *Barite*.

HEDGE Process that allows a metal customer to lock in both the price of a metal in their products and the price of a metal in their raw material purchases and not have their manufacturing margin eroded by metal price fluctuations. Accomplished by taking a forward position on the LME.

HEDGING Use of futures contracts, options on futures or cash forward positions to offset an anticipated cash market transaction in the future. By so doing, the hedger locks in the price of zinc for future sales or delivery and insulates himself from market variations during the time period of the futures contract.

HEGELER PRODUCER Furnace for the manufacture of producer gas.

HEGELER ROASTING FURNACE A mechanically raked, two compartment, gas-fired, muffle furnace first used to roast zinc sulfides in 1882 at LaSalle, IL. The first commercially successful mechanical blende roaster built in the United States. The Hegeler furnace also permitted the manufacture of sulfuric acid, and from the early 1880's to 1925, it was the only furnace used in the United States roasting blende for acid production. When zinc flotation concentrates became generally available in the late 1920's, American roasting practices underwent change, resulting in the discarding of the Hegeler-type furnaces in favor of the McDougall and Wedge types.

HELD IN COMMON Claim with more than one owner or in the case of more than one claim, the words *claims held in common* means that work on one of the claims is sufficient to meet the statutory requirements of all of them.

HEMATITE PROCESS A process to remove iron from the zinc-iron sulfate solution derived by hot sulfuric acid leaching of zinc ferrite in leach residues. In this process, hematite (iron oxide) is precipitated from the solution in an autoclave at about 200 degrees C in the presence of oxygen. See *goethite* and *jarosite* processes.

HEMIMORPHITE A common, secondary zinc-ore mineral. A hydrous zinc silicate.

HERRESHOFF FURNACE A mechanical, cylindrical, multi-deck muffle furnace of the McDougall type for roasting zinc ores.

HG Acronym for High Grade zinc metal.

HI ZINC Name given to the nonmagnetic product obtained in the processing of residues in electrothermic smelting. The product consists of coke and sinter containing appreciable zinc. Part of the hi zinc is returned to the raw mix for the sinter machine and part is passed through an air separator, in which the coke and sinter particles are separated for reuse. Also see *Lo Zinc.*

HIGH BRASS Copper-zinc alloy containing 34% zinc and possessing high tensile strength. Used for springs, screws, rivets, etc.

HIGH GRADE ZINC Specification grade of zinc initially approved by the ASTM in 1911. It is to be composed of at least 99.90% zinc with a maximum of 0.10% combined lead, cadmium and iron content. High Grade (HG) metal was used chiefly in special brasses and die castings before increased production of SHG in the late 1920's. Thereafter, it was used to make common brass, rolled zinc, and zinc oxide, and was used for galvanizing purposes.

HIGH-GRADER One who steals and sells, or otherwise disposes of high-grade or specimen ores. A common practice especially in the early days of gold mining.

HIGH-GRADING [1] Theft of valuable pieces of ore. [2] Said of a mining operation when higher-grade ores are mined preferentially outside the mine plan, often as a result of falling profitability or in the final stages of a mine's life.

HOEPFNER PROCESS An electrolytic process to produce zinc metal from sulfide ores. The ore was given a sulfate roast, the sulfate leached and converted to chloride by salt, which was then electrolyzed. The process was not a commercial success as regards zinc metal production but was successful as part of operations at chlorine-producing plants in England and Germany.

HOLE In Joplin, Missouri, A local term for a mine shaft.

HONEY Code word for yellow brass scrap. See *Apple.*

HOPEITE Hydrous zinc phosphate mineral.

HORN TIFF In Missouri, calcite stained with carbonaceous material; sometimes dark enough to be mistaken with sphalerite.

HORSE RUN Device with which horses draw loaded vehicles up an incline from excavations.

HORSE WHIM Horse-powered winding drum for raising ore from a mine. A horse gin.

HORSE Mass of country rock lying within a vein.

HORWOOD PROCESS A flotation process in which a mixture of iron, copper, lead, and zinc sulfides is roasted, the surfaces of the three former sulfides can be changed to oxide from sulfide at a comparatively low temperature, whereas the blende is practically unaltered. The partly roasted material is then subjected to a heat-acid-oil-flotation process by which the zinc can be floated, the other metals staying behind.

HOT BRIQUETTING Making smelter-feed briquettes on briquetting rolls using oxidic zinc materials that were preheated to 350-450 degrees C.

HOT CHAMBER MACHINE A die casting machine in which the metal chamber under pressure is immersed in the molten metal in the furnace. The chamber is sometimes called a gooseneck and the machine a gooseneck machine.

HOT QUENCHING Quenching in a medium at an elevated temperature.

HOT ROLL To roll while hot, as a metal.

HOT ROLLING Rolling zinc strip between 350 and 400 degrees F. See *Cold Rolling*.

HOT-DRAWN Signifies the product of drawing, when the operation is preformed on material that is hot.

HOT-SPOT PROTECTION Special cathodic protection designed and applied locally to overcome a particularly bad corrosion condition, worse than encountered elsewhere in the installation.

HUASCOLITE Variety of galena in which part of the lead is replaced by zinc.

HUFF SEPARATOR Electrostatic device for the separation of sized, electrically conducting and non-conducting minerals. The ore is passed over a charged roller in which the constituents take various charges according to their conductivity and are separated by their deflection or non-deflection when falling past charged electrodes. Huff separators were first used to separate sphalerite and marcasite in Wisconsin beginning in 1908, and the following year at Bingham, Utah to separate sphalerite and pyrite.

HUNDRED-WEIGHT Commonly reckoned in the United States, and for many articles in England, at 100 pounds avoirdupois; but commonly in England, and formerly in the United States, at 112 pounds. There is also an older hundredweight, called the long hundredweight, of 120 or six-score pounds.

HUNG SHOT See *hangfire*.

HUNTINGTON-HEBERLEIN PROCESS Sink-float process employing a galena medium and utilization of froth flotation as a means of medium recovery.

HURDY-GURDY WHEEL Water wheel operated by the direct impact of a stream upon its radially-placed paddles.

HURDY-GURDY Dance house in a mining camp.

HUTCH Bottom compartment of an ore-dressing jig.

HYDE PROCESS Flotation process patented in 1911, in which a small amount of sulfuric acid, with or without the use of copperas, is used to give the slimy portion of the ore a preliminary coagulation before flotation. Sulfides are then floated off, yielding an impure concentrate, which is re-treated in a second machine. The process was used mainly on complex western sulfide ores.

HYDRAULIC FILL; MINE FILL Waste material transported underground and flushed into place by use of water. When tailings are used, cement is sometimes added to firm the material.

HYDRAULIC MINE-FILL-ING Filling a mine with material transported by water. Commonly, mill tailings, often with cement added, are pumped into mine workings for fill.

HYDRO T METAL Trademark for a fine-grained zinc sheet.

HYDROFRANKLINITE *Chalcophanite*.

HYDROGEN EMBRITTLEMENT Low ductility in metals resulting from the absorption of hydrogen.

HYDROGEN OVERVOLTAGE In electroplating, associated with the liberation of hydrogen. The theoretical voltage to deposit zinc from a zinc sulfate solution is about twice the voltage necessary to decompose water, however in the cell there is considerable resistance (overvoltage) to the production of hydrogen, such that a narrow margin exist between hydrogen generation and the deposition of zinc. This phenomenon is the primary reason that zinc can be recovered in the electrolytic process. In zinc battery systems, mercury is sometimes added to further increase hydrogen overvoltage.

HYDROMETALLURGY Treatment of ores, concentrates, and other metal-bearing materials by wet processes, usually involving the solution of some component and its subsequent recovery from the solution.

HYDROTHERMAL SOLUTION Hot-water solutions originating within the earth and carrying dissolved mineral substances.

HYDROTHERMAL Pertaining to hot water, especially with respect to its action in dissolving, transporting, and redepositing, and otherwise producing mineral changes within the earth's crust. Hydrothermal systems are believed to have been instrumental in the formation of many zinc deposits, including those in the Tri-State district, southeastern Missouri, and the Appalachian Province.

HYDROZINCITE Basic zinc carbonate. The mineral is commonly found as an alteration product of zinc minerals. Also caller zinc bloom.

HYPOGENE Applied to ores or ore minerals that have been formed by generally ascending waters, as contrasted with supergene ores or minerals.

I

IGNITER A device to relight safety lamps by friction, such as an attached sparking or igniting match device.

ILZRO International Lead Zinc Research Organization.

ILZSG International Lead Zinc Study Group.

IMMERSION PLATING Depositing a metallic coating on a metal immersed in a liquid solution, without the aid of an external electric current.

IMPERIAL SMELTING PROCESS (ISP) A blast furnace process to produce both lead and zinc. In the process, zinc- and lead-containing sinter or briquettes and metallurgical coke are charged into a blast furnace, somewhat similar to a standard lead furnace. The charge is blast heated to about 1,150 degrees C., volatilizing zinc and liquefying slag components and lead. The lead is tapped periodically; the slag is also tapped and discarded; and the zinc vapor leaves the top of the furnace and is captured by a lead splash condenser. The hot lead containing the zinc is cooled to about 450 degrees C., at which point excess zinc separates out and, after treatment is cast into slabs. The first ISP plant started up in England in 1950 and since then, 13 other plants have come on stream around the world, but none have been built in North America.

IMPORT QUOTAS Limits placed on the quantities of zinc concentrates, metal, and scrap that could be imported into the United States in the period from October 1958 through November

1965. In 1957 a domestic lead/zinc producers group petitioned the U.S. Tariff Commission for restrictive regulation of imports, claiming serious injury to the domestic industry owing to trade and tariff concessions made by the Government under the GATT agreements. As a result of an ensuing Trade-Act, escape-clause investigation, President Eisenhower proclaimed the imposition of import quotas on zinc (and lead) metal and ore beginning October 1, 1958. The quotas were set at 80% of the average annual competitive rate in the years 1953 through 1957. Allocations were on a quarterly basis, and major exporting countries received individual quota allowances whereas other countries were included in a group quota. The annual import limits for zinc were fixed at 379,840 short tons of zinc in ores and 141,120 short tons of metal. Government imports were entered ex-quota. Based on a 1965 Tariff Commission report on the condition of the lead and zinc industry, the import quotas were terminated effective October 22 for ores and November 21 for metal.

IMPURITIES Elements or compounds whose presence in a material are undesired.

IMS PLASMA-BASED PRO-CESS A process in which zinc, lead, and cadmium is extracted from steelmaking, electric-arc-furnace dusts in a plasma reaction vessel wherein a dc plasma arc and coke breeze perform a carbothermic reaction which results in the vaporization of the zinc, lead, and cadmium in the dusts. The metal vapors are captured in a splash condenser.

INCIPIENT FUSION Beginning fusion, as in the sintering process.

INCLINE Shaft not vertical; usually on the dip of a vein.

INDIAN TIN Spelter from India in 17th century.

INDIANA FURNACE A simple Belgian-type zinc furnace in which natural gas is fired under the lowest row of retorts.

INDICATED ORE Unmined ore for which the tonnage and grade are computed partly from specific measurements, samples, and production data, and partly from projection for a reasonable distance in three dimensions on geologic evidence. Essentially the same as *Probable Ore*.

INDIRECT PROCESS Manufacture of zinc oxide by burning the zinc vapor emitted from boiling zinc metal. The French Process. Compare *Direct* or *American Process*.

INDUCTIVELY COUPLED PLASMA EMISSION SPECTROMETRY Analytical technique that utilizes the spectra of elements in the sample for the analysis. fluxes. The sample in solution is introduced into a radio-frequency excited plasma (about 8000 degrees Kelvin). Each element in the solution produces a characteristic spectrum. The intensity of the spectral lines are proportional to the quantity of the element present.

INERT ANODE An anode which is insoluble in the electrolyte under the conditions obtained in electrolysis. In zinc electrowinning, the lead or lead-silver anode is the inert anode. In electrogalvanizing wire, lead, lead-silver, cast magnetite, etc. may be the inert anode.

INFERRED ORE Quantitative estimates are based largely on broad knowledge of the geologic character of the deposit and for which there are few, if any samples or measurements. The estimates are based on an assumed continuity or repetition for which there is geologic evidence. Used essentially in the same sense as *Possible Ore*.

INFILLING Material used for filling in; filling. Compare with *Back Filling*.

INGATE Same as *Gate*.

INGOT Casting or block of metal suitable for working or remelting.

INHIBITING PIGMENT An agent, such as zinc chromate, which is added to paints, normally in relatively high proportions to retard the corrosion of metals.

INITIATION Process of causing, by a fuse, primer or detonator, a high explosive to detonate.

INJECTED HOLE Borehole in which a cement slurry has been forced into by high pressure and allowed to harden. The reasons for cementing a hole are many. In diamond drilling, such a process may be used to correct excessive hole deviation, surmount a hole blockage, provide access through a weak or fault zone, etc.

IN-SITU RESERVES Same as *Geologic Reserves*.

INSOLUBLE ANODE See *Inert Anode*.

INSTROKE The right to raise or take ore from a leased mine through the shaft or tunnel of an adjoining mine.

INTERBEDDED Occurring between beds or lying in a bed parallel to other beds of a different material; interstratified.

INTERCALATED A body of rock interbedded or interlaminated with another body of different rock.

INTERGRANULAR CORROSION Corrosion occurring preferentially at grain boundaries.

INTEGRATED MARKET Mines, mills, smelters, and/or fabricating plants that are financially linked.

INTERMEDIATE GRADE Specification grade of zinc approve in 1911 by ASTM mainly for the manufacture of specialty alloys and brasses. Intermediate is composed of 99.5% zinc with a maximum of 0.5% cadmium, iron and lead. It was dropped as a specification metal in 1967.

INTERMETALLIC COMPOUND An intermediate phase in an alloy system, having a narrow range of homogeneity and relatively simple stoichiometric proportions, in which the nature of the atomic binding can vary from metallic to ionic.

INTERSECTION SHOOT A mineral deposit localized along vein intersections or cross fissures; they are among the oldest known and commonest types.

INTERSTRATIFIED Interbedded; strata laid between or alternating with others.

INVESTMENT CASTING [1] Casting metal into a mold produced by surrounding or investing an expendable pattern with a refractory slurry that sets at room temperature after which the wax, plastic or frozen mercury pattern is removed through the use of heat. Also called precision casting, or lost-wax process. [2] A casting made by the process.

INVESTMENT Outlay of capital for a period in the expectation of receiving it back at the end of the period and during the period, of receiving regular income from it.

INWALL Interior walls or lining of a shaft furnace.

IOLA FURNACE A natural-gas-fired furnace used at Iola, Kansas beginning in 1895 for the distillation of zinc. It was a direct adaptation of the Hegeler furnace.

ION EXCHANGE Interchange of different ions in a liquid or solid.

ION Atom, or group of atoms, that has gained or lost one or more outer electrons and thus carries an electric charge. Positive ions, cations, are deficient in outer electrons. Negative ions, anions, have an excess of outer electrons.

IRISH BUGGY Wheelbarrow.

IRISH DIVIDEND Assessment on mining stock.

IRON HAT A weathered ironstone capping; a gossan. Often an indication of an underlying ore body containing sulfides.

IRON JACK In the Missouri zinc region, solid flint rock with disseminated specks of black jack (zinc blende).

IRON PURIFICATION STEP Removal of iron by the formation and precipitation of ferric hydroxide as a purification step in preparing the zinc sulfate leach solution for electrolysis. In the process, ferrous iron is first oxidized to the ferric state followed by precipitating the ferric sulfate by a neutralizing agent, most often zinc calcine. The precipitation of ferric hydroxide not only gets rid of the iron but also greatly assist in the removal of other impurities.

ISOTOPE Atom having the same atomic number as all other atoms of the same number (a chemical element) but having a different atomic weight. The nuclei of the atoms have the same number of protons but different numbers of neutrons. Zinc, for example, has an atomic number of 30 (30 protons), but has 5 stable isotopes (different numbers of neutrons). Because each zinc isotope is represented in nature at differing relative abundance, the atomic weight of zinc is the proportional average of the 5 isotopes or 65.38.

ISRI Acronym for Institute of Scrap Recycling Industries, Inc.

J

JACK *Zinc blende.*

JACKHAMMER Hammer type, non-reciprocating rock drill worked without a tripod and provided with an automatic rotating device. It uses hollow steel through which air passes and blows the cuttings from the drill hole.

JACKKNIFING Collapsing of square-set timbers by wall pressure or through poor erection.

JACKROLL A windlass worked by hand.

JACOBY CONVEYOR A long, rotatable cast iron tube, 14 inches in diameter and up to 150 feet long, that is used to move hot roaster calcine to a calcine storage bin. Calcine is cooled as it is moved through the tube by a helical screw.

JAGGING BOARD An inclined board on which ore slimes are washed, as in a buddle.

JARDINIERE GLAZE Former type of unfritted lead glaze containing lead, zinc, aluminum, potassium, calcium, and silicon oxides.

JAROSITE PROCESS A process used to increase zinc recovery in the roast-leach-electrowin pro-cess whereby additional zinc is recovered from ferrite in the the residue remaining after the primary leach. Any zinc ferrite present, which forms during the roasting of zinc concentrates, remains in the leach residue with other insolubles. Although the ferrite can be readily dissolved by hot (95 degrees C) sulfuric acid and the iron precipitated if the solution is neutralized, the resulting precipitate is difficult to filter and wash. In the jarosite process, the iron is precipitated by an ammonium salt as an ammonium jarosite which is readily filterable from the zinc in solution, which then can be recovered. The process was named after the structurally similar, naturally occurring mineral, jarosite, a basic sulfate of potassium and ferric iron. See also *hematite* and *goethite* processes.

JEFFERSONITE A zinc-bearing manganese pyroxene.

JET PROCESS A process whereby jet streams of air or superheated steam wipes excess zinc off the surfaces of the emerging sheet in continuous hot-dip galvanizing to yield a more uniform coating and/or different coating thicknesses on the two sides of the sheet. See *Air Knife*.

JIG BED An agent used in a jig which consist of the heavy minerals in the ore that behaves in some respects like a dense fluid. Pulsation of the water and /or the motion of the screen keeps the bed open and in suspension during part of the cycle so the heavy minerals entering the jig can settle into the bed.

The lighter minerals can not penetrate the jig bed and so are forced to remain on top and are eventually discharged. Other agents used are lead shot, iron shot and punchings, pyrite, and magnetite.

JIG [1] A machine or apparatus in which ore minerals are concentrated on a screen or sieve in water by a reciprocating motion of the screen, or by the plusion of water through the screen. [2] To separate heavier from lighter materials, such as ore from gangue, by agitation in water.

JIGGING Separation of the heavy and light minerals in an ore from each other by means of a jig.

JINNY ROAD A gravity plane underground.

JINNY Stationary engine for hauling on a jinny road, when not operated by gravity.

JOINT VEINS Small veins confined to one bed of rocks that give no signs of displacement, or at least so slight that they can not be noticed.

JOLLY BALANCE Delicate spring balance used especially for the determination of densities of solids by the method of weighing in water and air.

JOPLIN ORE SCALE A sliding-scale contract to market Joplin ore, developed by the Zinc Ore Producers' Association in 1910. Contracted prices were based on the average published zinc price of the preceding week, with adjustment for zinc content and impurity levels. The Scale lasted only a few years owing to dissatisfaction of producers, especially during periods of rising prices when they received less for their concentrate than they would have on the open market.

JUMBO CATHODES Cathodes of large surface area, 2 to 3 square meters, used in jumbo zinc electrolytic cells. The term super jumbo is sometimes used to denote cells with surface areas over 3 square meters.

JUMBO CELLS Large electrolytic cells for the recovery of zinc made possible to a large extent by mechanization in the handling and cleaning of electrodes and in the stripping of zinc from the aluminum cathodes. Jumbo electrodes and cathodes are employed in such cells.

JUMBO [1] In mining, a drill carriage on which several drills of drifter type are mounted. [2] A mobile scaffold to assist drilling in large headings.

JUMPER One who jumps a claim. A claim jumper, one who takes possession of a claim or mine by stealth, fraud or force.

JUNK BOND High-yielding unsecured debt used in U.S. buy-outs.

JUVENILE WATER Water from the interior of the earth which is new or has never been part of the general system of groundwater circulation. See *Magmatic Water*.

K

K061 WASTE An Environmental Protection Agency hazardous waste category. Includes steel-making electric arc-furnace dusts.

KADMIA Alchemist term for zinc oxide.

KALAMEIN [1] An anticorrosive alloy of lead, tin, bismuth, and nickel for coating iron. [2] To coat in a manner similar to galvanizing, but using kalamein.

KAMMERLING FURNACE A modification of the Belgian zinc smelting furnace with two combustion chambers separated by a central longitudinal wall.

KARST TOPOGRAPHY Irregular topography developed in limestone regions by the solution of carbonate rocks by surface and ground waters. See *Karst*.

KARST A limestone region with many sink holes, abrupt ridges, caverns, and underground streams.

KEEPS; KEPS Wings, catches, or rests, to hold the cage when it is brought to rest at a landing. Also called dogs, fans, chairs and shuts.

KEEVE A tub used in collecting grains of heavy ore or metal; a dolly tub. A keeve of rich slime is stirred with water, and then struck on the side, which causes the heavy mineral to settle to the bottom. Also spelled *Kieve*.

KEIL FURNACE A gas-fired furnace containing one or more vertical retorts for the distillation of zinc.

KELLER FURNACE A multiple-deck roasting furnace for sulfide ore. It is a modification of the Spence furnace.

KIBBLE A steel bucket used during shaft sinking.

KICKER Ground left in first cutting a vein, for support of its sides.

KIEVES Strong tubs with sides flaring upward, in which separation is affected by mechanical agitation in a deep mass of thick pulp. Stirring is used for preliminary mixing and hammers or heavy striking bars are used for the final separation. Kieves are used to finish the concentration of fine products or slimes that are nearly rich enough to ship.

KILLED STEEL Steel deoxidized with a strong deoxidizing agent such as silicon and aluminum in order to reduce the oxygen content to such a level that no reaction occurs between carbon and oxygen during solidification.

KILN [1] A furnace for the calcination of coarsely broken ore or stone; also, an oven for drying, charing, etc. [2] A large furnace used for baking, drying, or burning fire brick or refractories. [3] To dry in a kiln.

KILN-DRY To dry in a kiln. Said of a material dried in a kiln.

KILNHOLE Mouth or opening of an oven or kiln.

KINGBOLT Bolt with which the cage is attached to the hoisting cable.

KING-POT Large central pot or crucible in a brass-melting furnace.

KISH Dross on the surface of molten lead.

KIVCET PROCESS A process for smelting complex sulfide concentrates with simultaneous production of zinc, lead and other metals. Kivcet is an acronym of Russian words meaning oxygen-flash-cyclone-electrothermic pro-cess. In the process, fine dry concentrate is fed into a cyclone furnace where it is autogeneously roasted and smelted with oxygen. The concentrate particles are flash smelted in suspension and pass into a distribution chamber where the melt and sulfur dioxide gases are separated. The melt flows into an electric resistance furnace where settling of the metals and volatilization of zinc takes place under reducing conditions. The zinc is either captured by a zinc splash condenser or is oxidized and collected as the oxide.

KLEEMAN CONDENSER Rectangular clay pipe in which distilled zinc is condensed. Used exclusively in Silesian retort smelting. Replaced by Dagner-type condensers in the 1890's.

KNITS; KNOTS Small particles of ore.

KNOCK-OFF To stop work for the day or part of a day.

KNOWN TO EXIST A vein or lode is known to exist when it could be discovered by anyone making a reasonable and fair inspection of the premises for the purpose of a location.

KNURLING Impressing a design into a metallic surface, usually by means of a hard roller that carries the corresponding design on its surface.

KOMSPELTER Proposed trade name for spelter from the Kansas, Oklahoma, and Missouri fields.

KREMNITZ WHITE A pure white lead made by treating litharge and lead acetate with carbon dioxide and formed into tablets. It is used in fine painting.

KYACK Pack sack to be swung on either side of a pack saddle.

L **LABEL** Scrap code word for new brass clippings. See *Apple.*

LABORER [1] Man hired by the contract miner to assist him. [2] Mine laborer; a man working for day wages in and about a mine; a company man as distinguished from digger or contractor.

LACE Scrap code word for brass shell cases without primers. See *Apple.*

LACING Timber or other material placed behind and around the main supports. Lagging.

LADDERWAY; LADDER ROAD Particular shaft, or compartment of a shaft, containing ladders.

LADLE Receptacle used for transferring and pouring molten metal.

LAGGING Planks, slabs, or small timbers placed over the caps and behind the posts of the timbering, not to carry the main weight, but to form a ceiling or a wall, preventing fragments of rock from falling through.

LAKE Scrap code word for brass small arms and rifle shells. See *Apple.*

LALANDE CELL An alkaline primary wet cell in which the negative electrode is zinc with some mercury and the positive pole is copper oxide; the electrolyte is either potassium or sodium hydroxide. Because of its reliability, it was widely used for railway signaling and fire alarm systems.

LAMINABLE Capable of being rolled or hammered into thin sheets.

LANA CALAMINARIS Calamine.

LANA PHILOSOPHICA Zinc oxide. To alchemists same as *philosopher's wool.*

LAND DISTRICT Division of a State or Territory created by law in which is located the land office for the disposition of public lands therein.

LAND SURVEYING The locating of boundaries, area, characteristics, of tracts of land.

LANDING [1] Level stage in a shaft, at which cages are loaded and discharged. [2] Platforms at regular intervals constructed in the manway or ladderway of raises to assist mining operations and safety. [3] Top or bottom of a slope, shaft or inclined plane. [4] Platform from which to charge a furnace.

LANYON SHIELD An iron curtain, stiffened by ribs of angle iron, suspended from trolley wheels running on a rail parallel with and in front of a zinc retort furnace. Its main purpose is to protect the worker from the furnace heat during retort charging, metal extraction, and retort discharging.

LARSENITE A lead-zinc silicate mineral found at Franklin, New Jersey.

LARVIK FURNACE A modified chambered muffle furnace that is heated electrically via graphite resistors (glow bars), which, in the zinc industry, is used for the processing of zinc scrap to yield high purity zinc metal or zinc dust.

LATENT HEAT [1] Thermal energy released or absorbed when a substance undergoes a phase change. [2] Thermal equivalent of the energy expended in melting a unit mass of a solid or vaporizing a unit mass of liquid; or conversely, the thermal equivalent of energy set free in the process of solidification or liquefaction.

LATERAL SECRETION Theory that the contents of a vein or lode are derived from the adjacent wall rock.

LATTEN BRASS Metallic compound into which brass and scrap-brass enter, and which is rolled or hammered into thin plates. Formerly much used for church utensils.

LATTEN Metal in thin sheets, especially (and originally) brass, which in this form is also called *Latten-brass*.

LAUNDER [1] Trough, channel, or gutter, by which water is conveyed; specifically in mining, a chute or trough for conveying crushed ore, or for carrying water to or from crushing apparatus. In metal making, a channel for the conveyance of molten metal. [2] A box conduit conveying particles suspended in water.

LAY-BY At Joplin, Missouri, an underground siding at or near a shaft for storing empty mine cars.

LAY-DOWN SYSTEM See *Pit System*.

LEACHING Extracting a soluble material from a heterogeneous solid by dissolving in a suitable liquid.

LEACH To dissolve minerals or metals out of an ore, calcine, residue, etc., by the use of water, acids, bases, and other solvents.

LEAD DAYS Early expression at Joplin for the time when only lead was mined; zinc minerals were ignored or cussed as being in the way of extracting the lead.

LEAD FUME Fume escaping from lead furnaces containing both volatilized and mechanically suspended metallic compounds. Because zinc was almost always present in lead concentrate, zinc not trapped in furnace slag generally was emitted from the charge as fume. Early practice released the fume to the air, but virtually all was captured thereafter.

LEAD PAINT Ordinary paint, so called because white lead is used as a base.

LEAD SLAG The slag resulting from the smelting of lead ores, concentrates, and scrap. Lead slag from the smelting of ore and concentrate, often contains substantial amounts of zinc and, between 1927 and 1987, up to 100,000 tons of zinc was extracted annually from such slags in the United States by the process called slag fuming.

LEAD TREE Crystalline deposit of metallic lead on zinc that had been placed in a solution of acetate of lead.

LEADED ZINC OXIDE Zinc oxide that contains white lead (basic lead sulfate) in amounts ranging from 3% to 50%. These oxides are made by fuming a charge containing lead sulfide ores admixed with oxidized zinciferous material or by physically blending components that were prepared in separate processes. Sometimes the terms high-leaded and low-leaded were used to indicate the lead content because of uncertainty of the materials composition. High-leaded oxide generally contained 35% to 37% white lead and low-leaded oxide, 3% to 12%. Leaded zinc oxides have been mainly used as paint pigments. See *cofumed, blends* and *white lead*.

LEAD-ZINC PROCESS Conventional, cut-sheet galvanizing process in which the pickled sheet is passed through a flux blanket into a lead bath where the sheet is preheated without coating. The sheet then passes through a shallow zinc bath floating on top of the lead and is coated with zinc. Compare with zinc-bath process.

LEAN Applied to poor ores, or those

containing a lower proportion of metal than is usually worked.

LEASE [1] An instrument through which interests are transferred from one party to another, subject to certain obligations and consideration. [2] A contract between landowner and another granting the latter the right to search for or produce minerals upon payment of an agreed rent, bonus, or royalty.

LECLAIRE'S PROCESS Developed in 1847, this was the process that eventually became known as the French process for the production of zinc oxide. LeClaire, a Frenchman, produced the oxide by burning the vapor issuing from a horizontal retort and collecting the oxide fume in a settling chamber. The metallurgy of the process has not changed since LeClaire's time, and although a number of furnacing techniques have been developed, the horizontal retort approach is still probably in use.

LECLANCHE CELL Zinc-carbon primary cell invented by George Leclanche of France in 1866. It was the first of the zinc-carbon cells. The original cell was known as a *wet cell* because the anode, an amalgamated zinc rod, and the cathode, crushed manganese oxide mixed with graphite, were immersed separately in a solution of ammonium chloride which was the electrolyte. Improvements in the cell resulted in development of the dry cell in the early 1880's. The dry cell increased versatility and eliminated the hazard of spilled electrolyte. In the dry cell, the electrolyte and cathode materials were formed into a paste, the zinc anode became the actual can that contained the cell, and the current collector was a solid carbon rod in the center of the cell. Both types of cells entered growing markets, being used to ring doorbells, power railroad signals, supporting the growing telegraph systems, and beginning in 1876, became the major source for telephone current. *In 1900 the flashlight came into being; it was so named because the dry cells of that day had very short lives, so to conserve cell life, the light was flashed on and off when used.*

LEGAL DESCRIPTION The description of a particular parcel of land according to the official plat of its cadastral survey (a survey by the Bureau of Land Management to establish units of land for management purposes).

LEG A prop of timber supporting the end of a stull or cap of a set of timber.

LEGS Uprights of a set of mine timbers.

LENS A body of ore or rock thick in the middle and thin on the edges; similar to a double convex lens.

LENTICULAR An orebody shaped like a double convex lens.

LETTER NAME A single letter or combination of letters used to designate a specific range, size, group, and/or design of diamond drill fittings, such as casing, drill rods, core barrels, etc. Terms such as EX, BX, NX, etc . indicate the the hole size and diameter of core that is attainable by a given drill set.

LEVEL [1] A horizontal passage or drift into or in a mine. It is customary to work mines from levels at regular depth intervals; often levels are known by their depth below the adit, shaft or surface, for example the 2000 level. [2] An instrument for finding a horizontal plane or line, or adjusting with reference to a horizontal line.

LEVERAGED BUY-OUT Acquisition of a publicly traded company by a small group, often including the company management, which takes the company private. Much of the purchase price is borrowed, with the debt repaid from company profits or by selling company assets.

LIBERATION SIZE The particle size at which substantially all of the valuable minerals are detached from the gangue.

LIBERATION Freeing, by comminution or crushing and grinding, particles of a specific mineral from their interlock with other constituents in the ore.

LIFE OF MINE Time, generally in months or years, that the ore reserves of a mine will last at scheduled mining rates.

LIFTER HOLES Any of the boreholes for blasting that are drilled horizontally or nearly so about floor level for insuring rock breakage to floor level.

LIGHT METAL ALLOYS Low-density metals and alloys, especially aluminum and magnesium. A relatively small number of these are zinc-containing.

LIMESTONE SINK See *Sinkhole*; *Karst topography*.

LIMESTONE General name for sedimentary rocks composed essentially of calcium carbonate.

LINKED VEIN A step-like vein in which the ore follows one fissure for a short distance, then passes by a cross fissure to another nearly parallel, and so on.

LIQUATION [1] Partial melting of an alloy. [2] Differentiation in which two immiscible liquids, such as molten lead and zinc, separate. The term also applies in cases where a phase crystallizes out resulting in the separation of the residual liquid.

LIQUID ASSETS Cash or other items easily converted to physical money.

LITHOPONE A pigment composed of zinc sulfide and barium sulfate, prepared by double decomposition of barium sulfide and zinc sulfate or by physically mixing the two end components. Most lithopone was prepared by the former method. It was invented in 1874 by John Orr of Scotland. Only small amounts were produced in the United States before 1900 as uses were few; however, lithopone became a great success in the linoleum and window shade industries and, thereafter, gained as a pigment in the paint industry through the invention of ready-mix flat wall paints in 1906. U.S. lithopone consumption rose from less than 1,000 in 1900 tons to 187,000 tons in 1929, its highest level, finding uses in rubber and paper industries as well as those mentioned above. By far the major use was in paint. Consumption in paint fell after 1929, as did that of zinc oxide, owing mainly to substitution by titanium dioxide pigment. U.S. consumption has since fallen to only a few thousand tons per year.

LIXIVIATION Separation of a soluble from an insoluble material by means of washing with a solvent.

LME OFFICIAL PRICE The last quoted price for each metal at the close of the second morning ring on the London Metal Exchange. Zinc prices are quoted in US dollars for cash, 3 months, 15 months, and 27 months.

LME London Metal Exchange.

LO ZINC A high-iron, strongly magnetic residue extracted from recycled spent sinter in the electrothermic process. Minus 3/4 inch sinter residue is subjected to magnetic separation, yielding three products of low, medium and high zinc content. The lo zinc product may partially be utilized to maintain the iron balance in electrothermic smelting, but most is discarded. Interestingly the oversized fraction is not reprocessed but is discarded or sold. It had been found from experience that the largest particles contain little zinc and enlarged owing to the loss of zinc resulting in the lowering of melting points, permitting slagging and agglomeration. Also see *hi zinc*.

LOCAL ACTION Corrosion due to the action of local cells, that is, galvanic cells resulting from inhomogeneities between adjacent areas on a metal surface exposed to an electrolyte.

LOCAL PLANT INDICATOR Plants which selectively grow in soils that contain high concentrations of a specific metal or those that have an affinity for absorbing certain metallic elements from the soil and which can be used in geochemical exploration. For example, the mock orange is an indicator plant for zinc In Washington.

LOCATABLE MINERALS Minerals that may be acquired under the Mining Law of 1872, as amended.

LOCATION NOTICE A written notice prominently posted on a claim, giving name of locator and a description of the claims extent and boundaries.

LOCATION WORK Labor required by law to maintain or establish the mining claim.

LOCATION [1] The act of fixing the boundaries of a mining claim, or the mining claim itself. [2] Perfecting the right to a mining claim by discovery of a valuable mineral, monumenting the corners, completing the discovery work, posting a notice of location and recording the claim.

LODE Strictly a fissure in the country-rock filled with mineral; usually applied to metalliferous deposits. In general miners' usage, a lode, vein, or ledge is a tabular deposit of valuable mineral between definite boundaries. The word should not be used for a flat or stratified mass or for a placer deposit. As used by miners, before being defined by any authority, the term lode simply meant that formation by which the miner could be led or guided. It is an alteration of the verb lead; and whatever the miner could follow, expecting to find ore, was his lode. Some formation within which he could find ore, and out of which he could not expect to find ore, was his lode. Lode, as used by miners, is nearly synonymous with the term vein, as employed by geologists.

LOG WASHER A slightly slanting trough in which revolves a thick shaft or log, carrying blades obliquely set to the axis. Ore is fed in at the lower end, water at the upper. The blades slowly move the lumps of ore upward against the current while any adhering clay is disintegrated and floated out the other end. Devices of this type were commonly used in the early years of zinc mining in the United States to upgrade the quality and zinc content of calamine ores.

LOHMANNIZING A process by which a protective zinc coating is amalgamated to a base-metal sheet.

LOISEAU FURNACE A gas-fired furnace for the distillation of zinc.

LONDON WHITE White lead or basic lead carbonate, used as paint pigment and pottery glazes.

LOST-WAX PROCESS Executing bronze casting by casing a wax model with plaster and afterwards melting out the wax.

LOW-GRADE ORE Ore relatively poor in the metal for which it is being mined. Sometimes used to indicate material below ore grade and would not normally be mined, except when it is possible to blend it with high-grade ore to normalize the mill head feed.

LUMEN BRONZE An alloy of 86 percent zinc, 10 percent copper, and 4 percent aluminum. It is especially valuable for high-speed bearings which do not carry a heavy load.

LUMINOUS PAINT A paint containing a phosphorescent oxide or sulfide in oil. For example zinc sulfide.

LUSTER The character of the light reflected by a mineral.

LUTE [1] A mixture of fireclay used to seal cracks between crucible and cover when heat is to be applied. [2] To seal with clay or other plastic material.

LYNEN FURNACE A zinc-distillation furnace with a common condensation chamber.

LYSTER PROCESS A flotation process that separates galena and sphalerite by treatment, at a low temperature, with eucalyptus oil and other frothing agents, and with agitation or aeration in a neutral or alkaline, but not acid, solution of the sulfates, chlorides, or nitrates of calcium, magnesium, sodium, potassium, or mixtures of these substances.

M

MACQUISTEN PROCESS A flotation process developed about 1905. The original process was a simple water flotation method that later provided for the treatment of tailings with oil. No acid or gas entered the flotation process. The machinery consisted of a near horizontal, spirally grooved tube, 1 foot in diameter and 4 to 6 feet long, that was rotated. Finely ground pulp, thinly diluted with water, was charged into the tube. The sulfide particles were caught and held on the surface by surface tension and superficial viscosity of the water and separately captured at the discharge end. The process was first used to float chalcopyrite at a mine in Nevada in 1906, and in 1910 was used at the Morning Mine in Idaho to separate sphalerite from sphalerite-barite-siderite table tailings. This was one of the first successful attempts to float zinc minerals in the United States. Also known as the *Mcquisten-Tube Process*.

MACROSHRINKAGE A casting defect, detectable at magnifications not exceeding ten diameters, consisting of voids in the form of stringers shorter than shrinkage cracks. This defect results from contraction during solidification where there is not an adequate opportunity to supply filler material to compensate for the shrinkage. It is usually associated with abrupt changes in section size.

MAGGOT ORE Lustrous, irregular-surfaced hemimorphite found at the Sterling Mine in New Jersey, so named because of its similarity to a mass of maggots.

MAGMATIC WATER Water derived from cooling magma.

MAGNETIC SEPARATOR A device used to separate magnetic from less magnetic or nonmagnetic materials. The crushed material is conveyed on a belt or vibrated down a chute or carried in a slurry past a magnet or electromagnet.

MAKE Amount produced; yield. For instance: The make from a furnace was 10 tons.

MALLEABILITY The characteristic of metals which permits plastic deformation in compression without rupture.

MALM PROCESS A process for treating complex ores, especially those of zinc, with dry chlorine in a tube mill at about 160 degrees F; followed by heating at about 750 degrees in a multiple-hearth furnace. The metals, dissolved in water, are precipitated out with zinc dust; Zinc remains in solution and is evaporated out. The resulting zinc chloride is fused and zinc recovered by electrolysis; the chlorine is returned to the process

MAN CAGE A special cage for raising and lowering men in a mine shaft.

MAN CAR A car with seats for transporting miners up and down steeply inclined shafts.

MANAGEMENT BUY-OUT Purchase of a business by its existing management with the help of a group of financial backers. Buy-outs are funded largely by loans secured on the assets of the company itself.

MANAGER An official who has control and supervision of a mine, mill or smelter.

MANHEIM GOLD A brass alloy resembling gold.

MANHOLE Small cut in the wall of a haulageway that allows miners to be safe from passing locomotives and cars. A safety hole.

MANTO Blanket-like replacement of rock (commonly limestone) by ore. In some districts the term has been modified to designate a pipe-like deposit confined within a single stratigraphic unit.

MANWAY A small passage used as a traveling way for the miner, and often used as an airway or chute or both. In a multi-compartment timbered raise, the laddered section with landings used by miners for access up and down the raise and associated workings.

MARCASITE A white iron pyrite, the orthorhombic diamorph of pyrite. It is less common in sulfide-ore deposits than pyrite, but is a prominent constituent in the Wisconsin lead/zinc deposits, where to some extent it caused serious problems for zinc producers because marcasite and sphalerite have similar specific gravities and could not be easily separated by gravity methods, resulting in the production of low-grade zinc concentrate. In 1904 a process was developed whereby the gravity concentrates were given a slight roast which made the marcasite slightly magnetic, and thereby easily separable by magnetic means.

MARMATITE A ferrous variety of sphalerite, containing more than 10% iron.

MASS NUMBER Sum of the neutrons and protons in the nucleus of an element.

MASTER ALLOY An alloy, rich in one or more desired addition elements, that can be added to a melt to raise the percentage of the desired element.

MATING SEASON General term for the period in the fall of the year when zinc miners and smelterers agree on treatment charges for the next year. Terms typically are arranged on an individual basis rather than a group effort.

MATRIX [1] Finer-grained material between the larger particles of a rock or the material surrounding a fossil or mineral. [2] Gangue; the carbonate material in zinc ore, for example. [3] The binding agent in briquetting. [4] Principal phase or aggregate in which another constituent is embedded.

MATT BLUE Color for a pottery decoration depending on the formation of cobalt aluminate; zinc oxide is usually added. Quoted recipe is 60% aluminum oxide, 20% cobalt oxide, and 20% zinc oxide.

MATTE A metallic sulfide mixture made in smelting sulfide ores of copper, lead and nickel.

MATTOCK A miner's pickaxe.

MATURATION In alchemy, the conversion of base metal into gold. A concept that metals took time to form, to mature, as brass into gold.

MAZAK Trademark name for die casting alloys. See *Zamak*.

MCDOUGALL ROASTING FURNACE A circular multi-hearth sulfide roasting furnace that contains up to 16 hearths, each of which have moving rabbles driven by a central rotating shaft. The rabbles advance the roasting concentrate across each hearth dropping the material onto the next lower hearth in succession until it is discharged at the bottom.

MEASURED ORE An ore in which the tonnage is computed from dimensions revealed in outcrops, trenches, workings, and drill holes and for which the grade is computed from detailed sampling. The block of ore is well established such that the size, shape, and the mineral content are judged to be accurate within limits. See proved ore.

MECHANICAL CLASSIFIER Machines such as the *Dorr classifier*, that are commonly used to classify a ball mill or rod mill discharge into finished product and oversize, which is returned for regrinding.

MECHANICAL LOADER A power machine for loading pay mineral and dirt.

MECHANICAL PLATING Plating wherein fine metal powders are peened onto the work by tumbling or other means.

MECHANICAL RABBLE A rabble worked by machinery.

MECHANICAL SAMPLING A process for sampling a steam of liquid, slurry or dry material in which a portion of the stream is systematically removed by mechanical means for analysis or testing.

MEGASCOPIC Large enough to be seen by the naked eye. The antithesis of microscopic.

MELTING POINT The temperature at which a pure metal, compound or eutectic changes from solid to liquid; the temperature at which the solid and the liquid are in equilibrium.

MERESTONE A stone used as a boundary; figuratively, a boundary.

MERRILL FILTER PRESS A variation of the plate and frame press.

MERRILLITE Trademark for a high-purity zinc dust.

METABOND A zinc-iron phosphate coating for application to iron and steel.

METAL PICKLING The immersion of metal objects in a hot acid bath to remove scale, oxide, tarnish, etc., leaving a chemically clean surface for galvanizing or painting.

METAL SPRAYING Coating metal objects by spraying molten metal upon the surface with gas pressure.

METAL [1] An opaque lustrous elemental chemical substance that is a good conductor of heat and electricity and, when polished a good reflector of light. Most elemental metals are malleable and ductile and are, in general, heavier than the other elemental substances. [2] As to structure, metals may be distinguished from nonmetals by their atomic binding and electron availability. Metallic atoms tend to lose electrons from the outer shells, the positive ions thus formed being held by the electron gas produced by the separation. The ability of these free electrons to carry an electric current, and the fact that the conducting power decreases as the temperature increases, establish one of the prime distinctions of a metallic solid. [3] From the chemical viewpoint, an elemental substance whose hydroxide is alkaline. [4] An alloy.

METALLIC ORE Of or belonging to metals, containing metals. Term applied to minerals having the luster of metals.

METALLIZING The forming of a metallic coating by atomized spraying with a molten metal or by vacuum deposition.

METALLOGENIC PROVINCE A large area of the earth's surface characterized by an unusual abundance of ores of a particular metal or of a particular type. Examples of metallogenic provinces are the lead-zinc-silver Coeur d'Alene, Idaho and the zinc-lead Tri-state (Kansas, Missouri, and Oklahoma) districts.

METALLURGICAL COKE A coke used as a fuel and reducing agent but also one possessing very high compressive strength at elevated temperatures such that it can provide essential support to the charge in the furnace.

METALLURGICAL SMOKE Gases, vapors, and the fine dust emitted from the stacks of, or where possible from, blast furnaces, reverberatory furnaces and roasting furnaces. Consists of three distinct substances: gases (including air), the flue dust and the fume.

METALLURGY The science and art of preparing metals for use from their ores by separating them from mechanical mixtures and chemical combinations. Includes various processes, as smelting, amalgamation, and electrorefining. Metallurgy, as generally understood, is concerned with the production of raw materials, the manufacture of which into finished articles, belongs to other arts.

METEORIC WATER Water that previously existed as atmospheric moisture or surface water, and then enters from the surface into the voids of the lithosphere.

MIDDLINGS [1] Second quality ore obtained by washing. [2] Ore particles in a concentrator operation incompletely separated from the gangue and/or another type of ore mineral. [3] Product intermediate between concentrate and tailings and containing enough of a valuable mineral to make retreatment profitable. Used in plural form.

MILL HOLE An auxiliary shaft connecting a stope with the level below.

MILL [1] Any establishment for reducing and/or beneficiating ores by means other than smelting. An ore concentrator or beneficiating plant. [2] In mineral processing, a machine or group of machines used in comminution. This group can include crushing, wet grinding, and ore reduction by other means. Representatives of mill machines are the ball, rod, hammer, stamp, and tumbling types.

MILLING ORE Any ore that contains sufficient valuable minerals to be treated by any milling process.

MILL-SITE ENTRY Nonmineral public lands to be used as a mill site under the General Mining Law of 1872, as amended, for the beneficiation of ore or in the development of a lode claim.

MINE DEVELOPMENT The process of preparing the mine for ore production. Involves efficient entry by shaft, adit or stripping, developing a network of drifts, crosscuts, raises, etc., and installation of ore transportation, ventilation, and water pumping systems. May also include construction of surface facilities, such as a tailings dam and concentrator.

MINE DUST Dust formed by rock drilling, blasting, or handling rock.

MINE RUN The entire unscreened ore output of a mine. Also called run-of-mine.

MINEABLE A mineral or material that can be mined under present day mining technology and economics.

MINERAL DEPOSIT Any valuable mass of ore. See ore deposit.

MINERAL DRESSING The treatment of natural ores or partly processed products derived from such ores in order to segregate or upgrade some or all of their valuable constituents.

MINERAL MONUMENT A permanent monument established in a mining district to provide for the accurate description of mining claims and their location.

MINERAL PAINTS Minerals used as paint pigments.

MINERAL [1] Naturally occurring homogeneous inorganic substance of definite or fairly definite chemical composition and with characteristic physical properties. [2] Naturally occurring substance with characteristics and economic uses that bring it within the purview of U.S. mineral laws and which make it obtainable under applicable laws from public lands by purchase, lease or preemptive entry.

MINERS' ELBOW Swelling on the back of the elbow due to inflammation of the bursa over the olecranon, so-called because it is often seen in miners.

MINERS' NEEDLE A long, slender, tapering, metal rod left in a hole when tamping and afterwards withdrawn, to provide a passage, to the blasting charge, for the squib.

MINERS' RULES Rules and regulations proclaimed by the miners in any district relating to the location, recording, and the work necessary to hold possession of a mining claim. It was the miners' rules of the early days of the mining industry that were the basis of present mining laws.

MINER'S SELF-RESCUER A small pocket form of gas mask carried on the belt of a miner for protection against carbon monoxide. Affords about 30 minutes of protection while the miner escapes from the contaminated area.

MINING CLAIM A portion of the public mineral lands which a miner, for mining purposes, takes and holds in accordance with mining laws.

MINING DISTRICT A settlement of miners organized after the plan that, in the first years of mining in the western United States, the miners, in the independence of all other authority, devised for their own self-government. [2] A section of country usually designated by name and described or understood as being confined within certain natural boundaries, in which ore minerals are found in paying quantities.

MINING WIDTH The minimum width necessary for the extraction of the ore regardless of the actual width of the ore-bearing rock.

MINING [1]Act or business of making mines or working them. [2] Processes by which useful minerals are obtained from the earth's crust. [3] Art and practice of operating mines profitably.

MISFIRE An explosive charge in a drill hole which has partially or completely failed to explode.

MISSISSIPPI VALLEY TYPE (MVT) Generic term for a general class of epigenetic zinc-lead deposits in carbonate rocks, so named because of the numerous and widespread occurrence of these deposits in the Mississippi River drainage basin of the United States. Many of the great zinc-lead deposits of the world are of this type. Ore mineralization is often extensive and deposits may occur over large areas. Deposits characteristically have simple mineralogy consisting of zinc, lead and iron sulfides, and associated jasperoid and dolomite. Ores occur as replacements of host rock and/or as joint, fault, fracture or breccia fillings. MVT deposits tend to occur at moderate to shallow depth in specific horizons, are marginal to cratonic and sedimentary basins, and are very often without any obvious mineralization source. Increasingly, there is evidence that many MVT deposits are genetically associated with large, interconnected mineralizing hydrothermal systems that were driven, to a large extent, by orogenic forces.

MOCK LEAD Cornish term for zinc blende; also called wild lead. See *Mock Ore*.

MOCK ORE Term used for sphalerite, so named because the mineral may look like galena but yields no lead. The term *blende* is similarly derived, being taken from the German, meaning blind or deceiving.

MOCK PLATINUM An alloy of 8 parts of common brass and 5 of zinc.

MOCK SILVER A white alloy of copper, tin, nickel, zinc, etc., of the same class as Britannia metal; pewter.

MODIFYING AGENTS; MODIFIERS In flotation, chemicals which increase the specific attraction between collector agents and particle surfaces, or conversely which increase the wettability of those surfaces.

MOLD [1] The form made of sand, metal or other material which contains the cavity into which molten metal is poured to produce a casting of definite shape and outline. [2] In powder metallurgy, it is the same as die.

MOLECULE The smallest part of a substance that can exist separately and still retain its composition and characteristic properties; the smallest combination of atoms that will form a chemical compound.

MOND PRODUCER A furnace used for the manufacture of producer gas.

MONHEIMITE A variety of smithsonite containing iron carbonate.

MONTEFIORE FURNACE A small furnace used for the recovery of zinc from blue powder by liquidation.

MORTAR MILL A mixing and stirring machine for combining lime, sand and other materials to make mortar. A form of pug mill.

MOSSY ZINC Granulated zinc obtained when the molten metal is poured into cold water.

MOTE; MOAT A straw filled with gunpowder for igniting a shot. A fuse.

MUCK [1] The broken rock or other material from a mine excavation. [2] To excavate and remove muck.

MUCKER One who loads mine cars, and in some mines, is also a trammer moving the cars to an ore chute, the shaft or the adit mouth.

MUCKING The operation of loading broken rock by hand or machine in shafts, tunnels and headings.

MUD CAP A charge of dynamite, or other high explosive, fired in contact with the surface of the rock after being covered with a quantity of wet mud, wet sand, or earth, no bore hole being used. The slight confinement given the dynamite permits part of the energy to be transmitted to the rock in the form of a blow. Also called adobe, dobe, and sandblast.

MUDCRACKING A surface condition characterized by cracks, like the surface of a dried mud puddle, that extend into the body of a zinc-rich primer.

MUFFLE FURNACE [1] A furnace devised to shield its contents from direct contact with the flames. [2] A furnace in which heat is applied to the outside of a refractory chamber containing the charge.

MUNDIC Miner's term for pyrite and/or marcasite.

MUNTZ METAL An alloy of copper and zinc, 3 parts to 2.

NASCENT Coming into existence, beginning to exist or to grow. A term used in the flotation process.

NAVAL BRASS An alloy of 62% copper, 1% tin, and 37% zinc. Resistant to sea-water corrosion.

NEEDLE A piece of copper or brass about 1/2 inch in diameter and 3 to 4 feet long, pointed at one end and turned into a handle at the other end. Thrust into a charge of blasting powder in a bore hole and while in this position the bore hole is tamped solid, preferably with moist clay. The needle is then withdrawn, leaving a straight pathway through the tamping for the miners' squib to shoot or fire the charge.

NEOGEN An alloy resembling silver, containing 58 parts copper, 27 parts zinc, 12 parts nickel, 2 parts tin and small amounts of bismuth and aluminum.

NEST A small isolated mass of any ore or mineral within a rock.

NET UNIT VALUE The difference between the gross unit recoverable value and the cost of mining, treating, marketing, and transporting the ore.

NEUTRAL LEACH SOLUTION The resultant solution from leaching zinc calcine with spent electrolyte and in which all excess sulfuric acid has been neutralized by the subsequent additions of excess zinc calcine. It is necessary to completely neutralize all sulfuric acid to insure the removal of most of the impurities from the solution in subsequent purification procedures. With a single leach process, milk of lime or ground limestone is often used to neutralize any acid remaining; however with a double leach system, excess calcine is used to utilize available acid to dissolve additional zinc. Both leach processes can be used in continuous operations.

NEUTRON ACTIVATION ANALYSIS An analytical technique which is dependent on measuring primarily gamma radiation which is emitted by radioactive isotopes produced by irradiating samples in a nuclear reactor. Each element that is activated will emit a characteristic fingerprint which can be measured and quantified.

NEUTRON An uncharged elementary particle with a mass that nearly equals that of a proton.

NEVADA SYSTEM A square-set stoping system.

NEW CLIPPINGS Scrap consisting of new pure zinc sheets, faulty stampings and clips from manufacturing processes, free of corrosion and foreign material or attachments.

NICHOLSONITE A variety of aragonite, containing up to 10% zinc, from Leadville, CO, and the Tintic district, UT.

NIHIL ALBUM; NIL ALBUM An old term for zinc oxide.

NILL Sparks of brass during manufacture.

NITRO An abbreviated form for nitroglycerin or dynamite.

NITROGLYCERIN A product of the action of nitric acid and sulfuric acid on glycerin. It is an oily substance about one-half times as heavy as water, is almost insoluble in water, and is the principal active ingredient in dynamite.

NOBLE METALS Metals that have little affinity for oxygen; those metals that are least liable to oxidize under normal conditions. Includes gold, silver, mercury, and the platinum-group metals. The term is of alchemistic origin.

NOMAD Scrap code word for yellow brass turnings. See *Apple*.

O'BRIEN FURNACE A roasting furnace of the Herreshoff type with a central vertical shaft carrying stirring arms.

OFFICE OF MINERALS EXPLORATION (OME) Office established under the Department of the Interior in 1958 to assume the functions of the Defense Minerals Exploration Administration (DMEA). See *Defense Minerals Exploration Administration*.

OIL FLOTATION A process in which oil is used in ore concentration by flotation. The term developed as a consequence of the early development of flotation in that virtually all of the early methods were based largely on the use of oils. As flotation development progressed, the process became complex involving a number of reagents, of which oils became only a part; and as a result the term, oil flotation, is little used today.

OILER In flotation, oil which provides a film around a mineral particle.

OLD DOMINION A leaded brand of spelter sold by the Bertha Mineral Co. from 1879 to about 1911. See *Bertha Pure Spelter*.

OLD ZINC Scrap term for zinc-products that have been discarded because of wear, damage, or obsolescence after serving a useful purpose.

OLEIC ACID A fatty acid derived from animal tallow and vegetable oils. An oily substance that functions as oil in flotation operations.

OLIVER FILTER A continuous-drum-type of pressure filter for separating liquid from solids in a mineral pulp.

ONE-PIECE SET Applied to a single timber support. When used vertically, it is called a post or prop and when used horizontally or inclined, it is called a stull.

ONE-SIDE GALVANIZING Processes or treatments where only one side of a sheet or strip is galvanized. It was developed mainly to provide a high quality surface (the uncoated side) for painting and a side highly resistant to corrosion. One sided is mainly used in the manufacture of automobiles panels.

OPEN HEARTH FURNACE A reverberatory melting furnace with a shallow hearth and a low roof. The flame passes over the charge on the hearth, causing the charge to be heated both by direct flame and by radiation from the roof and sidewalls of the furnace.

OPENCAST A working in which excavation is preformed from the surface. Commonly called *Open-Cut* and *Open-Pit*.

OPENING [1] Entrance to a mine. [2] A short heading driven between two or more parallel headings or levels for ventilation.

OPEN-PIT MINE; OPEN-CAST MINE A mine working or excavation open to the surface.

OPEN-PIT MINING The mining of metalliferous ores by surface-mining methods. Commonly the term open-pit mining is used to refer only to the mining of metal ores to distinguish it from coal or strip mining and from nonmetallic mining or quarrying.

OPEN-STOPE METHOD Stoping in which no regular artificial method of support is employed. The walls and roof are self supporting, although random pillars and occasional props or cribs may be used to hold local patches of insecure ground.

OPEN-TANK METHOD A way of treating mine timbers to prevent decay in which the timber is immersed in a tank of hot preservative and then into a tank of cold preservative. Preservatives used are creosote, zinc chloride, and other chemicals. More recent processes for timber impregnation use vacuum and pressure methods.

OPERATOR The company, person, owner or lessee that actually operates the mine or facility.

OPTION Privilege secured by the payment of a certain consideration for the purchase, or lease, of mining or other property, within a specific time period, or upon the fulfillment of certain conditions set forth in the contract.

OPTIONS In mineral exploration, options contain an exclusive right to explore a property or orebody over some period of time. Options are designed to keep front-end costs low, to provide adequate time to evaluate the property, and to lock-in a purchase or lease price at the end of the option period.

ORE [1] Any mineral or mineral aggregate having value on extraction. [2] At Joplin a lead, zinc, or lead-zinc concentrate obtained from milling and gravity separation. The crude ore was called *dirt*.

ORE BLOCKED OUT Ore exposed on three sides within a reasonable distance of each other. Essentially proved or measured ore reserves.

ORE BODY Generally a solid and fairly continuous mass of ore, which may include low-grade and waste as well as pay ore, but is individualized by form or character from adjoining country rock.

ORE CHUTE; ORE PASS An inclined or vertical passage for the transfer of ore to a lower level, generally for car loading and hoisting purposes. Gates are used at the bottoms of chutes to control or stop the flow of rock at discharge points.

ORE CONTROL Geologic features, such as an algal reef, fault, etc., that has influenced the deposition of ore.

ORE DEPOSIT General term applied to rocks containing minerals of economic value in which they can be profitably exploited and/or have a reasonable chance of profitability by a change in the economic circumstances that control their value.

ORE DRESSING Same as *Mineral Dressing*.

ORE GRADE A common term for average content of a metal or metal in an orebody.

ORE HOPPER A temporary storage bin for broken ore loaded from the top and discharged through the bottom through a door or chute at the bottom.

ORE IN SIGHT Term used to indicate two factors in an ore reserve estimate: (a) ore blocked out and (b) ore that reasonably may be assumed to exist, though not actually blocked out. (a) is synonymous with measured or proved ore reserves and (b) with indicated and inferred or with probable and possible ore reserves.

ORE MILL A stamp mill; a concentrator.

ORE PARTLY BLOCKED OUT Ore bodies that are only partly developed and the values of which can only be approximately determined. Synonymous with indicated or possible ore reserves.

ORE POCKET [1] An excavation near the hoisting shaft into which ore is moved and stored prior to hoisting. [2] Also used in such a phrase as *a rich pocket of ore* to describe an unusual concentration in the lode.

ORE RESERVES Usually restricted to ore of which the tonnage and grade have been established with reasonable assurance by drilling and other means and which can be mined and processed at a profit. Often, the reserves at a mine are the combined measured/proved ore reserves and all or part of the indicated/probable ore reserves. Because ore reserves definitions have varied and are not used in precisely the same way by those in the minerals industries, care is required in evaluating and comparing ore reserve data. [2] That part of the reserve base, which could be economically extracted or produced at the time of determination. See also *Demonstrated Resources* and *Reserve Base*.

ORE RESERVE Strictly, the term refers only to ore, the grade and tonnage of which is reasonably assured by drilling and other means, that is economically viable. Compare *Geologic*, *In-situ*, and *Demonstrated Ore Reserves*.

ORE RUNS A local term in Tennessee for zinc ore that occupies two or more stratigraphic horizons that generally are separated by non-commercial intervals.

ORESHOOT A large and usually rich aggregation of mineral in a vein. It is a more or less vertical zone or chimney of rich vein matter extending from wall to wall, and has a definite width laterally.

ORE TRADING The selling or trading of ores and concentrates on the open market.

ORICHALCEOUS Having a color between brass and gold, of or pertaining to orichalc.

ORICHALC Under the Roman empire, an alloy of copper and zinc, resembling gold in appearance; brass. There was also a white orichalc.

ORIGINAL CONTOUR The pre-mining surface configuration of the land, not necessarily the pre-mining elevation.

ORMOLU A brass made to imitate gold and used in mounts for furniture, cheap jewelry, and other decorative items. Also ground or powdered for pigment use on objects to be gilded. Also called *Mosaic Gold*.

OROGENY Process of mountain building.

ORTON CONES Pyrometric cones for heat recording, similar to Seger cones.

OUTCROP An exposure of rock formations, veins, etc. at the Earth's surface. When used in connection with a vein or lode as an essential part of the definition of apex, it does not necessarily imply the visible presentation of the vein on the surface of the earth, but includes those deposits that are so near the surface as to be found easily by digging.

OVERBURDEN [1] To charge in a furnace too much ore and flux in proportion to the amount of fuel. [2] Material of any nature, consolidated or unconsolidated, that overlies an orebody that is mined or is planned to be mined by open-pit methods.

OVERHAND CUT-AND-FILL In this method, two level drives are first connected by a raise or two raises located apart along the vein. Mining begins at the bottom and proceeds upward in horizontal slices some practical distance out from the raise or between the two raises. The ore is completely removed and replaced with fill, except that working space is left to make the next slice if necessary. At the Sterling Mine in New Jersey, the ore was of such high grade that the surfaces of the fill were cemented to provide for complete removal of the ore and to prevent dilution. Compare shrinkage stoping.

OVERHAND STOPE One in which the ore above the point of entry to the stope is attacked, so the broken ore tends to gravitate toward discharge chutes and is self emptying.

OVERHAND STOPING Working a block of ore from a lower level to a level above.

OVERHEAD CHARGES General charges or expenses which cannot be charged up as belonging exclusively to any particular part of the work or product.

OVERRIDE A royalty or percentage

of the gross income from production deducted from the working interest.

OVERVOLTAGE The difference between the actual electrode potential when appreciable electrolysis begins and the reversible electrode potential.

OXIDATION A reaction in which there is an increase in valence resulting from a loss of electrons. Contrast with reduction.

OXIDE SHAKES See *Brass Founder's Ague*.

OXIDIZED ORES Ores resulting from alteration of metalliferous minerals by weathering and the action of surface waters, and their conversion, partly or wholly, into oxides, carbonates, and sulfates.

OXIDIZED ZONE That portion of an ore deposit which has been subjected to the action of surface waters carrying oxygen, carbon dioxide, etc. That zone in which sulfides are altered to oxides, carbonates and hydrous minerals. See *Secondary Minerals*.

OXIDIZING FLAME Gas flame produced with excess oxygen.

P

PACHUCA TANK A cylindrical tank or vat, tall in proportion to its diameter, with the bottom ending in a 60 degree cone. Within the tank is a hollow column extending from the bottom to less than a foot of the top. The apparatus works on the air-lift principal, aerated pulp in the tube flows upward and discharges at the top while more pulp enters at the bottom. Essentially a pulp agitator useful for leaching purposes. Also known as a *Brown Tank*.

PACK TRAIL A path or narrow road for the passage of pack trains only.

PACKER A person who transports goods by pack animal; a carrier; a pack animal. Common in early mining districts.

PACK [1] A wall or pillar built of gob to support the roof. [2] To fill in stopes and old mine workings with waste rock to support the roof.

PACKFONG Chinese term for German silver, a silver-white alloy of copper, zinc and nickel.

PACK-ROLLING METHOD Rolling mill, sheet-reduction method in which as many as 8 to 18 rough-rolled zinc sheets are stacked together and rolled simultaneously.

PAINT MILL A machine for grinding mineral paints. The chief purpose of grinding is to amalgamate the pigment, intimately with the oil or vehicle, so as to make a homogeneous mixture. In early practice, painters handground the zinc oxide pigment, drier (if used), and linseed oil to make paint, whereas, manufacturers developed machines to mix the pigment and oil first in a pug mill, followed by rolling to obtain the right characteristics.

PAN MILLS See *Chilean mill*.

PAN A shallow, concave steel or porcelain dish in which drillers or samplers wash drill sludge to concentrate heavy minerals to obtain a quick visual of valuable minerals in the rock being drilled.

PARAGENESIS General term for the order of formation of associated minerals in time of succession, one after the other. To trace out in a rock or vein the secession in which the minerals developed.

PARAMAGNETIC A property of many substances, related to ferromagnetism, by virtue of which, when placed in a non-uniform magnetic field, they tend to move toward the strongest part.

PARKES PROCESS [1] A process to de-silverize lead by adding zinc to molten argentiferous lead. Zinc and the silver rise to the surface of the bath as a scum, which is then taken off and afterwards, the silver is recovered by distilling off the zinc. In 1851, Alexander Parkes, an Englishman, patented the process which made use of the fact that silver has a greater affinity for zinc than lead. In 1864 Edward Balbach, Jr. from New Jersey patented a lead desilverizing process using zinc, but within a few years the Balbach process had been modified to a large extent, such that it closely resembled the Parkes process. To a large extent the Parkes process used today is fundamentally the same as originally developed. [2] Process to soften (refine) lead and recover precious metals from lead, based on the principal that if 1% to 2% zinc is stirred into molten lead, a compound of zinc, gold and silver separates out and can be skimmed off, followed by recover of the above metals.

PARSON'S WHITE BRASS One of the first zinc alloys used in die casting. It is composed of 65% tin, 30% zinc and 5% copper. Although this was not a satisfactory die casting alloy, it encouraged further work on high-zinc-content die cast alloys.

PASSIVATION An alteration of the chemically active surface of a metal to a much less reactive state. Contrast with activation.

PASSIVE METAL Metal on which oxide film helps prevent further corrosion. Zinc, aluminum, and nickel are examples of metals that attain surface passivity.

PASSIVITY Condition in which a piece of metal, because of an impervious covering of oxide or other compound, has a potential much more positive than where the metal is in the active state.

PATENT PAINT Used for all liquid or ready-mix paint from the 1870's into the 1930's, so called because of Averill's patent of a ready-mix paint formula in 1867.

PATENTED ROPE Galvanized steel rope.

PATTINSON PROCESS A process for separating silver from lead, in which the molten lead is slowly cooled so that crystals poorer in silver solidify out and are removed, leaving the melt richer in silver.

PAY DIRT; PAY ROCK Earth, rock, veins, etc., which yields a profit to the miner.

PAY ORE Parts of an ore body which are both rich and large enough to work for a profit.

PEAVY A stout lever like a cant hook, but having the end armed with a strong and sharp spike.

PEBBLE JACK Zinc blende in small crystals or pebble-like forms not attached to rock but found in clay openings in the rock.

PEENING A mechanical working of metal by hammer blows or shot impingement.

PEG To mark out a miners' claim by pegs at the four corners, each bearing the claimant's name. To peg a claim.

PELLETIZING [1] A method in which finely divided material is rolled in a drum or on an inclined disc, so that the particles cling together and roll up into small, spherical pellets. [2] Moulding, by the application of pressure, of various materials into artifacts smaller than briquettes.

PENALTY In connection with contracts for the purchase of concentrates and scrap metals, a penalty is a deduction from the agreed price for failure to attain the agreed assay value, level of contaminants, or some quality of the delivered material. In construction, a penalty is assessed if the work is not completed on time, fails to meet specifications, etc.

PER DESCENSUM By descent, as distillation of zinc vapor in a retort downward as in the Indian method in the 14th century or the English method developed by Champion about 1740. As opposed to *per ascensum* or rising distillation vapors as in the case of all other zinc retort methods.

PERCENTAGE DEPLETION A depletion allowance for Federal tax purposes. It is a fixed statutory percentage of the gross income of the property. The income must be based entirely on the depleting mineral deposit. The statutory depletion rate for zinc is 22%, but in contrast to cost depletion, the percentage depletion allowance cannot exceed 50% of the taxable income of the property, calculated without allowance for depletion.

PERCUSSION CAP See *Detonator; Primer.*

PERCUSSION TABLE An inclined table, agitated by a series of shocks, and operating at the same time like a buddle. It may be made self-discharging and continuous by substituting for the table an endless rubber cloth, slowly moving against the current of water, as in the frue vanner.

PERCUSSIVE DRILLING A form of drilling in which rock is penetrated by the repeated impact of a reciprocating drill.

PERMANENT MOLD A metal mold, other than an ingot mold, of two or more parts that is used repeatedly for the production of many casting of the same form. Liquid metal is poured in by gravity.

PERMANENT MONUMENT A monument of lasting character for marking a mining claim. It may be a mountain, hill, butte, gulch, lake, stake, post, shaft, tunnel, or well known adjoining patented claims.

PERMISSIBLE A machine or explosive that has been approved by the U. S. Bureau of Mines for use underground under prescribed conditions.

PERSUADER Common term for a crowbar, lever, or some such article used as a manual aid in moving heavy objects.

PETER; PETER OUT To fall gradually in size, quantity, or quality, as in the mine has petered out.

PETROLEUM COKE The residue obtained by the distillation of petroleum. On account of its purity, it has found application in metallurgical processes and in making battery carbons.

PHASE A physically homogeneous and distinct portion of a material system.

pH The negative logarithm of the hydrogen ion activity; it denotes the degree of acidity or alkalinity of a solution.

PHILOSOPHER'S STONE An imaginary stone, or solid substance or preparation, believed to have the power of transmuting the bases metals into gold or silver, and hence much sought by the alchemists.

PHILOSOPHER'S WOOL A type of purified zinc oxide, first thought to have been produced by alchemist from cadmia obtained from the flues of copper smelters. The alchemists recognized cadmia as a way of producing immature gold but for medicinal purposes, they devised a special process for purifying cadmia. The furnace consisted of a hearth and a separate settling chamber; zinc vapor resulting from cadmia heated by charcoal and aided by air from a bellows, passed into a settling chamber, depositing mainly as fluffy zinc oxide powder on the chamber floor but some adhered to the walls and ceiling, increasing in size until it appeared like skeins of wool. To this latter material, they gave the name, philosopher's wool.

PHOSPHATIZING Treating an electrogalvanized surface immediately after coating to pacify the surface and to provide a surface that is readily paintable. In the process the coated work is thoroughly water rinsed followed by immersion in, spraying or brushing the surface with a phosphatizing agent, often a proprietary chemical, which chemically forms a very thin coating of insoluble tertiary zinc phosphate crystals. Phosphatized, electrogalvanized flat-rolled materials often find uses in refrigeration, trailers, truck, automobile, office, cabinet, etc. applications

PHOSPHORESCENCE The continued emission of light by a substance (not incandescent) produced especially after heating, exposure to light or radiation, or to electrical discharge. Zinc sulfide and to a lesser extent, zinc oxide are important phosphors. Zinc sulfide can be made to fluoresce or phosphoresce in many colors by preparing the pure material with traces of an activator, such as copper or silver or by partial substitution of zinc by cadmium or sulfur by selenium. Specially prepared zinc oxide can be made to phosphoresce when activated by X-rays and ultraviolet rays. A copper activated zinc sulfide commonly makes the bright green color in cathode ray tubes and on radar screens, where the image remains for a second or more, providing the operator with a better visual of the overall scan.

PHOTOZINCOGRAPHY A process for reproducing pictures, etc., by using a zinc plate on which the design had been photographically reproduced.

PICK TONGS Tongs for handling hot metal.

PICKER [1] A small tool used to pull up the wick of a miner's lamp. [2] A laborer who removes high-grade ore, wood and/or iron from ore as it passes on a conveyor belt to the crusher.

PICK A pickax; a heavy iron or steel tool pointed at both ends and often curved, fitted with a wooden handle inserted in an eye between the ends. Used in various forms in mining and many other occupations.

PICKING (THE EYES) To extract over a prolonged period an undue portion of the richest ore, thus lowering the average grade of the remaining ore reserves.

PICKLE LIQUOR A spent acid pickling bath.

PICKLING Removing surface oxides from metals by chemical or electrochemical reaction.

PIEL An iron wedge for piercing stone.

PIG An oblong mass of metal that has

been run, while molten, into a mold.

PIGTAILER In Joplin, Missouri, the term was used to denote one employed by a mining company to assist trammers in long-distance haulage, where tramming was done by men. An assistant trammer.

PIKE A pick or pickax.

PILE A fortune. A miner who has made money has made his pile.

PILLAR A piece of ground or ore left to support the roof or hanging wall in a mine.

PINCH [1] Narrowing of a vein or deposit. [2] A crowbar with a short projection and a heel or fulcrum at the end; used to pry forward heavy objects; a pinch.

PINCH-OUT See *Pinch [1]*.

PINTO PROCESS A process to remove iron from continuous galvanizer-grade top and bottom dross by treatment in a Pinto furnace or reactor, a cylindrical vessel that can be agitated by a multi-bladed axial turbine. In the process the dross is melted in an induction furnace and transferred to the Pinto furnace. Excess aluminum is added beyond that needed to form a solid iron-aluminum intermetallic having a lower density than zinc and higher melting point (1,150 degrees C.). The crystallized intermetallic rises in the melt (kept at about 650 degrees C.) and is vigorously stirred to repel adhering zinc, followed by addition of air to oxidize its surfaces to make it unwettable by molten zinc. The refined zinc is poured into a holding furnace for alloying back to specifications.

PIPE DOG A hand tool that is used to rotate a pipe whose end is accessible.

PIPE METAL An alloy of tin and lead, sometimes with zinc, for making organ pipes.

PIPE An elongate body of mineral; a narrow portion of rich ore extending down the lode.

PIPE-SHAFT SYSTEM A method of mining introduced in 1890 to mine the calamine deposits at Bertha, VA., and named from the form and diminutive appearance of the small circular shafts used. The shafts, which were constructed of numerous steel skinned, circular sections connected together, had an average diameter of 3.5 feet, which was said to bear a close relationship to the horizontal extent of a miner's spinal column when in a position for digging in one of these shafts. These shafts were sunk as deep as 125 feet and were located so they would penetrate the principal basin and zones between the limestone chimneys where the bulk of the ore was located. Ore was hoisted up the shaft in an iron bucket capable of holding 1400 pounds of ore. When the shaft was no longer needed the wood linings and steel plates were removed to be used elsewhere.

PIT SYSTEM A method of preparing the charge for retort smelting. In the process, roasted blende or calcined carbonate ore and reducing fuel, separately crushed and coarsely sized (2 to 6mm), were spread in one or two pits in superposed layers. Water was added in all cases to assist intimate mixing, counteract dusting and help packing in the retort. The pit mixture was either shoveled directly into a car for retort loading or moved by conveyor to improve mixing. Also known as the *Lay-Down System*.

PITCHER One who picks over dumps for pieces of ore.

PITCHING BAR A pick used by miners in beginning a drill hole.

PITCHING FERRULES Short lengths of galvanized steel pipe set into a reinforced concrete mass and used for handling it.

PITCH Inclination of an orebody or vein from the horizontal; the rake or dip.

PLANCHET Metal disk with edges milled ready for coining.

PLASMADUST PROCESS Patented process in which zinc, lead and other metals are recovered from waste materials such as EAF dusts. In the process dust, powdered coal, and superheated gas are injected into cavities formed in a coke-filled shaft furnace. The superheated gas is produced by a plasma generator. The zinc and lead compounds are reduced, the metals are vaporized and collected in a condenser outside the furnace.

PLASTIC AND SEMIPLASTIC EXPLOSIVES Explosives that can be shaped by moderate pressure; some have viscosities that make it possible to produce cartridges by extrusion.

PLASTIC BRONZE A zinc-containing, high-lead bronze used for bearings.

PLASTICITY The property possessed by clay of forming a plastic mass when mixed with water.

PLATE-AND-FRAME FILTER PRESS Filter press consisting of plates with a gridiron surface alternating with hollow frames, all of which are held by means of lugs on the press framework. The corners of both frames and plates are cored to make continuous passages for pulp and solution. Filter cloth is placed over the plates to effect separation of solids and liquid. Well known types were the Dehne and the Merrill.

PLATE A flat-rolled metal product of some minimum thickness and width arbitrarily dependent on the type of metal.

PLATING Art or process of covering anything with plates or with a coating of metal.

PLENUM Mode of ventilating a mine or heading by forcing fresh air into it.

PNEUMATIC CONCENTRATOR A gravity jig, shaking table, and other device in which suitably ground minerals are separated by gravity during their exposure to a continuous or pulsating current of air.

PNEUMATIC FLOTATION CELL Machine in which the air used to generate a mineral froth is forced into the cell at or near the cell bottom.

PNEUMATIC JIG A jigging machine in which an air blast performs the work of separation of minerals.

POCKET A small body of ore; an enlargement of a lode or vein; an irregular cavity containing ore.

POINT OF THE HORSE Point where a lode splits or divides into two parts.

POLARIZATION In electrolysis, the formation of a film on an electrode such that the potential necessary to get a desired reaction is increased beyond the reversible electrode potential.

POLING; POLLING [1] Poles are used instead of planks for lagging. [2] The process of protecting the face of a level, drift, cut, etc., by driving poles or planks along the sides or back of yet unbroken ground.

POLYGONAL METHOD An ore reserve computation method in which the assumption is made that the area of influence of each drill hole extends halfway to the neighboring drill holes.

POLYMORPH A substance that crystallizes in several distinct forms; also any one of these forms. Sphalerite and wurtzite are polymorphs of zinc sulfide.

PONY SET A small timber set or frame incorporated in the main sets of a haulage level to accommodate an ore chute or other equipment from above or below.

POP A BOULDER To place and explode a stick of dynamite on a boulder so as to break it for easy removal from the mine.

POP SHOT In mining, a shot fired for trimming purposes or one by which a boulder in a mine is broken up by drilling a hole and blasting or by placing a stick of dynamite on top of it and exploding it.

POPPET; PUPPET Pulley frame or the headgear over a shaft. A headframe.

PORTAL TO PORTAL Term used to measure the time a miner or shift is underground. At many mines it defines the hours a miner works; that is, pay begins on entering the portal and ends when leaving the portal; hence the term portal-to portal pay or 8-hour shift portal to portal.

PORTAL [1] Surface entrance to a drift, tunnel, adit, or entry. [2] Concrete or masonry arch, retaining wall, etc., erected at the opening of a drift, adit, or tunnel.

POSITIVE ORE Ore exposed and sampled on four sides in reasonable blocks or size, having in view the nature of the deposit as regards uniformity and value per ton. Essentially the same as proved or measured ore reserves.

POSSESSIO PEDIS The actual possession of a mining claim by the first arrival.

POSSIBLE ORE Ore which may exist below the lowest workings or beyond the range of actual vision. Essentially the same as *inferred* reserves (resources).

POST JACK A jack for pulling posts. See *Post Puller*.

POST PULLER A lever and chain device for safely removing and recovering posts from worked out portions of a mine.

POST A mine timber. Commonly used in metal mines instead of leg, which is the coal miner's term.

POTATO STONE A potato-shaped geode of quartz having a central cavity lined with crystals.

POTTER-DELPRAT PROCESS The original Potter process (1902) was one of flotation in a 1 to 10 per cent acid solution. The mixture was 1 : 1 of ore and acid solution; this was agitated freely and heat applied, with the generation of carbon dioxide from the carbonates present. This caused the sulfides to rise where they were allowed to flow off or were skimmed off. This was clearly a surface tension process. Delprat, also in 1902, accomplished the same thing with acid salt-cake solution. Both processes were tried out at Broken Hill, Australia. Later patents indicated that oil assisted these processes. These inventors worked independently, became involved in litigation, and eventually pooled their interests.

POTTERY UNIT The group that

makes new retorts or place where retorts are manufactured.

POUNDER An ore-stamp mill.

POUR Used in founding: [1] the amount of material, as melted metal, poured at a time; [2] the act, process, or operation of pouring melted metal; as make a pour at noon.

POURING-GATE A channel in a mold, through which to pour molten metal.

POWDER BOX A wooden box in a miner's chamber in which were kept black powder, cartridge paper, cartridge stick, squibs, lampwick, chalk and tools.

POWDER MAN The person in charge of explosives in an operation. A *Powder Monkey*.

POWDER MONKEY A person who distributes powder, dynamite and fuse to the miners at the working face.

POWDER SIZE In powder metallurgy, material ranging from 0.1 to 1000 microns.

POWDER Miner's term for explosives.

PRECIPITATE [1] To cause a solid to form from a solution by chemical means, by lowering the temperature, or by evaporation. [2] Ejection of a solid substance from a solution as a result of insolubility that has developed, usually from changes in chemical or physical conditions. [3] Solid product that was precipitated.

PREFERENTIAL FLOTATION Name applied to a special type of differential flotation in which a mixture of two flotatable minerals is given a slight roast in order that one will be slightly oxidized, and therefore not float, and the other remain unchanged.

PREMIUM PRICE PLAN U.S. Government program initiated in WWII to maintain and expand production of copper, lead, and zinc and to maintain price stability in these metals. During the summer of 1941, it became evident that zinc and copper production could not be brought to the desired levels at the prices agreed to earlier between the Government and leading producers, chiefly because of increasing labor and material costs. In February 1942 a subsidy system, the Premium Price Plan, began whose primary purpose was to obtain production of critically needed non-ferrous metals by compensating for abnormal costs mainly in the production of marginal ores. Premium payments were made on production above specific quotas based on the estimated maximum production which could be obtained at ceiling prices at any given time. The Premium Price Plan was ended on June 30, 1947.

PRESSED DENSITY In powder metallurgy, the density of the unsintered compact. Also called green density.

PRESSURE CASTING [1] Making castings with pressure on the molten or plastic metal, as in injection molding, die casting, centrifugal casting, and cold chamber pressure casting. [2] Casting made with pressure applied to the molten metal or plastic.

PRESSURE FILTER A filter machine in which the liquid slurry to be filtered is forced through filtering material by a pressure greater than its own weight.

PRESSURE LEACHING A process to produce zinc sulfate in solution directly from zinc sulfide or bulk lead-zinc concentrates. Concentrate and spent electrolyte are fed into a pressure leaching autoclave, where the zinc is hot leached under oxygen pressure. The discharge slurry undergoes classification and flotation to separate out solid elemental sulfur from a slurry of zinc sulfate solution and lead jarosite residue. This slurry in piped to the calcine leach step in the main plant. An advantage of the sulfur separation is the non-necessity of an acid plant or the handling of liquid acid.

PRESSURE TANK A pressurized tank into which timber is inserted for impregnation with creosote, zinc chloride, or other preservatives.

PRIMARY CELL A cell or battery designed to deliver its rated capacity once and be discarded; not designed to be recharged.

PRIMARY METAL Metal extracted from ores, natural brines, or sea water. Also called virgin metal.

PRIMARY ORE Ore that has remained practically unchanged from the time of original formation.

PRIMARY ZINC Zinc metal obtained directly from ores by smelting. Also called virgin zinc.

PRIME WESTERN ZINC (PW) A specification grade of zinc composed of 98.0% zinc with a maximum of 1.4% lead, 0.5% iron, 0.2% cadmium, and 0.05% aluminum. PW was the principal product of retort smelters and was the most common grade until electrolytic production became prominent. PW is mainly used for galvanizing purposes.

PRIMER A dynamite cartridge, or package of any explosive, which contains the detonator, whether blasting-cap, or electric blasting-cap.

PRIMING HORN Miner's powder horn.

PRIMING POWDER A detonating fulminating powder for firing a charge.

PRIMING TUBE A tube containing fulminating powder for firing a charge. A detonator.

PROBABLE ORE Any blocked-out ore not certain enough as to limits, uniformity, and continuity. May include any undiscovered ore of which there is a strong possibility of existence. More or less equivalent to indicated reserves including in some cases inferred reserves.

PROCESS FLOWSHEET A basic diagram or plan showing the main operational steps within a plant, the movement of various materials between the steps, and the final products obtained, and often, also the quantities of materials with which the plant must be capable of dealing at various points.

PRODORITE A proprietary asphaltic product similar to naturally-occurring gilsonite used to line zinc electrolyzing cells.

PRODUCER GAS The gas obtained by the partial combustion of coal or coke in air. It consist mainly of carbon monoxide and nitrogen, with a small portion of hydrogen, methane, and carbon dioxide.

PRODUCTION The yield or output of a mine, metallurgical plant, or mill.

PROLONG Generally a simple cone or canister of sheet iron 2 to 3 feet long which is slipped over the nose of retort condenser mainly to collect the blue powder that passes off from the retort. In American practice, prolongs were not generally used because it hindered the furnace man from observing the working of the charge and because the metal recovered in the prolong was not considered sufficient to cover the time or cost involved.

PROMOTER [1] Person who alone or with others arranges or takes the preliminary steps in a scheme or undertaking for the organization of a company, the floating of bonds, stock, etc., or the carrying out of any business project. A mine promoter. [2] A reagent used in the flotation process; usually called the collector.

PROOF SPIRIT Old standard of alcoholic content of liquid, based on the strength of the alcohol to water mixture which, when poured on gunpowder, just permitted ignition. Now 57.1% ethanol. Therefore, 70 proof is 70 x 57.1 / 100 or 40% alcohol.

PROP Timber set upright or at right angles to the dip (a raking prop or a raker).

PROSPECT [1] Name given to any mine workings the value of which has not been determined. [2] To examine land for the possible occurrence of valuable minerals by sampling, drilling, etc.

PROSPECTING Searching for new ore deposits; preliminary exploration.

PROSPECTOR A person engaged in exploring for valuable minerals.

PROTORE [1] Low-grade material which by natural processes of enrichment was convertible into ore; as, for example, the formation of high-grade oxidized zinc deposits from low-grade primary zinc sulfides. [2] Any primary material to low in tenor to be ore but from which ore could be formed through enrichment.

PROVE UP To show that the mining or ore deposit requirements for receiving a patent for government land has been fulfilled.

PROVED ORE Ore where there is practically no risk of failure of continuity. See positive ore; same as measured reserves.

PRUDENT-MAN (PERSON) TEST The most durable and famous test of a mineral discovery on Federal lands. The basis of the test was first laid down in 1894, in which the Secretary of the Interior stated:

> "... where minerals have been found and the evidence is of such a character that a person of ordinary prudence would be justified in the further expenditure of his labor and means, with a reasonable prospect of success, in developing a valuable mine, the requirements of the statutes have been met."

PUDDLER The system of small pipes admitting compressed air to a tank of water and zinc chloride to affect a thorough solution for use as a timber preservative.

PUFFER BOY A person employed to operate an engine for hauling loaded mine cars through haulageways. Also an operator of any small stationary hoisting engine.

PUG MILL A mill for kneading or mixing clay.

PUG [1] Pug mill. [2] To mix and stir clay when wet. [3] To fill or stop with clay by tamping.

PUGGING The mixing and working clay for bricks, retorts, etc.

PULLEY FRAME A gallows frame or head frame.

PULP DENSITY In flotation, the

amount of solids in a pulp, typically ranging from 10 to 25%, by weight.

PULP DILUTION The ratio of water to solids by weight. It is expressed as a ratio; for example, a pulp dilution of 3 to 1 means that a pulp contains 3 tons of water for each ton of solids.

PULP Pulverized ore mixed with water. Also applied to dry crushed ore.

PULVERIZATION The reduction of a material to powder by mechanical means. Synonymous with comminution.

PULVERIZE To reduce or be reduced to a fine powder or dust by crushing, grinding, or the like.

PUNCH The member of a tool that forces the metal into the die during blanking, coining, drawing, embossing, forging, powder molding or similar operations.

PURIFICATION In the electrolytic zinc process, impurities in the neutral leach have to be removed to extremely low levels for successful electrolytic zinc production. Early attempts to produce zinc electrolytically failed because of high impurity levels in the solution. Arsenic, antimony, germanium, nickel, iron, copper, cobalt and selenium lower the hydrogen overvoltage, reduce current efficiency, and cause poor plating. Impurities are commonly reduced or cemented out of solution in stages by additions of zinc dust aided by reagents such as arsenic, antimony and copper. The solid residues resulting from purification, known as filter cakes, are generally treated to extract zinc and cadmium, and if of sufficient content, are shipped for recovery of copper, cobalt and nickel.

PW Prime Western zinc metal.

PYRITE A hard, shiny, yellow, iron sulfide mineral, common in most base-metal deposits, including zinc deposits. Also called iron pyrites and fools gold.

PYROMETALLURGY Metallurgy involved in winning and refining metals where heat is used, as in roasting and smelting.

PYROPHORIC SPHALERITE A variety of sphalerite that gives off sparks or glows when abraded. Some is so sensitive that the effect is obtained by scratching with a fingernail.

Q

QUALITATIVE TEST A test that determines the nature or presence of the constituents in a compound or material, not necessarily their amount.

QUANTITATIVE TEST A test that determines the amounts of the constituents present in a compound or mixture.

QUILL A slow burning fuse made formerly of the quill of a feather filled with powder.

R

RABBIT EAR A recess in the corner of a die to allow for wrinkling or folding of the blank.

RABBLE [1] An iron scraper serving for a rake in removing scoria from the surface of melted metal in a reverberatory furnace. [2] A mechanical stirrer used to stir the ore charge in a roasting furnace.

RABBLING Stirring molten metal, ore or other charge, using a hoe-like tool, arms, rakes or other devices.

RACKING TABLE A table on which to wash ore slimes.

RACKING The process of separating ores by washing on an inclined plane.

RAIL HAULAGE SYSTEM A materials transportation system consisting of steel rails on which cars are moved about by powered traction units or locomotives.

RAIL HAULAGE Transport of ore, waste, or mill products in rail cars on steel rails by locomotives or powered traction units.

RAILROAD WHITE A paint developed for U.S. railroad use around 1880, containing equal volumes of zinc oxide and barites. When colored, it was known as railroad colors.

RAISE A mine shaft driven from below upward. An opening, like a shaft made in the back of a level to reach the level above.

RAKE CLASSIFIER A type of mechanical classifier utilizing reciprocal rakes on an inclined plane to separate coarse from fine material contained in a water pulp, overflowing the fines and discharging the coarse by means of the inclined raking system.

RAKE See *Pitch*.

RAKING PROP An inclined prop in timbering.

RAM Moving member to which the forming punch is fastened.

RANDOM WALK A system for exploration drilling of large tracts of land that possibly contain hidden ore deposits, which for the most part, are only suggested by the geology of the region. The premise for this type of exploration is that a series of widely-spaced drill holes on the order of miles apart and random into the favorable rock horizon, will find ore. This technique was used to some extent in exploring for uranium on the Colorado Plateau, but rarely for base metals. The most famous random walk exploration program in search of zinc was that of The New Jersey Zinc Co. in middle Tennessee which resulted in the finding of the Elmwood zinc deposit in 1967. On the 79th hole drilled some 38 miles from their initial hole, they hit the Elmwood orebody at 1384 feet. Successful random walk exploration requires a certain amount of luck. One year later, New Jersey Zinc drilled more holes at the Elmwood site, the first of which was 100 feet north of hole 79 to obtain better core recovery. This hole hit only weak mineralization, and interestingly, probably would not have attracted enough special attention to warrant further drilling in the area if it had been drilled first.

RANGE POLE A long wooden staff, usually painted alternately red and white at one-foot intervals, used by surveyors for long sights.

RANGE [1] One of the north-south rows of a township in a U. S. public-land survey that are numbered east and west from the principal meridian of the survey. [2] Orderly arrangement of diamond drill fittings, such as casing, core barrels, drill rod, etc., with diameters appropriately related to each other and intended to be used together. Ranges commonly are designated by letter names, using letters such as E, A, B, and N individually or as the first letter in two and three letter names. [3] Difference between the greatest and least of a number of results. [4] Term used in Wisconsin and Utah for more or less horizontal runs of zinc ores that extend for long distances.

RAPPER A lever or hammer at the top of a shaft or inclined plane for signals from the bottom.

RAW ORE Ore that is not roasted or calcined.

The U.S. Zinc Industry

REAGENT A chemical or solution used to produce a desired chemical reaction; a substance used in assaying or in flotation.

REAMER A tool for enlarging the diameter of a borehole.

RECESSION A mild decrease in economic activity marked by a decline in real GDP, employment, and trade.

RECIPROCATING Having a straight back-and-forth or up-and-down motion.

RECLAMATION The process of restoring mined land to its former or other productive uses.

RECORD TABLE A heavy duty shaking table used to treat relatively coarse sands. Shaking is by double-link eccentric motion, with longer and slower throw than with a Wilfley type of table.

RECOVERABLE GRADE Mill-head grade less metallurgical losses.

RECOVERY [1] The proportion or percentage of ore mined from the original deposit. [2] General term to designate the percentage of a valuable metal in an ore, calcine, matte, slag, etc., extracted by a mill, smelter, refinery, or process, or an entire processing sequence, i.e. from mine to refined metal. For example, the recovery of zinc at the concentrator is 85%, indicating that of the zinc in ore fed to the mill, 85% was captured in concentrates, whereas 15% was lost to tailings. [3] Proportion of the desired component obtained from ore in processing, usually expressed as a percentage.

RECRYSTALLINES Early term used to describe crystallized dolomite formed by the process of dolomitization. Sometimes used in describing dolomites associated with zinc mineralization in the Tennessee and Virginia zinc districts.

RECTIFICATION The process by which electric current is transferred from an alternating-current to a direct-current circuit.

RECUPERATION The process of recovering sensible heat from hot gases from a furnace and using it to dry or heat incoming charge or fuel gases.

RECUPERATIVE FURNACE A furnace that permits the recuperation or recovery of heat from the waste gases of combustion, typically used for preheating furnace gases, drying roaster feed, etc.

RED BRASS A copper-zinc alloy containing less than 15% zinc.

RED OXIDE OF ZINC See *Zincite*.

RED SEAL See *Seals*.

RED ZINC ORE Synonym for *Zincite*.

REDISTILLATION A process used to upgrade a spelter to high purity by utilizing the difference in boiling points of zinc and the impurities. In early practice, retort furnaces were used for redistillation of spelter. This process became especially important during WWI, owing to the demand for fine zinc for ordnance brasses. In the 1930's, The New Jersey Zinc Co. developed a continuous vertical redistillation process capable of refining zinc to Special High Grade specifications. This process involved the use of two vertical distillation columns, one to separate zinc from lead, iron and other high boiling-point impurities and the second to separate cadmium from zinc. In the process, molten metal is fed into the first column and heated to drive off zinc and cadmium vapor, which is condensed and fed into the second column. This melt is again heated to boil off the cadmium, leaving high purity zinc.

REDUCING AGENT A substance that causes reduction.

REDUCING FLAME A gas flame produced with excess fuel.

REDUCING ROAST Reduction of certain metallic oxides by heating in contact with coal or other reducing agents.

REDUCTION WORKS A works for reducing metals from ores, as a smelting plant.

REDUCTION [1] In cupping and deep drawing, a measure of the percentage decrease from blank diameter to cup diameter, or of diameter reduction in redraws. [2] In forging, rolling and drawing, either the ratio of the original to the final cross-sectional area. [3] Reaction in which there is a decrease in valence resulting from a gain in electrons. Contrast with oxidation.

REEDING The operation of forming serrations and corrugations by coining or embossing.

REEF See *Bioherm*.

REFINE To free from impurities; to free from dross or alloy; to purify as metals.

REFINING The purification of crude or impure metallic products. Zinc can be refined by distillation, electrolytic processes, and liquidation. Zinc compounds can be purified by a number of chemical processes.

REFRACTORY ORE Ore difficult to treat.

REFRACTORY [1] A material of very high melting point with properties that make it suitable for uses as furnace linings and kiln construction. [2] Quality of resisting heat

REFUGE HOLE A small opening or place made in an underground haulageway in which a miner can take refuge during the passing of a train or when shots are fired.

REGENERATION In mineral leaching, reconstitution of barren leach solution after it has completed its chemical attack on mineral and its leached values have been removed.

REGRIND CIRCUIT A circuit employing pumps and grinding and classifying equipment to reprocess mill and flotation tailings and low-grade products to increase the recovery or to upgrade the product.

REJECT [1] Material extracted from the feed during cleaning for retreatment or discard. Includes middlings, residues, tailings, etc. [2] Minerals in ore removed and discarded at any stage of treatment.

RELEASE MESH The grain size or mesh required to free a specific mineral from associated but different minerals in an ore. The complete release of a specific mineral may or may not be the optimum grind economically owing to factors related to costs, sliming, affect on the recovery of other ore minerals, flotation characteristics, materials handling, etc.

RELIEF HOLES Bore holes loaded and fired for the purpose of relieving or removing part of the burden of the charges to be fired in the main blast.

REMELT ZINC Melted metallic zinc

scrap too contaminated to be sold for slab zinc or for making end products. Serves as raw material feed for primary and secondary smelters.

REMILLING PLANT A centralized mill set up in a mining district or region to treat and upgrade crude sulfide gravity concentrates produced at nearby mining operations.

RENN-WALZ PROCESS A smelting process for reclaiming iron and other metals from the waste materials produced in the smelting of zinc and lead ores.

REPLACEMENT ORE BODY; REPLACEMENT VEIN A mass of ore formed by the dissolution of previous minerals and their replacement by others.

RESERVE BASE That part of an identified resource that meets specified minimum physical and chemical criteria related to current mining and production practices, including those for grade, quality, thickness, and depth. The reserve base is the in-place demonstrated (measured and indicated) resource from which reserves are estimated. It also includes those resources that are currently marginally economic (marginal reserves) and some of those that are currently sub-economic (sub-economic resources).

RESERVE That which is held back or kept in stock for future use.

RESIDUAL ORE DEPOSIT An in-place accumulation or concentration of valuable minerals, formed by removal of other constituents in the rock or lode by weathering or leaching.

RESIDUE [1] Solid matter remaining after a liquid has been filtered or evaporated. [2] The waste or final product from a hydrometallurgical plant which, at the time of operation, may or may not be valuable with further processing. e. g., The residue resulting from the acid leaching of roasted zinc concentrate (calcine) in the roast-leach-electrowin process often contains economic quantities of lead and silver which can be recovered. [3] Discharged materials from retorts, traveling grates, and Waelz kilns after extraction of zinc.

RESIN JACK See *Rosin Jack*.

RESIN TIFF Light colored sphalerite.

RESOURCE A concentration of naturally occurring solid, liquid, or gaseous material in or on the Earth's crust in such a form and amount that economic extraction of a commodity from the concentration is currently or potentially feasible.

RESUE To mine or strip sufficient barren rock to expose a narrow but rich vein, which is then extracted without dilution.

RETORT A vessel used for the distillation of volatile material, as in the separation of certain metals and in the destructive distillation of coal.

RETREAT To treat over again. Said of tailings from ore dressing plants.

RETURN ACID Sulfuric acid regenerated during the electrolyzing step.

RETURNING CHARGE A charge made per unit of ore or concentrate treated by a smelter in custom smelting based mainly on the smelter schedule.

RETURNS A term used for the water and sludge that overflows the collar of a borehole.

REVERSE BRASS An alloy widely used before the development of die cast alloys. It is composed of about 90%

zinc, 7% copper, and small amounts of lead and iron. Also known as *Fontainemoreau bronze*.

REVERSED LOADER A front-end loader mounted on a wheel tractor having the driving wheels in front and the steering at the rear.

RHENISH FURNACE A zinc distillation furnace that is mainly a modified type of Belgian furnace but which had some features of the Silesian furnace. It differed from the Belgian furnace in that it had three tiers of oval retorts instead of the usual six in the latter. The furnace was used largely in Europe superseding Silesian-type furnaces, although two plants in the United States were built using Rhenish furnaces.

RID UP To clean out rubbish or waste from a mine, metallurgical plant, etc.

RIDING Said of mine timbering when the sets are thrust out of line or leaning.

RIFFLE [1] A device used to reduce the volume or weight of a sample. It consists of a thin metal plate on which is mounted a series of metal strips to guide or deflect a small portion of the sample into a separate container. Sample splitter. [2] Groove or indentation set in the bottom of a trough for arresting heavy mineral in sands and gravels.

RIGG PROCESS A sintering process, developed by G. Rigg at Port Pirie, Australia in 1917. It was the first commercial zinc sintering process. In the process ore roasted down to 9% sulfur is mixed with green ore and is fed to a Dwight-Lloyd sintering machine on which it is ignited to effect almost complete desulfurization and at the same time, fritting the ore into a cake.

RILL STOPING A method of stoping, such as overhand, inclined, or pyramidal, in which the miners can rise on an inverted pyramidal heap of broken ore, its apex in a winze or mill hole through which ore gravitates to the tramming level below.

RING CRUSHER A hammer mill with a high-speed horizontal shaft upon which a series of steel rings are swung.

RING The place where trading sessions at the LME are is held 4 times daily for 5 minutes each per metal. The ring is an open exchange session whereby all prices are shouted or cried out.

RINGER A crowbar.

RIPARIAN RIGHTS The rights of a person owning land containing or bordering on a watercourse or a body of water. Such rights may include reasonable use of the water, preventing the diversion or misuse of the water upstream, and ownership of the stream bed.

RISE To dig or work upward in mining, in opposition to sink. A raise.

RISER The reservoir of molten metal connected to the casting to provide additional metal to the casting, required as the result of shrinkage before and during solidification.

ROAD METAL A rock suitable for surfacing macadamized roads and for roadbeds.

ROASTER A reverberatory or a muffle used in roasting ore.

ROASTING Heating an ore to effect some chemical change that will facilitate smelting.

ROAST-LEACH-ELECTROWIN PROCESS A common term used for

the typical electrolytic-zinc process, in which the major elements of the process are expressed.

ROB To extract pillars previously left for support; or, in general, to take out ore from a mine for immediate gain rather than for longer-term production.

ROCK BOLT; ROOF BOLT A bar, usually steel, that is inserted into a pre-drilled hole and secured for the purposes of ground control or mounts for electric cables, fan line, etc. Rock bolts are classified according to the means by which they are secured or anchored in rock. There are four main types: expansion, wedge, grouted and explosive.

ROCK BURST The sudden yielding, sometimes with explosive violence, of the rocks in pillars, at the bottoms of shafts, and in mine workings, due largely to the release of accumulated strain energy.

ROCK CORE A cylindrical column of rock cut out by a rotary core drill.

ROCK DRILL A machine for boring rock either by percussion, affected by reciprocating motion, or abrasion, affected by rotary motion. Compressed air is the usual motive power, but steam and electricity are also used.

ROCK FILLING Waste rock, used to fill up worked-out stopes to support the roof. See sand filling.

ROCK OXIDE Furnace accretions of zinc oxide.

ROCK PRESSURE BURST See *Rock Burst*. Also called a rock bump.

ROCKER SHOVEL LOADER A mechanical mucking machine. Digging and loading machine consisting of a bucket attached to a pair of semicircular runners which when rolled, lifts and dumps the bucket load into a car behind the machine.

ROD MILL [1] A mill for rolling rod. [2] A mill for fine grinding somewhat similar to a ball mill but employing long steel rods instead of balls to affect the grinding.

RODMAN One who uses or carries a surveyor's leveling rod.

ROLLED ZINC Zinc in the form of sheet, strip, plate, rod and wire.

ROLLING MILL An establishment in which metal is made into sheets, rods and rails by working it between pairs of rolls.

ROLLING The act of reducing the cross-sectional area of metal stock, or otherwise shaping metal products, through the use of rotating rolls.

ROLL-THEIR-OWN Term used by paint grinders in the mid-1800's for the mixing of dry pigments or the colors in oil and paste white into paint on the job.

ROOF BOLTING A system of roof support in mines, in which a pattern of rock bolts, their ends protruding, are affixed with end plates or tied together to support the roof

ROOM AND PILLAR A system of mining best suited to extensive, relatively flat ore bodies or bedded deposits. The ore ideally is mined in parallel rooms separated by narrow ribs or by pillars. The irregular nature and thicknesses of zinc orebodies generally requires that modifications or combined mining methods be used in mining flat, tabular zinc deposits.

ROPE General term for steel cable used in hoisting and winding.

ROPP FURNACE A long, reverberatory furnace over the hearth of which a series of rakes or plows is drawn by a continuous cable, moving the ore steadily from the feed to the discharge end. Difficulties in handling flotation zinc concentrates beginning in the mid 1920's, and non-capture of sulfur in the roasting process led to rapid replacement of Ropp and also Hegeler-type roasters by more efficient McDougall and Wedge roasters at many U.S. smelters.

ROSIN BLENDE Iron-free sphalerite. The term was used in the early days of zinc mining in Colorado and New Mexico to distinguish iron-free from ferruginous varieties of sphalerite and marmatite.

ROSIN JACK A yellow variety of sphalerite.

ROSIN ZINC Sphalerite with a rosiny appearance.

ROTARY CONVERTERS An early term for machines used in electrolytic zinc plants for the conversion of alternating current to direct current.

ROTARY DRILL Broadly, various types of drill machines that rotate a rigid, tublar string of rods to which is attached a bit or cutting rock to produce boreholes.

ROTARY SHEARS A sheet-metal cutting machine with two rotating-disk cutters mounted on parallel shafts driven in unison.

ROUGHER CELLS Flotation cells in which the bulk of the gangue is removed from the ore. Usually the first bank of cells to receive the crushed and ground ore.

ROUGHER CONCENTRATE The concentrate from the first group of cells treating lower-grade ores, which is seldom rich enough for a final product. Retreated rougher concentrate yields a cleaner concentrate. In some cases the rougher tailings may be reground and refloated forming a low grade scavenger concentrate, which is returned to the first rougher cell.

ROUGHING The upgrading of run-of-mill feed either to produce a low-grade concentrate or to reject valueless tailings at an early stage. Preformed by gravity on roughing tables or in flotation in a rougher circuit.

ROYALTY [1] Payment to a lessor by a lessee of a mineral lease for a share of the production of minerals from the leased property; it is generally a percentage of the value of the minerals mined. [2] The amount paid by a lessee or operator to the owner of the land, the mineral rights, or the mine equipment, based on a certain amount or percentage per year, per ton, per total production, etc. The term had its origins in medieval times when a tax was paid to the monarch. Also see advance royalty.

ROYALTY INTEREST The right to receive a cost-free share of any mineral actually produced or a right to a share of the production.

RUBBING ROPES Special guide ropes to prevent possible collision between cages or skips at the passing point.

RUBY ZINC A popular name for transparent sphalerite of a deep red color, and also for zincite with the same characteristics.

RUN [1] Term employed in the Central United States for a lead-zinc deposit that follows a certain line in a plane of stratification parallel to a joint or fault system. [2] Length of time a reduction works or facility is kept in operation without stopping to clean up, make repairs, or for other purposes. Same as campaign.

RUNNER BOX A distribution box that divides the molten metal into several streams before it enters the mold cavity.

RUNNER [1] A channel through which molten metal flows from one receptacle to another. [2] Portion of the gate that connects the downgate, sprue or riser with the casting.

RUN-OF-MINE Raw ore as it comes from the mine.

RUNOUT [1] An unintentional escape of molten metal from a mold, crucible or furnace. [2] A defect in a casting caused by the escape of metal from the mold.

RUNS Term employed in the central United States for lead-zinc deposits (in Paleozoic limestone and dolomite) following a certain line in the plane of stratification parallel a joint or fault system.

RUST A corrosion product consisting of hydrated oxides of iron. A zinc corrosion product is known as *white rust*.

S.E.G. Society of Exploration Geologists.

SACRIFICIAL ANODES Anodes used for cathodic protection against corrosion of metallic structures buried in soils, in contact with sea water, and other electrolytes. Fundamentally, cathodic protection consists of establishing an electromotive force on a structure to make it cathodic. Zinc anodes in an electrolyte, such as sea water, provide current of sufficient magnitude to oppose the galvanic currents on the structure to prevent the loss of metal on the structure. In doing so, the anodes are consumed or sacrificed to protect the structure. Cathodic protection can be traced back to 1824, when Sir Humphry Davy reported his successful use of zinc protectors to prevent corrosion of copper sheeting on ships hulls exposed to se water.

SACRIFICIAL PROTECTION The act of reducing the extent of corrosion of metal in an electrolyte by coupling it to another metal that is electrochemically more active in the environment.

SAFETY FUSE; BLASTING FUSE A train of powder enclosed in cotton, jute yarn, and waterproofing compounds; used for firing cap which sets off the explosive charge. The fuse burns at rate of 2 feet per minute.

SAL SKIMMINGS Skimmings from the surfaces of galvanizing baths when fluxes are used. This material often contains chlorine in excess of 4 percent.

SALT Sodium chloride. Sometimes added to the charge in horizontal retort smelting to dissolve the oxide film occurring on freshly condensed zinc droplets so they will coalesce into pool of liquid zinc rather than as blue powder.

SAMPLE A representative fraction of body of material removed by approved methods. In mining and metallurgy, bulk samples of ore and products of ore treatment are taken for the purpose of developing and testing suitable processes. Smaller samples: channel samples, grab, cores, chips, panning, etc., are primarily made to establish the character and value of the ore.

SAMPLER [1] A mechanical device for selecting certain fractional part of ore to be used as an assay sample. [2] One whose duty it is to select samples for assay or to prepare samples by grinding, splitting, etc., for assay or analysis.

SANDER'S PROCESS A flotation process used in 1907 to separate sphalerite and fluorite middlings in Kentucky. The middlings concentrate was placed in a bath of basic aluminum sulfate (alum) and heated to about 90 degrees C., rising bubbles of hydrogen sulfide (?) floated off the sphalerite. In 1908, the Tri-Bullion Smelting and Development Co. in Kelley, NM installed a 100 ton-per-day plant using the process to separate pyrite and sphalerite.

SANDMAN In metal mining, a laborer who switches the flow of sand in pipes and flumes from one mined-out stope to another so they are properly sand filled to provide ground support for nearby mining areas.

SANDS The coarser and heavier portions of the crushed ore in a mill.

SAPONIFICATION A process in which fatty substances form soap, by combination with an alkali. Used in the flotation process.

SAPROLITE A disintegrated, somewhat decomposed rock that lies in its original place and to some extent appears to be typical rock.

SAUCONITE [1] Artificial zinc hydroxide which has one amorphous and six polymorphic phases. [2] A naturally-occurring zinc smectite (clay) mineral found and named from its occurrence in the Saucon Valley near Allentown, PA.

SAVES Scrap code word for old zinc die cast. See *Apple*.

SCABS Scrap code word for new zinc die cast. See *Apple*.

SCAFFOLDING Encrustation on the inside of blast furnace.

SCALE DOWN Removing fragments of rock threatening to fall or break from the roof or walls. See *Bar Down*.

SCALE [1] A crust of metallic oxide formed by cooling hot metals in air. [2] To get rid of film of oxide formed on the surface of metal, by brush, abrasion, pickling, etc. [3] Rate of wages to be paid depending on various contingencies such as type of job. [4] The *Joplin ore scale* was related to the rates paid for lead and zinc concentrate based on grade and quality of the ore, relative to the prevailing price.

SCARFING Splicing timbers, so cut that when joined the resulting piece is not thicker at the joint than elsewhere.

SCAVENGER CELLS Secondary cells for the retreatment of rougher and cleaner cell tailings.

SCAVENGER CONCENTRATE Low-grade concentrate produced in retreating the tailings from rougher and cleaner flotation cells. Typically scavenger concentrates are fed back into the first rougher cells.

SCAVENGING In mineral processing, the final stage of froth flotation before discarding of tailings. The cells are so worked as to remove for retreatment as much low-grade rising mineral as possible under the given working conditions. The scavenger concentrate is typically fed back into the rougher cells.

SCHNEIDER FURNACE A distillation furnace for the reduction of zinc ores containing lead, with recovery of the latter metal as well as the zinc.

SCHOOP METALLIZING A zinc-coating process by which atomized zinc is sprayed upon cleaned or sandblasted iron or steel surface at high pressure to form galvanized surface. Zinc wire is fed through spraying apparatus in which it is atomized by oxygen-hydrogen blowpipe flame containing excess hydrogen so as to be reducing. Air, under pressure, in central tube atomizes the molten zinc and

The U.S. Zinc Industry

blows it at high velocity onto the object to be galvanized.

SCHORI PROCESS A method of metal spraying in which zinc dust is the feed material for applying protective coating of zinc on steel surface.

SCOBS The dross of metals.

SCONCE Protection, cover, shelter, or screen. A metal cover and holder combined for holding miner's candle, especially for hanging on wooden timbers.

SCOOT Scrap code word for zinc die cast automotive grilles. See *Apple*.

SCOPE Code word for newly-plated zinc die cast scrap. See *Apple*.

SCORE Code word for old zinc scrap. See *Apple*.

SCOURING Having the quality of eroding the furnace hearth, as some kinds of slags and cinders.

SCRAM To search for and extract ore in mine that is apparently worked out.

SCRAMMER One who scrams.

SCRAP Defectively made or discarded metallic material from whatever source that can be reclaimed through melting and refining.

SCREEN APE Used at Joplin, Missouri for one who attends the grizzly or screen. He breaks the large pieces of ore and picks out such waste rock as he can as it passes over the screen.

SCREENED MATERIAL Material which has been size separated by being passes through screens of various mesh size.

SCREEN [1] At Joplin, Missouri, the grizzly near the top of the head frame. [2] Scrap code word for new zinc clippings. See *Apple*.

SCREENING See *Screened Material*.

SCRIBE Code word for crushed clean fragmentizers die cast scrap, as produced by automobile fragmentizers. See *Apple*

SCRIP Credit slips or tickets issued by mining company to its employees before pay day in lieu of cash. The scrip drawn is charged against the pay of the employee, but is exchangeable for goods and food at the company or local store at its face value.

SCROLL Code word for unsorted zinc die cast scrap. See *Apple*

SCRUBBERS Special apparatus for cleaning waste gasses.

SCRUB Code word for hot dip galvanizer's slab zinc dross (batch process). See *Apple*

SCULL Scrap code word for zinc die cast slabs or pigs. See *Apple*

SCUM Impure or extraneous matter that rises or collects at the surface of liquids or as dross on bath of molten metal.

SEAL Scrap code word for continuous line galvanizing slab zinc top dross. See *Apple*.

SEALS A term for specified grades of French-process zinc oxide mainly as indicators of brightness and density for pigment-type uses. White Seal is the brightest and least dense; Green Seal is less bright; and Red Seal is the least bright and most dense.

SEAM [1] A ridge in casting, marking the place where the mold parted. [2] Scrap code word for continuous-line galvanizing bottom dross. See *Apple*.

SEASON CRACKING Cracking resulting from the combined effects of corrosion and internal stress. The term is usually applied to stress-corrosion of brass.

SECOND CLASS In the Joplin district, a term used to indicate relatively rich, small deposits, typically 100,000 tons or less, that could be exhausted in few years. Second class deposits largely were in the form of runs, lenses and irregular bodies as opposed to the large sheet-ground deposits. Until about 1901 nearly all of the Joplin district production was derived from deposits of the second class. Also see *first class*.

SECONDARY CELL [1] A cell which receives its electrical energy from charging operation, then stores the energy until it is required. The lead-acid battery is the most common type of secondary cell. [2] Group of flotation cells in which the product from the primary cells is retreated.

SECONDARY CRUSHER Crushing and pulverizing machines next in line after primary crushing to further reduce the particle size.

SECONDARY FERTILIZER COMPONENT Fertilizer components, other than primary, but essential for proper plant growth, including trace elements such as zinc, copper, boron, etc.

SECONDARY METAL Metals recovered from scrap by remelting and refining.

SECONDARY MINERAL Minerals resulting from the alteration of primary mineral. Thus sulfides by oxidation change to sulfates, oxides, and carbonates, and by hydration they become hydrous forms of the same. Many zinc deposits originated through surficial weathering and oxidation, giving rise to mineable near-surface orebodies of smithsonite, hemimorphite, hydrozincite and, rarely, descloizite. The oxidation process generally stops at the water table. For the most part, secondary zinc orebodies are underlain by primary (sulfide) zinc ores.

SECONDARY SHAFT A shaft that extends downward from the bottom of the primary shaft but offset from the primary shaft.

SECONDARY ZINC Slab zinc made by refinning zinc ashes, skimmings, drosses and old scrap.

SECONDARY ZINC OXIDE Zinc oxide produced by wet or chemical means from zinc solutions obtained from chemical processes or from the dissolving of galvanizers dross or fume from brass mills. The zinc is precipitated, generally as a hydroxide or carbonate, and calcined to make the oxide.

SECOND-CLASS ORE Any ore that must receive treatment before it's a marketable grade. Same as *Seconds*. Compare with *Second Class*.

SECONDS Ore that requires dressing. See *Firsts*.

SEDIMENTARY ROCKS Rocks formed by the accumulation of sediments transported by water (aqueous deposits) or wind (eolian deposits).

SEDIMENTATION In powder metallurgy, classification of powder particles by settling in gas or liquid.

SEISMIC PROSPECTING Method of geophysical prospecting based on the fact that the speeds of transmission of shock waves in rocks vary with the elastic constants and the densities of the rocks through which they pass. Explosive charges provide the shock waves and the travel times of selected waves are picked up by sensitive recorders. The data reveal information on the nature, structure, and extent of the underlying strata.

SELECTED ASTM specified metal composed of 98.75% zinc and maximum of 0.8% lead, 0.04% iron, and 0.75% cadmium, together not to total over 1.25%. Selected was used mainly to make the cheaper brasses such as rod suitable for machining to make screws, bolts, tire valves, etc. Dropped as specification metal by ASTM in 1967.

SELECTIVE FLOTATION Generally understood to refer to the surface or froth selection of valuable minerals rather than the gangue. Sometimes used to mean differential flotation.

SELECTIVE MINING [1] A method of mining whereby ore of unwarranted high value is mined in such manner as to make the low-grade ore left in the mine unprofitable. Frequently called robbing mine. [2] Process of mining whereby low-grade or sub-grade ore is mined to dilute high-grade ore so as to maintain uniform mill head grade for efficient mill recoveries.

SELF-ANNEALING Applied to metals such as lead, tin and zinc, which recrystallize at air temperature and in which little strain hardening is produced by cold working.

SELF-RESCUER Small gas-mask, filtering device carried by a miner underground, either on his belt or in his pack, to provide the miner immediate but temporary escape protection against carbon monoxide, other gases, or smoke in case of mine fire or explosion.

SEMIKILLED STEEL Steel that is incompletely deoxidized and contains sufficient dissolved oxygen to react with the carbon to form carbon monoxide to offset solidification shrinkage.

SEMIPLASTIC EXPLOSIVES See *Plastic Explosives*.

SENDZIMIR PROCESS Continuous in-line sheet galvanizing process in which the annealing or heat treating of cold-rolled strip and cleaning are carried out continuously in the same operation as the galvanizing process. This process was the first to attain commercial success as a continuous sheet galvanizing process. The first Sendzimir line was installed in 1936 at Armco's mill at Butler, PA.

SET OF TIMBERS The timbers which compose any framing, whether used in shaft, stope, level, raise, or drift. Also called *set* or *timber set*.

SETTLER A separator; tub, pan, vat, or tank in which mineral separation can be effected by settling.

SEVERENCE AND PRODUCTION TAXES A tax, often by the State in which the deposit occurs, based on the value of the mineral extracted or on each unit of mineral extracted.

SHAFT FURNACE A high furnace, charged at the top and tapped at the bottom.

SHAFT An excavation of limited area compared with its depth made for finding or mining ore. The term is often specifically applied to vertical shafts, as distinguished from an incline or incline shaft.

SHALE DRILLING Exploration term used in Tri-State by drillers when trying to locate the contact and contour of the Cherokee shale and the underlying Boone limestone owing to the fact that zinc-lead orebodies frequently are found adjacent to shale basins or slumps.

SHAVEN LATTEN Very thin sheet brass.

SHEET DEPOSIT A mineral deposit extending in length and breadth and having relatively small thickness, thus including both lodes and beds as distinguished from irregular masses. Sometimes applied in a more limited sense to deposits (blanket veins) occurring in an approximately horizontal plane.

SHEET Flat-rolled metal product of some maximum thickness and minimum width arbitrarily dependent on the type of metal. It is thinner than plate.

SHELF Scrap code word for prime zinc die cast dross. See *Apple*.

SHERARDIZE; SHERADIZING A galvanizing process well suited for zinc coating articles such as tubing, conduit, nuts, nails, and small castings. In the process, cleaned or pickled items and fine dust are placed in a drum that is sealed, rotated and heated to 700-800 degrees F. for several hours. Zinc sublimates and alloys with the items forming an adherent coating consisting of an outer zinc surface overlying an inner zone of zinc-iron alloys. The process was discovered in 1900 by an Englishman, Sherad Cowper-Coles.

SHG Special High Grade zinc metal.

SHINES A general term used by exploration drillers in the Tri-State district that indicated the approximate percentage of zinc sulfide in the drill cuttings. The following terms; thin zinc shines, fair zinc shines, zinc shines, and good zinc shines, represented zinc sulfide values of 0%, 0.5%, 2.0%, and 2.5%, respectively in some interval of the drill cuttings. These estimates and others of higher values by the drillers were generally accepted without assay and the logs were often used as principal guides in deposit development.

SHIPPING ORE Any ore of greater value when broken than the cost of freight and treatment. Direct shipping ore. Calamine ores were generally shipped directly to smelter, although those that underwent some treatment, such as washing, drying and even calcining to upgrade the zinc content and lower shipping weight were considered direct shipping ore by some.

SHOE [1] Piece of iron or steel attached to the bottom of a stamp for crushing ore. [2] Steel pieces fastened to the ends or sides of cages, which slide on guides when the cage is in motion. [3] Trough to convey ore to the crusher.

SHOOT [1] A shot in a single operation of blasting. [2] Elongated body of ore; an ore shoot.

SHORE-UP To stay, prop up, or support by braces.

SHORT FUSE Any fuse that is cut too short.

SHORT LEG One of the wires on an electric blasting cap, which has been shortened so that when placed in bore hole, the two splices or connections will not come opposite and short circuit.

SHOT DRILL A type of core drill employed in rotary-drilling boreholes of less than 3 inches to more than 6 feet in diameter in hard rock using chilled-steel shot as a cutting medium. The bit is an annular-shaped, flat faced, steel cylinder with diagonal slots cut in the bottom edge. As the bit and attached core barrel are rotated, shot is fed at intervals into the drill stem with water. The shot works its way under the flat face of the bit and wears away the rock. The core is removed similar to that as in diamond drilling.

SHOT A charge or blast.

SHOVEL LOADER A car loading machine for removing blasted rock at a drift face. A bucket hinged to the chassis, scoops up the material in front of the machine; the load is elevated over the machine and discharged into a trailing ore car. Another type has a conveyor built into the loader that conveys the material back into the cars.

SHOWER ROASTING See *suspension roasting*.

SHRINKAGE CAVITY A void left in the cast metal as result of solidification skrinkage.

SHRINKAGE STOPING Also known as *Back Stoping, shrinkage with waste fill, overhand stoping with shrinkage and delayed filling, and overhand stoping with shrinkage and no filling*. The method is a modification of overhand stoping and its main characteristic is the use of part of the ore for the purposes of support and as a working platform. As applied to small orebodies two modifications are used: stoping without ore passes (chutes) and stoping with ore passes (to remove excess ore). As applied to large orebodies the stopes are separated by pillars or ribs and the name used is *shrinkage stoping with alternate pillar and stope*.

SIDE PLATE In timbering, where both cap and sill are used and the posts act as spreaders, the cap and sill are spoken of as sideplates.

SIEVE FRACTION That portion of a material in which its particles fall within a certain size range, generally attained by sieving through a screen or series of screens.

SILESIAN FURNACE Rectangular combustion chamber containing about 20 large muffles for the distillation of zinc. Furnaces commonly were built in pairs with chambers between each for the calcination of the ore. The furnace was developed in Silesia in the early 1800's and accounted for most of the production in that region up to World War I. By the mid-1930's the furnace had largely passed out of use. In 1836 zinc distilling using hybrid furnaces using features of both the Silesian and Belgian furnaces began in western Germany. This furnace eventually became known as the *Rhenish furnace*, but by this time, the Rhenish furnace was more Belgian in character than Silesian.

SILICA RETORT A retort composed of plastic fire clay, grog, and 20 to 25 percent silica flour. The retort was developed in 1919 by the American Zinc Company to improve retort life which was decreased by plastic flow under the combination of temperature and unsupported span which were encountered in furnaces then. Silica retorts were used in most U.S. horizontal retort smelters thereafter.

SILICATE ORE Miner's term in Joplin district for deposits of smithsonite and hemimorphite.

SILICATE Used in the Joplin, Missouri district for zinc carbonate.

SILLS Strong timbers laid horizontally to support posts or other drift timbers.

SIMILOR Golden-colored variety of brass. Also called *Mannheim gold; Prince Rupert's metal*.

SINGLE JACK Light single-hand hammer used in manual drilling, especially in metal mines. The hammer is used in one hand while the drill is held by the other.

SINGLE LEACHING In the electrolytic process, two general types of calcine leaching processes are used, single or double. In single leaching, sufficient spent electrolyte is used with single batch of calcine to dissolve all of the soluble zinc in single step. Lime or limestone are added to neutralize the solution and cause precipitation of soluble impurities and aid filtration of the pulp, yielding a clear solution. The Bunker Hill smelter in

Idaho employed a continuous single leach system that involved five steam-heated tanks. Leaching was carried out sequentially in the first two tanks followed by neutralization in the other tanks.

SINGLE SHOT The firing of the charge in one drill hole only as contrasted with multiple shots where charges in number of holes are fired at the same time.

SINKER A miner who sinks mine shafts.

SINK-FLOAT PROCESSES Processes that separate particles in a liquid heavy media on the basis of specific gravity. In the zinc industry, the process is used to separate coarse ore-rich and ore-poor rock fragments in a liquid heavy media. The medium usually consists of a thickened water suspension containing either fine galena, magnetite, or ferrosilicon, resulting in a controlled density (2.6 to 3.0 generally) to affect good separation without undue loss of zinc and/or lead to tailings. Coarse crushed and sized material ranging from one-quarter to 1 inch is fed into the heavy-media tank; fragments lighter than the medium remain on top, whereas the heavy zinc-ladened fragments sink. Both heavy and light products are washed to remove adhering heavy media grains, which are later recovered by flotation (galena) or by magnetic separator (magnetite and ferrosilicon). The sink product and the original crusher run undersize are further processed to concentrate the zinc and/or lead. The float product is generally used for mine fill or sold for construction material.

SINKHOLE A vertical hole in limestone rock, caused largely by solution of the rock along joints and fractures. Caving of overlying rock is common; the resulting breccias often are the sites for commercial lead/zinc deposits.

SINKING FUND Money reserved for amortization of wasting assets.

SINKING The process of excavating shafts.

SINTER A mass of fine particles agglomerated by incipient melting resulting from heating for a prolonged time just below the melting point.

SINTERING The bonding of adjacent surfaces of particles in a mass (metal, ore, concentrate or other materials) with or without fluxes or other components by heating.

SINTER PLANT Where sintering is carried out.

SINTER-MACHINE OPERATOR In ore dressing, smelting, and refining, one who burns out sulfur and other volatile impurities in zinc and other ores prior to smelting.

SIROSMELT PROCESS A zinc recovery process from slags, concentrates, and waste materials, such as EAF dusts. The process utilizes a refractory-lined, cylindrical vessel, open at the top, through which fuel, air, and oxygen, if desired, can be blown into the molten bath through a standard steel pipe lance. Depending on requirements, oxidation or reduction reduction reactions can be carried out. Zinc is volatilized at suitable temperatures under reducing conditions and is captured in the form of zinc oxide.

SKIDOO BELL A bell placed near the bottom of the shaft to warn men of any impending danger, as falling material, fire, descending cage, etc.

SKIM GATE A gating arrangement designed to prevent the passage of slag or other undesirable materials into the casting or another work area.

SKIMMER A tool for removing scum, slag or dross from the surface of molten metal.

SKIN TO SKIN As close as possible. Timbers set up so close as to be touching each other are said to be skin to skin.

SKIP The bucket or tub in which ore or rock is hoisted up the shaft or incline. Used also to lower supplies and sometimes is adapted to carry men.

SLAB ZINC General term for commercial zinc cast in various sizes and shapes.

SLABBING [1] Close timbering between sets of timber. [2] Cutting a slice or slab from the side of a pillar.

SLACK Fine coal sorted by screening. The coarse coal was used for firing at zinc smelters whereas the slack was used mixed in the charge for reduction.

SLAG A nonmetallic product resulting from the dissolution of flux and impurities in smelting operations.

SLAG FUMING A process to extract zinc from lead smelter slags. Lead concentrates derived from lead-zinc ore, invariably contain zinc. In smelting the lead concentrate, most of the zinc reports to the slag which, not uncommonly, will contain zinc in amounts ranging between 5% and 20%. To extract the zinc, a continuous coal-air mixture is introduced via tuyeres into a molten slag bath for about 2 hours, which is generally sufficient to fume out the zinc. The zinc with lead and cadmium which are also fumed out, are captured as oxides in a baghouse. The captured product is deleaded and densified, and sent as feed material to a zinc refinery. The first commercial use of the process was at the Anaconda plant at East Helena, MT in 1927.

SLICE BAR A thin, wide iron tool for cleaning clinkers from the grate bars of a furnace.

SLICE In an ore body of considerable lateral extent and thickness, the ore is removed in horizontal layers, termed slices, which may be 6 to 40 feet thick.

SLICK Ore in state of fine division; synonymous with slimes.

SLIME [1] A material comprised of extremely fine particles encountered in ore preparation. [2] A fine, mudlike, mixture of metals and insoluble compounds that form on the anode in electrolysis; Known as anode slime or mud. [3] A product of wet ore crushing in which particles of valuable minerals are so fine as to be carried in suspension by water.

SLIME TABLE A table for the treatment of slime. A buddle.

SLIMING Another name for over grinding in ball mill.

SLUDGE A fine mudlike mixture of water and bore meal produced in drilling rock.

SLUG Mass of half-roasted ore.

SLUSH CASTING A casting made by pouring molten metal into a metal mold, allowing the metal to freeze sufficiently to form a wall of suitable thickness and then pouring out the remaining molten metal from the interior. With the advent of zinc die casting, this form of casting zinc became very limited.

SLUSHER A scraper loader, often moved back and forth by cables, that is used to scrape ore into chutes and/or move sand and rock fill into stopes.

SMALL MINES STABILIZATION PROGRAM A Government subsidy program that provided stabilization payments to small domestic producers of lead and zinc ores and concentrates authorized by Public Law 87-347 in October 1961, in part to provide protection against low tariffs on imports of zinc ores and concentrates. Stabilization payments were based on 55% of the difference between 14.5 cents and the monthly market price per pound for PW zinc for the month in which the sale was made. In the 8-year period (1962 through 1969) of the program, payments of $2.6 million were made to 91 producers in 11 States on sales of 35,830 short tons of lead and 68,860 short tons of zinc.

SMELT To reduce metals by processes that includes fusion.

SMELTER RETURNS In a contract, the returns from the ore or concentrate shipped, less smelting charges, without deducting transportation charges

SMELTER SCHEDULE That part of a contract between a seller of ore or concentrate and a smelter which deals with the factors to be used in ascertaining the money due the seller. The amounts to be deducted from the gross value of recoverable metals typically consist of two parts, smelting charges and marketing charges, the former consist of a base treatment charge, deductions for processing losses, penalties for undesirable compnents, and extra value for various byproduct constituents if over a certain amount.

SMELTING Thermal processing wherein chemical reactions take place to produce liquid metal from beneficiated ore.

SMITHSONITE A zinc carbonate mineral, formerly a major source of zinc. See also *calamine*. The mineral was named in honor of James Smithson (1754-1829), who founded the Smithsonian Institution in Washington, D.C.

SMOKE ZONE Area surrounding a smelting plant in which the smoke or fumes damage vegetation, or in which it may be classed as a public menace or nuisance.

SNAKING A term used in the Tri-State district for the practice of making blasting holes in ore by pushing an iron bar 5 to 6 feet into clay or limestone between chert and jasperoid layers to save on drilling costs.

SNOW WHITE Early European term for French-process zinc oxide that collects in the farthest collection chambers. It was the purest and the whitest.

SOAK PIT A pit where wet clay is allowed to soak preparatory to molding.

SOAP FLOTATION A flotation process where soaps are used as collectors. Soap collectors were used in zinc flotation but generally they are used to float minerals that do not have a metallic luster and appearance.

SOFT BRASS Brass which has been annealed after drawing and rolling; used where ductility is essential.

SOFT BUCKFAT See *Hard Buckfat*.

SOFT DRIVING See *Hard Driving*.

SOFT GALVANIZING See *Hard Galvanizing*.

SOFT GROUND [1] Heavy ground; bad ground. [2] Rock about underground openings that does not stand well and requires heavy timbering.

SOFT ORE Sometimes used in the Tri-State district for second class ore.

SOLDER An alloy for uniting metal. Brazing solders are alloys of zinc and copper, while soft solders are alloys of

The U.S. Zinc Industry

lead and tin.

SOLUBLE ANODE An anode which goes into solution during an electrolytic process.

SOLID SOLUTION Single solid homogeneous crystalline phase containing two or more chemical species.

SOLIDIFICATION SHRINKAGE The decrease in volume of liquid metal during solidification.

SOLUTE The substance dissolved in solution.

SONOROUS Giving sound when struck or bent, as sonorous metal. Some early U.S. spelters were particularly sonorous when bent, possible the result of high impurity levels. The so-called *cry of tin* when tin is bent, however is attributed to slippage along crystallographic planes.

SOREL'S PROCESS A process for making zinc oxide based on the direct oxidation of the fused metal. It consisted of heating zinc in large muffles to the fusion point and then inflaming it with current of air, which entrained the oxide into collection chambers. The air current carried only *snow white*; the oxide retained on the metal was raked off and recovered by sifting.

SORTER One who sorts or classifies ore by hand.

SOUTHERN A leaded brand of spelter sold by the Berth Mineral Co. from 1879 to about 1911. See *Berth Pure Spelter.*

SPAD A flat spike with a hole for threading a plumb line. A spad is used to mark an underground survey station.

SPANGLE The bright, sparkley pattern that develops on the surfaces of hot-dip galvanized products owing to the development of numerous crystallization centers in the coating as the metal cools. The spangle appearance and size can be controlled to some extent by the addition of small amounts of various metals, such as tin, antimony, and cadmium.

SPARATALITE An old name for zincite.

SPEAUTER; SPIALTER Names given to spelter by the Dutch in the 17th century. The Dutch were said to have first obtained the metal about 1620 from captured Portuguese ship carrying spelter from the Far East. The Latinized version of this name is thought to have given rise to the term, *spelter.*

SPECIAL HIGH GRADE ZINC (SHG) A specification grade of zinc first considered and recommended by the ASTM in 1929 for diecastings. SHG is composed of 99.99% zinc with maximum limit of 0.01% for lead, cadimum and iron. It was used mainly for die castings since its initial production; however in recent decades it has gradually usurped the uses of HG, and since the early 1980's, it has found extensive use in electrogalvanizing, especially for steel sheet for the automobile industry.

SPECIFIC GRAVITY The ratio of the weight of body to that of an equal volume of some standard substance, water in the case of solids and liquids. Numerically equal to the density.

SPECIFICATIONS Formulated, definite and complete statement of what the buyer requires of the seller. In the field of materials specifications for zinc metal, galvanized items, and rolled zinc, the ASTM standards are generally used, but depending on requirements additional specifications may be required.

SPECTROGRAPHY Qualitative or quantitative analysis by visual, photographic, or electronic graphing of the spectrum of a substance.

SPELTER SOLDER [1] A brazing filler metal of approximately equal parts of copper and zinc. [2] Hard solder containing zinc.

SPELTER [1] The zinc of commerce, more or less impure, in slabs, plates or ingots cast from molten metal. It does not include zinc dust. The term dropped out of favor after WWII in the United States, being superseded by the term, slab zinc. [2] Metallic zinc of varying purity obtained in smelting zinc ores.

SPENCE FURNACE A roasting furnace of the muffle or reverberatory type, the ore being supported on shelves and stirred mechanically.

SPIEGELEISEN A manganiferrous white cast iron used in the manufacture of steel by the Bessemer process. Called spiegel; spiegel-iron. The first spiegeleisen produced in the United States was produced in 1855 from franklinite from New Jersey. At that time it had little use, but within few years, the Bessemer process for making steel was developed and opened the market for all that could be produced. The early production of spiegel from zinc ore was carried out by The New Jersey Zinc Co. at South Bethlehem, PA, and in 1904, at its zinc smelter at Palmerton, PA. The spiegel was produced from the residuum from the company's zinc oxide furnaces and Waelz kilns that processed franklinite ore. The spiegel was made by smelting the residuum, coke and limestone in a blast furnace, operated somewhat the same as those in the iron industry.

SPIESSED When a retort has been filled, it is spiessed, i.e., a small iron rod is inserted toward the top of the charge to furnish an opening for vapors and gas to escape, and prevent the charge from being blown out.

SPIESSING The act of keeping a spiessed hole open during the firing cycle in the retort furnace. Also known as rodding.

SPILL-TROUGH A trough to receive melted brass that might be spilled in pouring from a crucible into a flask.

SPINTHARISCOPE An instrument having a screen coated with zinc sulfide or other fluorescing substances, on which scintillations are observable when bombarded by radioactive rays.

SPITTING Lighting the fuse for a blast.

SPLASH CONDENSER Device used to condense zinc vapor emitted from continuous zinc distilling processes. The condensing unit consists of motor driven rotor that extends through the condenser wall into a bath of molten zinc or lead, which is splashed, like rain, throughout the condenser. The zinc vapor condenses on the drops and on the splash-covered walls and flows into the bath, with the excess overflowed into a collecting pot. The minor amount of zinc passing through the condenser and remaining in the gas stream is scrubbed out by water streams and recovered.

SPOT The price of a metal for immediate delivery in the physical market. In the futures market, spot refers to the immediate or current trading month.

SPREAD Price differential between two specific dates. The spread can be expressed as either a *contango* or a *backwardation*.

SPREADER [1] A horizontal timber below the cap of a set, to stiffen the legs. [2] A bar used as a distance piece or as a cross bearer to support a line of rails in an adit.

SPRUE, DOWNSPRUE, DOWNGATE The channel that connects the pouring basin with the runner. sometimes used to mean all gates, risers, and runners.

SPUD Nail, resembling a horseshoe nail, with hole in the head. A *spad.*

SQUARE SET STOPING Same as *Square Set System.*

SQUARE SET A set of timbers composed of cap, girt, and post. These members meet so as to form a solid 90 degree angle, and are joined with 3 other sets, forming a solid rectangular unit.

SQUARE-SET SYSTEM A method of mine timbering in which heavy timbers are framed together in rectangular sets, 6 to 7 feet high, and 4 to 6 feet square, so as to fill in the mined out area and provide support as the orebody is removed. The system is generally used with overhand stoping.

SQUIB [1] A tapered paper tube, about 7 inches long, filled with fine gunpowder, one end of the tube being treated with chemicals so as to form a slow burning match, which when ignited, burns so slowly as to give the miner time to reach a place of safety before the explosion. [2] Slow-burning fuse used in blasting.

STAGE [1] Stage grinding is successive comminution. [2] Stage concentration is stage grinding repeated on concentrate produced between grinding stages. [3] Stage addition in flotation refers to deliberate use of insufficient reagent in the early part of the treatment, in order to increase selectivity of conditions, followed by further addition at later point in the process.

STAKE [1] Short for grubstake. [2] Short pointed piece of wood driven into the ground to mark a boundary, survey station, etc.

STAKING OUT The physical act of locating a lode mining claim.

STAMPING [1] The act of reducing ore to the desired fineness in a stamp mill. [2] General term covering almost all press operations. Includes blanking, shearing, hot and cold forming, drawing, bending and coining.

STANDARD ERROR The standard deviation of many samples from the mean. It can be used, for example, to show the amount of inconsistency between the sample and an average sample.

STANDARD ZINC DUST A mixture of zinc powder and zinc-oxide powder, used as a pigment and corrosion agent in protective paints for metals.

STANDARD ZINC-LEAD WHITE A mixture of lead sulfate and zinc oxide, used as a paint pigment.

STARTING SHEET A thin sheet of metal used as the cathode in electrolytic refining.

STARVATION [1] In flotation, the deliberate inadequate addition of a reagent in order to restrict its effect. [2] In precipitation of gold from pregnant cyanide, limiting the use of zinc powder so as to attract gold salts rather than copper ones in the first stage of precipitation.

STATION The open area adjoining the shaft at each level where workers and materials are removed or delivered. Also a surveying observation point, an underground drill site, a pumping station, etc.

STEARATES Stearates of aluminum, calcium, magnesium, and zinc are used chiefly as die lubricants for dry pressing certain ceramic products.

STEEL JACK Sphalerite.

STEEL SETS Steel mine supports used in place of timbers in main entries and shafts and in areas where extreme ground pressures preclude the use of timbers.

STEEL Slang for drill rod, including diamond drill rod, in mines.

STEMMING ROD; STEMMING STICK A nonmetallic rod used to push explosive cartridges into position in a shot hole and to ram tight the stemming.

STEMMING Mining term applied to the inert material used on top of charge of powder or dynamite. See *tamping*.

STENCIL METAL Metal from which stencils for the enamel shop are cut are usually sheet lead or zinc.

STERLING FURNACE A furnace in which zinc and certain accompanying metals are reduced from their oxides by carbon utilizing heat supplied by open electric arcs. A commercial-sized furnace was placed in operation at Palmerton, PA by The New Jersey Zinc Co. in 1951, mainly to test various ores. Two Sterling furnaces were later built in Peru; both discontinued operations in 1958.

STERRO METAL An alloy of copper, 3 parts; zinc, 2 parts; with some iron and tin. It is stronger than gun metal.

STINKDAMP A mining term for hydrogen sulfide which smells like rotten eggs, hence the name. The gas can be released from rocks and has been in shaft sinking in the Middle Tennessee zinc district. Without proper ventilation it can be dangerous.

STOKE HOLE A hole in a furnace for introducing a rabble or other tool for stirring.

STOPE HOIST A small portable compressed-air hoist for operating a scraper loader or for pulling heavy timbers into position, often used in narrow stopes.

STOPE An excavation from which the ore has been extracted in a series of steps, either above or below a level. Usually applied to highly inclined or vertical veins. Frequently used incorrectly as a synonym of room, which is a wide working place in near flat deposits.

STOPER A light percussive drill used in stope mining.

STOPING UNDERHAND Mining a stope downward in such a sequence that it appears like a flight of stairs.

STORAGE STAIN A white alteration product that stains the surface of galvanized sheet when closely piled in damp or outdoor storage. Also referred to as *white rust*.

STOWING A method of mining in which all the material of the vein is removed and waste is packed (stowed) into the space left by the working.

STRAKE Place where ore is sorted on the floor of the mine; a dressing floor.

STRATABOUND An ore deposit limited and contained within a specific bed or strata.

STRATEGIC MINERALS Minerals and metals essential to the national defense, the supply of which we are wholly or partially dependent upon sources outside the continental limits of the United States , and for which strict measures controlling conservation and distribution are necessary. Compare with *Critical Minerals*.

STRATIFIED Formed or lying in beds, layers, or strata.

STRATIGRAPHIC TRAP A type of trap caused by variation in the lithology of the ore-containing host rock. Such traps can be the loci where ore deposits form.

STRAWBERRY JACK Used in Wisconsin for berry-shaped sphalerite aggregates found in soft-clay pockets.

STREAK The color of a mineral powder

obtained by rubbing on a streak plate or by scratching with a file or knife.

STRIKE [1] The direction or bearing of a horizontal line in the plane of an inclined vein, stratum, fault, joint, etc. [2] Stopping of work by workmen to obtain or resist change in conditions of employment. [3] To find ore; to strike a vein. [4] Thin electrodeposited film of metal to be followed by another plated coating.

STRIKING Electrodepositing, under special conditions, a very thin film of metal which will facilitate further plating with another metal or the same metal under different conditions.

STRINGER A narrow vein or irregular filament of mineral traversing a rock mass of different material. A veinlet.

STRIP A thin sheet of metal in which the length is many times the breadth.

STRIPPING Removing a coating from a metal surface, as stripping the zinc from an aluminum cathode sheet.

STROMATOLITE The term is applied to Pre-Cambrian and Paleozoic laminated, reef-like structures that are generally attributed to blue-green algae. Commonly called algal structures. Characteristically laminated with varied gross forms, from near-horizontal to markedly convex, columnar and sub-spherical.

STULL The top piece of a set of mine timbers; timber prop supporting the roof of a mine opening.

SUBLEVEL An intermediate level opened a short distance below a main level.

SUBSTRATE The layer of metal underlying a coating, regardless of whether the layer is the basis metal.

SULFATE ROASTING Term commonly used in connection with roasting for the electrolytic zinc process. It is actually a misnomer in that it is not desirable to have more zinc sulfate in the roasted product than is required to offset the loss of acid in the leaching plant. The term was originally coined to apply to a set of roasting conditions necessary to supply acid for a plant using limestone or milk of lime for neutralization. Typically, less than 3% sulfate sulfur is desired in calcine that is to be leached.

SULFATIZE; SULPHATIZE To convert into sulfates, as by roasting sulfide ores. In the early practice for the manufacture of zinc sulfate, sphalerite was heap roasted to form zinc sulfate which was extracted by leaching, followed by evaporation and crystallization.

SUMP An excavation for the purpose of collecting or storing seepage water before pumping to the surface. The bottom of the shaft is sometimes used as the sump.

SUPERALLOY An alloy developed for very high temperature service where relatively high stresses (tensile, thermal, vibratory and shock) are encountered and where oxidation resistance is frequently required.

SUPERGENE Applied to ores or ore minerals that have been formed generally by descending waters. When ore deposits are exposed to weathering by erosion, the surface waters oxidize many minerals such that metals are solublized generally as sulfates. Zinc and some other metals, typically react with limestone or available silica to form secondary zinc minerals in the lower areas of the zone of oxidation, i.e. above the water table, forming mineable oxidized zinc deposits. Rarely, zinc is precipitated below the water table as a sulfide in the zone of supergene sulfide enrichment.

SUPERPLASTIC ZINC A 78% zinc

and 22% aluminum alloy that displays fine grain structure and easy formability at relatively low temperatures (480 to 520 degrees F.), but once a formed part is cooled slowly, the part has the strength of a die cast zinc part. The formability of superplastic zinc is akin to that of molten glass or heated thermoplastic in the temperature range above but loses that ability above and below that range.

SURFACE DRILLING Boreholes collared at the surface as opposed to those collared in mine workings or underground.

SURFACE RIGHTS [1] The ownership of the surface of land only, where mineral rights are reserved. [2] Rights to land exclusive of mineral rights.

SURFACE-ACTIVE AGENT An agent which modifies physical, electrical, or chemical characteristics of a solid's surface, also surface tensions of solids and liquids.

SUSPENSION ROASTING A form of sulfide roasting in which suspended particles are flash roasted as they pass through the hot zone in the furnace. Suspension processes to roast zinc sulfide evolved when large tonnages of flotation concentrates became available in the mid-1920's. Because of high capacity and economic advantages, suspension roasting displaced circular hearth furnaces entirely, but they in turn, have largely been replace by fluidized-bed processes.

SUTTON, STEELE, AND STEELE DRY TABLE A concentrator of the Wilfley type in motion, but instead of using water, stratification is by means of rising currents of air. The heavy grains are pushed forward by the head motion, while the lighter grains roll or flow down the slope toward the tailings side.

SWAMPER Rear brakeman in a metal mine.

SWEET ROASTING Complete roasting, or until arsenic and sulfur fumes cease to form.

SWINGING A CLAIM The adjustment of the boundaries of mining claim to more nearly conform to the strike of the vein. A reasonable time is allowed the discoverer to explore the vein or lode to find out its strike and thus enable him to properly lay out his claim.

SWITHER Used in Wisconsin zinc-lead regions to denote a crevice or crack branching from main lode.

SYMON'S CONE CRUSHER A modified gyratory crusher used in secondary ore crushing that consists of downward-flaring bowl within which is gyrated a conical crushing head. The main shaft is gyrated by means of long eccentric which is driven by bevel gears.

SYMON'S DISK CRUSHER A mill in which the crushing is done between two cup-shaped plates that revolve on shafts set on a small angle to each other. These discs revolve with the same speed in the same direction and are so set as to be widest apart at the bottom. Feed is from the center, and is gradually crushed as it nears the edge, and is then thrown out by centrifugal force.

SYNDICATE An association or group of persons, usually financiers or capitalists, who combine to carry out on their own account, financial or industrial project, as the underwriting of an issue of bonds, stocks, the carrying out of an industrial enterprise, etc.

SYNGENETIC DEPOSITS An ore deposit formed contemporaneously with the parent rock and enclosed within it. Contrasted with *epigenetic deposits*, which are of later origin than the enclosing rock.

TABLING Separation of two materials of different densities by passing a dilute suspension over a slightly inclined table having reciprocal horizontal motion or shake with a slow forward and fast return.

TAG [1] A numbered piece of metal or wood that miner places on a mine car loaded by him. When it is dumped, the car is credited to the miner. [2] A numbered metal disc assigned to miners and visitors to be carried by them in the mine for identification purposes in case of an accident.

TAIL The poor grade of ore slime at the lower end of the slime-box as it flows from the stamps.

TAILINGS POND An area closed at the lower end by a containing wall or dam in which mill effluents are dumped. Mill slurry is piped to the pond, the solids separate while the liquid may be treated and discharged, allowed to evaporate, or be reused by the mill.

TAILINGS In ore dressing, the term generally means that part of the ore being processed that is discarded as inferior or worthless, i.e., the gangue and other refuse materials resulting from the washing, concentration, or treatment of ground ore. Tailings, as used by the mining industry, is used in plural form.

TAINTON ALLOY Lead-silver alloy containing up to 1% silver, developed in the early 1920's for anode use in electrolytic zinc plants. These alloys reduced lead contamination in the zinc cathode, improving the purity so that Special High Grade zinc became a common product in electrolytic zinc refining.

TAINTON PROCESS High-density, strong-acid process for producing electrolytic zinc. The process, which was used at the Bunker Hill zinc plant in Idaho, was characterized by high-temperature, high-acid, single-leach system and high current density relative to most other electrolytic zinc processes. Because of the higher acid and temperatures used, zinc ferrite was readily dissolved in the leaching process.

TAKING UP STOPE Term used in the latter phases of mining in the Tri-State in Oklahoma for extraction of low-grade ore from beds below (less often above) the floor of the generally worked-out main orebody.

TALLOW CLAY In Arkansas, a commonly occurring mixture of calamine and clay, red, brown, or yellow in color, recognized to some extent by its peculiar feel. Also known as *buck fat*.

TALMI-GOLD Kind of brass made to resemble gold, sometimes plated. Called also *Abyssinian gold*.

TAMP To fill (usually with clay) the bore hole or other opening through which an explosive charge has been introduced for blasting.

TAMPING For a long time the word has been used to designate both the inert material used on top of a charge of powder or dynamite, and the operation of compressing it in place. See *stemming*, which is the term preferred for the inert material, while tamping more correctly is the act of compressing the stemming.

TAP HOLE The opening through which molten metal is tapped or drawn from furnace.

TAPPING [1] Opening the outlet of a melting furnace to remove molten metal. [2] Removing molten metal from a furnace.

TATHAM FURNACE A stationary crucible furnace for retorting zinc crusts.

TCHESA STICK An igniting stick used to light powder fuses when firing a round of shots. Also called *Fire Stick.*

TECTONIC BRECCIA DEPOSITS Breccias produced by folding, faulting, intrusion or other tectonic forces. In the Appalachian zinc deposits and some localized zinc deposits in the Tri-State and Arkansas zinc districts, such breccias have been variously referred to as crush, rubble, crackle and shatter breccias.

TENDERFOOT Newcomer in rough or newly settled region, especially when not inured to the hardship or rudeness of the life.

TENOR The percentage or average metallic content of an ore, matte, or impure metal.

TEST HOLE A drill hole or shallow excavation for testing an ore body.

TEST PIT A shallow exploration shaft, trench, or hole made to determine the existence, extent, and grade of an ore deposit or to obtain samples or information on the soil and rock for mining or construction purposes.

THAW HOUSE A small building, designed for thawing dynamite, of such size as to provide enough thawed dynamite for the day's work.

THAWING The warming of frozen dynamite until it becomes soft and plastic.

THE INDUSTRIAL REVOLUTION Began in the second half of the eighteenth century. Principal factors that led to it were the use of coal in the manufacture of iron about 1740 and James Watt's invention of the steam engine in 1769.

THE PRESIDENT Name given to the world's largest pumping engine (in early 1870's) to pump water out of the Ueberoth Mine at Friedensville, Pennsylvania. The engine was a single cylinder, walking-beam engine with a pair of flywheels. The stroke was 10 feet. In 1876 the mine was making up to 20,000 gallons per minute, and although the pumping engine was capable of handling the flow, the cost was too high and the mine closed.

THICKENER An apparatus or vessel for increasing the solids in a dilute pulp. A settler.

THIN-WALL ZINC DIE CASTING A technology that allows the designer to reduce the weight of die cast parts without sacrificing strength by adding in thin wall sections to support the same load in somewhat the same manner that honeycomb sections can support tremendous loads without failure. Thin Wall, a trademark, is not a specific thickness but one that will do the job. It was developed by ILZRO in response to the competitive challenge posed by plastic components.

THIOCARBONATES Strong collector agents used in the flotation process where xanthates fail.

THREE-IN-ONE A paint formula developed before 1880, containing one-third each zinc oxide, white lead and barytes. Despite the mix, it was generally referred to as white lead into the early 1900's and into the 1920's by some painters to whom all white paste paints were *white lead.*

THROUGHPUT The quantity of material passed through a mill or plant section in given time or at given rate.

THUM FURNACE A gas-fired furnace especially for the treatment of zinc ore high in lead.

TIE COAT An intermediate coat used to

bond different types of paint coats.

TIFF Common name for calcite in Wisconsin and Missouri zinc fields, and for barite in southeast Missouri.

TIMBERING The operation of setting timber supports in mine workings or shafts.

TOBIN BRONZE An alpha-beta brass or Muntz metal containing 0.5 to 5 percent tin, used when resistance to seawater corrosion is required.

TOLERANCE The specified permissible deviation from a specified nominal dimension, or the permissable variation in size of a part.

TOMBAC Any one of several copper and zinc alloys, as *Prince's metal, Mannheim gold*, etc. Also spelled Tambac; Tombak.

TON MILE In railroading, a standard measure of traffic, based on the rate of carriage per mile of each ton of freight.

TOPCOAT The coating intended to be the last coat applied in a coating system. Zinc-rich primers are often topcoated to improve durability for severe exposure, color, gloss, etc. Topcoat is also known as finish coat.

TRACE ZINC Exploration drill-log term in Tri-State district for drill cuttings that contain about 0.5% zinc sulfide content. Also see *shines*.

TRACTOR SHOVEL; FRONT-END LOADER; LOADER Excavating equipment that has a bucket supported from the front end of the tractor.

TRAIL A long tube or pipe at zinc oxide plant in which the furnace oxide is transported and cooled before entering the bag house.

TRAINROAD A temporary track in a mine, used for light loads.

TRAMMING The practice of pushing tubs, mine cars, or trams around by hand.

TRAMWAY A suspended cable system along which ore and rock is transported in suspended buckets.

TRANSITION ELEMENTS Elements characterized by incompleteness of the second outermost shell of electrons in their atoms. Zinc is transition element.

TRANSITION METALS See *Transition Elements*.

TREATMENT In metallurgy, the reduction of ores by any process whereby the valuable constituent is recovered.

TREEING In electrolytic refining, the tendency of zinc to deposit at localized points on the cathod rather than uniformly over the entire surface. To minimize treeing, a colloid, such as glue, goulac, gum arabic, agar agar, with sodium silicate and/or cresylic acid is generally added to the electrolyte.

TREES Visible projections of electrode-posited metal formed at sites of high current density.

TRENCH In geological exploration, a narrow, shallow ditch cut across a mineral deposit to obtain samples or to observe character.

TRIMMERS Shotholes drilled around the periphery of a shaft or drift which break or trim the sides of the excavation to the shape and size required.

TRIMMING In forging and die casting, removing the parting-line flash and gates from the part by shearing. In casting, the removal of gates, risers and fins.

TRI-STATE ZINC-LEAD DISTRICT Name given to the group of mining districts in southwestern Missouri, southeastern Kansas, and northeastern Oklahoma in 1917, as substitute for the more cumbersome name, Missouri-Kansas-Oklahoma mining district. Prior to 1917 the district generally was known as the Joplin mining district. The Tri-State district, which covers approximately 1,200 square miles, has been the principal zinc-producing area of the United States. Lead mining, which began about 1850 and zinc mining in 1872, was continuous in the district for more than 100 years, until all zinc-lead mining ceased in the late 1950's.

TROMMEL [1] A revolving sieve for sizing ore. Also called according to size, washing drum; washing trommel. [2] A revolving cylindrical screen used in grading coarsely crushed ore.

TROOSTITE A variety of willemite, in large reddish crystals, in which zinc is partially replaced by manganese.

TUBBER In mining, a double pointed pickaxe; a beele.

TUMBLING An operation where the work, usually castings, are rotated in a barrel with metal slugs or abrasives to remove sand, scale, or fins. It can be done either dry or wet. Sometimes called rumbling or rattling.

TUNGSTEN-CARBIDE BITS Drilling bits tipped with tungsten carbide; widely used in mining to drill blasting holes.

TUNNEL BORER Any boring machine for making a tunnel.

TUNNEL In mining, often used as a synonym for adit, drift, or gallery.

TURKEY FAT In Arkansas and Missouri, a local name for a variety of smithsonite, colored yellow by cadmium minerals. So called from its appearance. Similar to *Fat Back* in Arkansas.

TUTANEG Early name for spelter.

TUTENAG [1] Zinc or spelter, especially that from China and the East Indies. [2] A white, copper-zinc-nickel alloy like German silver, for making tableware, etc.

TUTENEAGUE An eighteenth century European term for spelter imported from China.

TUTIA Alchemist term for impure oxide of zinc.

TUTTY An impure zinc oxide obtained as a sublimate in the flues of zinc smelting furnaces. Once used as a polishing powder.

TUYERE; TWEER; TWYER; TWERE [1] A pipe inserted in the wall of a furnace through which the blast is forced into the furnace. [2] Tube or opening in a metallurgical furnace through which air is blown as part of the extraction or refining process.

TWIN-BELT CASTER A machine with which zinc strip is continuously cast directly from molten metal. In the process, two moving, parallel casting belts up to 80 inches in width, are fed with molten zinc, which rapidly solidifies between the belts as cast strip. The continuous strip is then hot rolled in line with one or more tandem rolling mill stands to the gauge desired, and rolled into coils.

U ULTRAVIOLET LIGHT Black light. Invisible light rays from the portion of the spectrum that has wavelengths shorter than visible light and longer wavelengths than X-rays. Some minerals, including some zinc minerals, respond to exposure to ultraviolet light by emitting visible light colors even though the exposure is carried out in complete darkness. Also called black light and UV light.

UMPIRE An assay made by a third party to settle a difference found in the results of assays made by the purchaser and seller of ore, concentrate, metal or scrap.

UNCONFORMITY Surface of erosion or nondeposition, usually the former, that separates the younger strata from older rocks. An unconformity indicates a break in deposition or a break in time of the rock record;

UNDERFLOW Oversized material leaving the classifier.

UNDERHAND STOPING Mining ore from an upper to a lower level, underhand. A modified version of the method is employed at the Lucky Friday mine in Idaho. Stoping is carried out in progressive steps downward, however after each cut across the stoped area, the area is sand-filled with cemented tailings before the next cut below is made. This method was developed to increase safety against rock burst.

UNIT DIE In die casting, a die block that contains several cavity inserts for making different kinds of castings.

UNITED STATES-CANADA FREE TRADE AGREEMENT (FTA) On January 1, 1989, the United States and Canada formed the world's largest free trade area by implementing the FTA, through which the two countries would reduce or eliminate tariffs on U.S. and Canadian goods traded between them by 1998. U.S. duties on Canadian imports of zinc metal, alloys, and dutiable chemicals were to be phased out at 10% a year over 10 years or over a shorter period if agreed upon.

UNIT Smelter contracts make use of the term, unit. A unit means 1 percent. Since a short ton contains 2000 pounds, a unit is equivalent to 20 pounds. The statement that zinc will be paid for at a rate of one dollar per unit means that one dollar will be paid for each 20 pounds of zinc in the ore.

UNPATENTED CLAIM Mining claims to which a deed from the U. S. government has not been received. All such claims remain subject to annual assessment work in order to maintain ownership rights.

UPDRAFT KILN A kiln in which the heat enters the chamber from the bottom and passes up through the work.

UP-OVER Designating a method of shaft excavation by drifting to a point below, and then raising instead of sinking.

USBM The United States Bureau of Mines.

USGS The United States Geological Survey.

V

VACUUM MIXER A machine for the simultaneous de-airing and moistening of dry prepared clay as it is fed to the pug.

VACUUM PUG Pug with a vacuum chamber in which clay is de-aired before it passes into the extrusion chamber.

VAN To separate, as ore from a vein, by crushing the ore and washing it on the point of shovel. A simple way of testing ore.

VANNER A machine for dressing ore; an ore separator in which the peculiar motions of the shovel in the miner's hand in the operation of a van are imitated.

VAPOR GALVANIZING A process for coating a metal (usually iron or steel) surface with zinc by exposing it to the vapor of zinc instead of, as in ordinary galvanizing, to molten zinc. Also known as *Sherardizing*.

VAPORIZATION The act or process of changing a substance from a liquid to a gaseous state.

VAT A vessel or tub in which ore is

washed or subjected to chemical treatment.

VEIN. In common usage, vein and lode are essentially the same, the former being rather the scientific and the latter the miner's name for it. See *Lode*

VELVET Profit; easily earned money. By analogy, term used for galena in the Wisconsin zinc field when it could be separated from the blende without difficulty and sold as a byproduct.

VENETIAN WHITE A pigment consisting of a mixture of equal parts of white lead and barite.

VENT Small opening in a mold for the escaping of gases.

VENTURE CAPITALIST A dealmaker who provides funds and advice to entrepreneurs, either starting a business from scratch or staging a buy-out.

VERTICAL RETORT PROCESS A process for the continuous distillation of zinc. The first commercial plant was placed in operation in May 1929. Basically the process, consists of three components: (1) making briquettes containing roasted zinc concentrate or other zinc-bearing material, coal which cokes and will retain its strength, and binders; (2) distillation of zinc from pre-coked briquettes; and (3) condensation of the zinc vapor. The vertical retorts, which are made of silicon carbide bricks, are 25 to 35 feet in height, 5 to 7 feet long and about 1 foot in width. Pre-coked briquets are fed into the top of the externally heated retort and as they progress downward, zinc is reduced and distilled, rising through the charge to a condensing unit, which in early practice was an inclined conduit to rapidly cool the gases and a sump in which the molten zinc metal was collected. In later practice, a splash condenser was employed to condense the vapor. This process had many advantages over the horizontal retort processes; it was highly mechanized, had high productivity, vastly improved worker safety, and long retort life.

VIEILLE MONTAGNE FURNACE A mechanical roasting furnace for zinc ores.

VIRGIN METAL Metal obtained directly from ore.

VOLATILE Easily wasted away by evaporation; readily vaporizable.

VOLATILIZATION The act of vaporizing.

VOLCANOGENIC Processes directly connected with volcanism. Said of mineral deposits, such as massive sulfides and exhalites that have been produced through volcanic agencies and are demonstrably associated with volcanic phenomenon. The Red Dog deposit in Alaska is said to be of stratiform, stratabound volcanogenic origin.

VOLTA'S LIST A list of elements that are essentially the same as the electromotive series.

VOLTA'S PILE; VOLTAIC PILE First true battery (1800) to convert chemical energy to electrical energy. It consisted of coin-sized pairs of zinc and silver metal with wafers of pasteboard or leather between each pair of discs, which were stacked like piles of poker chips. The wafers were soaked in salt water, lye or other alkaline solution to provide the electrolyte. The tip of a metal strip attached to each end of the pile were placed into separate cups of mercury, which when properly connected provided a continuous current of electricity. In 1807, Sir Humphery Davy used a voltaic pile to isolate (and discover) sodium and potassium by electrolysis; other early discoveries using voltaic piles were the laws of Faraday and Ohm.

VOLTZITE A mineral; an oxysulphide of zinc.

VUG A cavity in rock, usually lined with crystalline encrustation and crystals.

VUGGY LODE Lode in which vugs or drusy cavities are of frequent occurrence.

VULCANIZING Process used to modify properties of rubber (strength, elasticity, stretch) by combination with sulfur or suitable sulfur-based compound aided by heat and chemical accelerators. Zinc oxide is commonly used in rubber stock, mainly in tires and electrical insulation. In these areas it functions as an accelerator and/or as an activator of organic accelerators, provides mechanical reinforcement, resistance to deterioration by sunlight, and high hiding power in white walls. In rubber insulation, it has a number of desirable electrical characteristics.

WAELZ PROCESS A process for the recovery of zinc and other metals from ores, concentrates, and residues by the use of rotary kilns, some of which are as long as 180 feet and up to 12 feet in diameter. The process was developed in 1923 in Germany and named after the German word, *Waelzen*, which means trundling motion, which aptly describes the movement of the charge through the slowly rotating kiln. In the process, well mixed feed (the zinc-containing ore, residues, dust, scrap, etc., coke or coal, and conditioners, such as silica, limestone and iron to keep the feed from slagging or building accretions) are fed into the high end of the kiln, which is slightly tilted so the charge works its way through by gravity. The temperature gradually increases as the charge moves down the kiln gradually reaching the reduction temperature, in part owing to heat transfer from gases passing countercurrently over the traveling bed. Reduction takes place under the bed, whereas simultaneously oxidation takes place above the bed. An essential feature of the process is that the reducing and oxidizing zones do not follow each other but occur superimposed over the length of the reaction zone. Gases carry dust and oxidized metals out the upper end of the kiln to a dust-settling chamber to eliminate the dust, then through a fume cooling system into a baghouse for collection of the zinc-rich fume. The recovered fume is generally impure and undergoes further treatment to obtain commercial zinc. Beginning in the 1980's, the Waelz process has been used successfully to extract zinc from steel-making electric arc furnace (EAF) dusts, a material that has been classed as hazardous (and requires treatment) by the Environmental Protection Agency. Aside from recovering the zinc from EAF dusts, the process provides an environmental bonus in that the discharged kiln residues are classes as non-hazardous wastes and can be disposed of in a landfill.

WAELZ SLAG The incandescent material discharging from a Waelz kiln. It typically consists of a porous, semifused mass of particles and after quenching, it breaks into inch-size chunks. The formation of liquid slag in the Waelz process is undesirable and is inhibited because accretion buildup and slag attack the kiln refractories.

WALK OUT A labor strike.

WARING PROCESS A process in which zinc is precipitated as zinc sulfide out of mine waters or from solutions derived by leaching zinc from old tailings. The process was used before 1910 to make lithopone, utilizing zinc in mine water.

WASH HOUSE A building on the surface at a mine where miners can wash before leaving for home. A change house; dry house.

WEATHERING The degradation of near surface rocks by processes involving the chemical action of air and rain water, plants and bacteria, and changes of temperature. Rocks exposed to weathering change in character, decay, and finally crumble into soil and become more susceptible to erosion.

WEATON-NAJARIAN CONDENSER A large U-shaped tube filled with molten zinc through which furnace-generated zinc vapor is drawn by partial vacuum and in which the zinc vapor condenses. These condensers were developed and first used by the St. Joseph Lead Co. in 1936 at their electrothermic plant in Pennsylvania. They were a major factor in making the electrothermic plant a successful metal producer.

WEDGE ROASTER A multiple-hearth vertical furnace. Rabbles rotate on each circular horizontal hearth and work continuously fed material alternately across to the periphery and then on the next hearth below toward center, so that it gravitates through either a central or a peripheral opening and is at the same time exposed to rising heat or air blown through rabble arms.

WEIGHER-AND CRUSHER MAN One who weighs zinc ore and other materials to be sintered, and crushes sintered ore preparatory to further reduction.

WEIGHT PER CENT Percentage composition by weight. Contrast with *atomic per cent*.

WEIGHTS AND MEASURES (ADIT) Local term, which persisted for many years, for an opening in an outcrop of zinc ore at the Franklin orebody in New Jersey. Zinc ore was mined at the site to make brass for official U.S. standards of weights and measures ordered by Congress in 1872.

WELD A union made by welding.

WELDING Joining two or more pieces of material by applying heat, pressure or both, with or without filler material, to produce a localized union through fusion or recrystallization across the interface.

WESTERN WETHERILL FURNACE See *Wetherill Furnace*.

WET CELL Primary battery that has a liquid electrolyte. The original battery developed by Leclanche in 1866 was a wet cell. The Leclanche cell gradually evolved, and in the 1880's, the electrolyte was formed into paste, creating what then became the *dry cell*.

WET DOWN Spraying or hosing down with water a pile of newly broken rock to suppress dust formation.

WET ROT Timber decay set up when mine props have not been treated with zinc sulfate, etc., and are exposed to alternating dry and wet conditions.

WETHERILL PROCESS A process developed in 1851 by S. Wetherill to produce zinc oxide, which later became known as the direct or American process. In the process, ore and coal were mixed and thrown in layers several inches thick on a hearth of perforated cast-iron plates; the furnace was closed after ignition of the coal and cold air was blown under the grate to increase the temperature of the charge to affect reduction and vaporization of the zinc. The vapor was oxidized and drawn off where the products of combustion were cooled and passed through muslin bags capturing the oxide. Two types of Wetherill zinc oxide furnaces known as the *eastern* and *western* types evolved. All furnaces consisted of four or more individual furnace units built together in a block. The *eastern* furnaces worked simultaneously though single units that could be worked independently. The vapors from each were channeled separately to a system of combustion chambers where the vapor was oxidized. From there, they entered a common flue leading to the bagroom where the products were collected as a blende of all the furnace units. The *western* furnaces were different, in that, the vapors were conducted into a common overhead flue and oxidized together in the same combustion chamber.

WETHERILL'S MAGNETIC SEPARATOR An apparatus for separating weakly-magnetic minerals such as franklinite, from non-magnetic minerals, invented by Samuel Wetherill in 1896. Although it had long been recognized that magnetic minerals could be separated by magnet, weakly-magnetic minerals could not be, unless they were roasted first. Wetherill conceived the idea of concentrating the magnetic field by tapering the pole pieces of the electromagnet; this proved successful in attracting weakly-magnetic minerals in ores as well as those that became magnetic when roasted. The Wetherill separator became one of the most important inventions to come out of the U.S. mining industry in the 19th Century and was especially important in the development of the domestic zinc industry. Not only did it result in the efficient development of the zinc deposits at Franklin, NJ by affording a simple means for separating franklinite and willemite into high-grade concentrates, but it also was successful in separating sphalerite in complex ores at Leadville, CO, and in achieving high-grade zinc concentrates from mixed blende/marcasite concentrate in Wisconsin after a suitable roast to make marcasite slightly magnetic was developed.

WETHEY FURNACE A multiple-deck, horizontal furnace for calcining sulfide ores. Resembles the *Keller furnace*.

WETTABILITY Term used in connection with the flotation process to describe the extent to which a specific mineral's surface attracts or rejects water. The wettability of a mineral's surface can be altered by modifiers to increase or decrease flotability.

WHIM A large capstan or vertical drum turned by horse power or steam power, for raising water or ore from mine.

WHIP Horse gear once used in hoisting ore in which the load was raised by a rope passing over a pulley and pulled by draught animal

WHITE BRASS An inferior brass containing more than 49 percent zinc.

WHITE LATTEN An alloy of copper, zinc, and tin in thin sheets.

WHITE LEAD A term used for both basic lead carbonate and basic lead sulfate. Both compounds can constitute the lead component in leaded zinc oxide. The former in French-process oxide and the latter in American-process and blended zinc oxides. In French-process oxide production, entrained lead vapor is converted into white lead carbonate in hot air charged with carbon dioxide. See also leaded zinc oxide.

WHITE METAL A general term covering a group of white-colored metals of relatively low melting points, such as lead, antimony, bismuth, tin, cadmium and zinc and of the alloys of these metals.

WHITE RUST Same as *Storage Stain*.

WHITE SEAL See *Seals*.

WHITE TOMBAC A variety of brass made white by the addition of arsenic.

WHITE VITRIOL Zinc sulfate; the mineral, goslarite. Also known as salt of vitriol; zinc vitriol.

WHITE ZINC Old European term for the heaviest French-process zinc oxide;

that which accumulates in the nearest chambers. Compare with *snow white*.

WIDOWMAKER [1] A reference to stoping drills by reason of the unhealthy effect of dust on the miner's lungs. [2] A large piece of loosened rock that might fall from the back or walls of a mine, alluding to the fact that it could kill a miner.

WILD LEAD Zinc blende.

WILFLEY SLIMER A form of shaking canvas table which is given a vanner motion.

WILFLEY TABLE A side-jerk table used in ore dressing. It has a riffled surface which separates the light and heavy grains into layers by agitation, and the jerking action then throws the heavy grains toward the head end, while the light grains are washed down over the cleats into the tailings box. The table tapers toward the head end, and the riffles are progressively longer toward the tailings side. The Dodd, Cammett, Halett and Woodbury are similar types.

WILLEMITE The naturally occurring zinc orthosilicate; an ore of zinc. Named in honor of Willem I, King of the Netherlands.

WIND BOX A compartment in a furnace that receives the blast and delivers it to the tuyeres. With the development of downdraft sintering, the wind boxes pulled the air through the sinter charge and drew the combustion gases into themselves rather than being used in updraft fashion.

WINDLASS A roll or drum with handles, used in winding or hoisting from shallow pits.

WIN To extract ore; to mine; to recover (as metal) from ore.

WINZE A vertical or inclined opening, or excavation, connecting two levels in mine, differing from a raise only in construction. A winze is sunk underhand and raise is put up overhand. When the connection between the two levels is made, and one is standing at the top, the opening is referred to as a winze, and when at the bottom, as a raise.

WIPE; TIGHT WIPE In wire galvanizing, an apparatus to control the thickness of a wire coating after deposition of the zinc. This is generally affected by pulling the coated wire through asbestos held in place by mechanical fixtures.

WIRE ROPE A rope whose strands are made of metal wires, twisted or woven together.

WIRE-MESH REINFORCEMENT Heavy welded fabric, fencing, and chain link fencing used underground to support the back and walls and prevent falling rock; it is generally held in place by roof bolts.

WITNESS CORNER A post set near a corner of a mining claim with the distance and direction of the true corner indicated thereon. Used when the true corner is inaccessible.

WOLF The name of carbide and electric lamps.

WORK Objects which are to be, are being or have been treated, as in cleaning or finishing.

WORKED OUT A mine or section of a mine where all the mineable ore has been removed; exhausted.

WORM An exudation or sweat of molten metal forced through the top crust of a solidifing metal by gas evolution.

WUENSCH PROCESS A heavy suspension method for the concentration of ores where the waste has a specific gravity of 2.7 or more. Minerals having a specific gravity of 5.25 must be used, since a suspension containing over 40 percent solids by volume is too plastic to use. Galena and ferrosilicon have been used. A heavy-media separation process.

WURTZITE A zinc sulfide mineral of the same composition as sphalerite, but hexagonal in its crystallization. Named after the French chemist, Adolphe Wurtz

XANTHATE A common specific promoter used in the flotation of sulfide ores. A salt or ester of xanthic acid.

X-RAY DIFFRACTION ANALYSIS A method to determine the identity of a crystalline material based on the characteristic lattice structure of that material when irradiated by x-ray of a specific wave length. The x-rays are diffracted by the material's internal atomic structure yielding reflections at definite positions that can be recorded on photographic film or by gamma-ray detection devices on chart paper. Every crystalline compound whether mineral, alloy, or chemical compound has a unique x-ray diffraction fingerprint that can be used for identification purposes. In multi-compound samples, such as an ore or rock sample, a number of the component minerals can be identified as present and some idea of their relative abundances estimated.

X-RAY FLUORESCENCE SPECTRO-SCOPY An analytical method in which the material to be analyzed is subjected to an intense beam of x-rays emitted from a single element source, commonly tungsten, iron, cobalt, or copper. The x-ray beam causes the elements in the sample to emit element-specific secondary x-rays which can be separated and quantified by their intensity. The method is commonly used to analyze ores, slags, alloys, mill products, and refractories. On-line units are common in milling operations to monitor and/or control various aspects of the mineral concentration process.

YELLOW BRASS An alloy of 70 parts copper and 30 parts zinc.

Z REAGENTS A Dow series of xanthate flotation reagents.

ZALCON Trademark for zinc ammonium chloride fluxes.

ZAMAK Trade name given to series of zinc-based die cast alloys. Name was derived from the alloying components: zinc, aluminum, magnesium, and kopper (copper). *The first name originally suggested was Mazak, which was based on the fact that magnesium was the most important alloying component. The New Jersey Zinc Company, the developers, wished to emphasize the alloys as zinc alloys and chose Zamak. Interestingly, when a production licence was granted to a British company, they chose to call the alloy series, Mazak, apparently without the knowledge of New Jersey Zinc's prior consideration of this name.*

ZETAX Trademark for a rubber accelerator, zinc 2-mercaptobenzothiazole.

ZILLOY Trademark for zinc-base alloys.

ZINC ARSENATE An insecticide and wood preservative compound.

ZINC BACITRACIN A common antibacterial medication.

ZINC BLOOM See *Hydrozincite*; *Zinc Oxide.*

ZINC BORATE Used for medicine, textile fireproofing, fungicide, and ceramic flux.

ZINC BOX A box containing zinc powder or shavings for precipitation of gold from cyanide solutions.

ZINC CALCINE A roasted zinc ore or zinc concentrate in which most of the zinc occurs in the form of zinc oxide.

ZINC CAPRYLATE A fungicide.

ZINC CARBONATE A compound used in ceramics and as a fireproofing filler in rubber and plastics.

ZINC CHILLS See *Brass Founder's Ague*.

ZINC CHLORIDE A crystalline powder or solution used as an antiseptic, flux, wood preservative, dental cement, electrolyte, galvanizing agent, and as a mordant, mercerizing and sizing agent in textiles.

ZINC CHROMATE A pigment used in artist paints, road paints, and rust-resistant primers and paints. Similar to zinc dichromate.

ZINC COLIC A form of colic thought to be caused by zinc-oxide poisoning.

ZINC CRUST The zinc-silver scum that rises to the top in desilverizing lead in the Parkes process.

ZINC ETCHING Zincography; also called a zincograph.

ZINC FLASH The colored surface produced on red bricks by the introduction of zinc into fireboxes of a kiln at the conclusion of firing; zinc vapor deposits on brick in various shades from yellow to green.

ZINC FLUORIDE Compound sometimes used in ceramic glazes and enamels.

ZINC GLASS A glass in which zinc oxide replaces part of the calcium oxide in ordinary lime-soda glass.

ZINC GRAY [1] Zinc dust used as pigment. [2] Ground sphalerite used as pigment. [3] Mixture of zinc white with finely divided charcoal or with lithopone, chalk or other pigment.

ZINC GREEN Cobalt green.

ZINC GREENS Any of various green pigments that are essentially mixtures of zinc yellow (hydrated zinc chromate) and Prussian blue (any of number of ferric ferrocynide compounds).

ZINC HYDRATE PROCESS An early (1913-1914) electrolytic zinc process developed to process the zinc ores of the Bully Hill Mine in northern California. The process was based on the widespread notion that good zinc deposits could not be electrowon from zinc solutions that contained much sulfuric acid, and therefore, any acid formed in the cell during electrolysis would have to be neutralized as fast as it formed. In the hydrate process, lime was added directly to the operating cell neutralizing the acid formed, and resulting in suspended precipitates of zinc hydrate and calcium sulfate. Although some good zinc cathode was made using the process, it was not competitive with the then emerging process of direct electrolysis of zinc sulfate with acid regeneration, and was discontinued.

ZINC OINTMENT An ointment, consisting of 20% zinc oxide mixed with a petrolatum or lard base, used in the treatment of skin diseases.

ZINC ORANGE A color with a red-yellow hue of high saturation and high brilliance; also called cowslip.

ZINC PHOSPHATE A compound used as a dental cement and phosphor.

ZINC POLE The negative pole in the Voltaic cell.

ZINC POOL A Government program that allocated a portion of domestic zinc supply to defense preparations beginning in April 1941. Initially, 5% of the January 1941 zinc production was to be requisitioned from the April production to alleviate shortages in defense industries. The pool requirements increased throughout the remainder of 1941 and until May 1942, when amounts were set at 75% for high-grade and 50% for lower grades. On May 1, 1942 the War Production Board assumed control of *zinc pool* and placed zinc under full allocation.

ZINC PROCESS [1] A process for the recovery of materials from cobalt-cemented tungsten carbide scrap. In the process cobalt and molten zinc react to form a series of intermetallic compounds which breaks down the matrix of the scrap. The tungsten carbide particles are separated and the zinc is removed by vacuum distillation, leaving a spongy cobalt matte which can be reused. [2] A process to vastly speed up the recycle of superalloys by increasing their dissolution rate in acids from days to a few hours. Melting superalloy scrap with zinc forms brittle intermetallic compounds that can be readily crushed into small particles that can be rapidly dissolved in acid.

ZINC RETORT A form of *bottle retort*, either stationary or tilting, that can be charged with solid or molten zinc dross or die cast scrap, weighing up to 4,000 pounds. The retort is contained in a furnace chamber and externally heated by either oil or gas till the charge boils. The zinc fumes can be captured as dust or metal or converted to oxide. Generally only metallics are processed in the zinc retort; however zinc oxide materials plus a reductant has been used in these retorts to make salable zinc metal products.

ZINC SALT FORMATION Non-uniform white discoloration on the surface of zinc-rich paint produced by the reaction of zinc dust with atmospheric constituents. Also referred to as *white rust*.

ZINC SCUM The zinc-silver alloy skimmed from the surface of the bath in the process of desilverization of lead by zinc.

ZINC SENDER A sender, which is used on long telegraphic lines, especially submarine cables, which automatically sends a momentary reverse current into the circuit after every signal, to counteract retardation.

ZINC SKIMMINGS The zinc metal and oxide and flux materials that is skimmed off the top of a hot-dip galvanizing bath to prevent poor coating and contamination of the products being immersed. Zinc skimmings are an important source of scrap for the secondary industry and, depending on the type and content of contained fluxes, are recycled back to metal or dust, or to compounds, often zinc chlorides.

ZINC SMELTING Distillation of zinc. So called because the reduction of zinc by carbon proceeds simultaneously with vaporization of the zinc metal, owing to the large spread between the temperatures required to reduce zinc oxide to metal (1,127 degrees C) and the boiling point of elemental zinc (907 degrees C).

ZINC SPAR Synonym for smithsonite.

ZINC STANDARD CELL A Clark cell.

ZINC STEARATE A zinc soap. Used in many applications, including cosmetics, lacquers, plastics, mold-release agent, metallurgical bath demister, and dermatitis medicine.

ZINC SULFATE A compound with many uses, including precipitating agent for viscose rayon, plant fertilizer and animal feed nutrient, ore flotation agent, water purification, and galvanizing.

ZINC SULFIDE ROASTING The application of heat and oxygen, usually from air, to convert sphalerite concentrate into zinc oxide and to eliminate the sulfur. Complete conversion of the concentrate is never attained in practice because some zinc sulfide is not oxidized and some zinc is converted into zinc sulfate. For the horizontal retort process low sulfur content is desired owing to the fact that any sulfur in the charge retains about twice its weight of zinc in the retort residues. Sulfur, if in the form of zinc sulfate, is not a problem when roasting for leaching and electrowining because it does not reduce zinc recovery and may supply needed acid for the leaching process.

ZINC SULFIDE A compound of many uses, including pigments, phosphors, lithopones, dyeing, and fungicides.

ZINC WHITE A common term for zinc oxide used for pigment. It is the whitest of all pigments, permanent and not poisonous, but it lacks the opacity and covering power of white lead or titanium oxide. *During the 1800's, painters often mixed paint starting with the zinc oxide powder. They first mixed it with oil to obtain stiff paint, and then with more oil, made liquid paint. However, because zinc white has a great thirst (absorption) for oil, painters sought out ways to be more economical; they found that less oil was needed and the covering qualities improved if water was added to the stiff mixture. This led some manufacturers to market a hydrated zinc white product for paints.*

ZINC YELLOW A commercial pigment composed of zinc chromate.

ZINC-AMMONIIUM CHLORIDE Used as welding, soldering, and galvanizing flux.

ZINCATE Any of various compounds formed by the reaction of zinc or zinc oxide with solutions of alkalies.

ZINC-BATH PROCESS Conventional, cut-sheet galvanizing process in which the pickled sheet is passed through a flux blanket and through a bath made up entirely of zinc plus any additions. Compare with the *lead-zinc process*.

ZINCBLENDE See *Sphalerite*.

ZINC-CARBON CELL Leclanche wet or dry cell. See *Leclanche cell* or *wet cell*.

ZINCEX PROCESS Commercial process, developed in Spain, that recovers zinc from roasted pyrite ore by chloridization followed by solvent extraction and electrowinning.

ZINCIC Relating to, containing, or resembling zinc.

ZINCIFEROUS Containing zinc or yielding zinc.

ZINCIFICATION The act or process of zincifying.

ZINCIFY To coat or impregnate with zinc; to galvanize.

ZINCING The act or process of heating iron or steel plate with zinc or zinc salts; galvanization.

ZINCKENITE; ZINKENITE A lead-antimony sulfide mineral.

ZINCKIFEROUS Carrying zinc.

ZINCKY; ZINKY; ZINCY Related to, containing, or having the appearance of zinc.

ZINCOGRAPH A plate prepared for printing by zincography.

ZINCOGRAPHY The art or process of putting designs of any kind in the form of a printing surface on zinc plates, and of producing impressions therefrom; sometimes, a process in which a relief plate is made by etching away part of the zinc. A process, the same in principle as lithography, in which the zinc plate replaces the stone.

ZINCOID Of, relating to, or resembling zinc.

ZINCOLYSIS A chemical decomposition produced by electrolysis in which the action is referred to the zinc element.

ZINCOLYSIS Electrolysis, especially with zinc anode.

ZINCOLYTE A body or compound that is decomposed by *zincolysis*.

ZINCOU Synonym for *zincic*.

ZINC-RICH PRIMER An anticorrosive primer for steel incorporating zinc dust in concentration sufficient to give electrical conductivity in the dried film to provide cathodic protection.

ZINCROMETAL Registered trademark for cold-rolled steel with a two stage coating (typically 0.5 mils thick) usually on one side to protect that side against corrosion. The strip is first coated with a proprietary chromium corrosion inhibitor, which is oven cured, followed by a top coat of zinc-rich primer (generally 90% zinc dust in suspension), which is also cured.

ZINKEN Zinc metal (German).

ZINKENITE A lead antimony sulfide mineral.

ZINROS; ZINAR; ZIREX; ZITRO Trademarks for high-melting zinc resinates used in adhesives, inks, lacquers, and rubber compounding.

Zn Chemical symbol for zinc.

ZONE OF OXIDATION Upper zone of a mineral deposit that has become oxidized.

ZONE OF SECONDARY ENRICHMENT The zone in which descending surface waters re-deposit their metallic contents derived from the oxidized zone, with the formation in the upper part of this zone of native metals, oxides and carbonates, and in the lower part of secondary sulfide minerals.

ZONE OF WEATHERING Down to the level at which ground water stands, the rocks are full of fractures and are exposed to atmospheric agencies, such as moisture, carbon dioxide, oxygen, etc. Here the rocks tend to decay, to be converted into carbonates and hydroxides, and to form soils. This zone is an area of rock destruction and alteration.

INDEX

Abbe Dony, 2
accelerator, 179
agreements
    Basel Convention, 185
Agricola, 15
American process, 3, 4, 8, 12
American Zinc Association (AZA), 12
American Zinc Institute (AZI), 8
Anaconda, 8
ASTM, 183
Austinville, 76
Austinville-Ivanhoe district, 77
AZA, 183

baghouse, 184
Balmat Mine, 68
barter agreements, 83
Bartlett Table, 185
Basel Convention, 124
Beart, 218
Belgian furnace, 2
Belgian Method, 2
beneficiation, zinc ores, 100
Bergenpoint, 22, 23
Bergenpoint spelter, 90
Bertha, 23
Bertha spelter, 76, 90
Bevill Amendment, 187
Bingham district, 74
blue dust, 40
bonded warehouses
    establishment of, 113
bottle retort, 190
brass industry
    early America, 111
Broken Hill, Australia, 54

Brown horseshoe furnace, 5, 30
Bruckner, 4
Bully Hill, 54
Butte district, 63

Cadmium, 2
calamine, 1, 5, 21, 15, 16, 111, 193
calcine. *See* sinter
Canadian Spelter Bounty, 194
Canadian Trade Agreement, 9, 114
Canon City, 55
Cappeau furnace, 31
cartel
    European, 116
cat's whiskers radio, 8
cementation, 197
cementation process. See processes
Champion, 1, 15
Clark cell, 199
Clevenger. *See* electrolysis
codes
    apple, 182
Coeur d'Alene district, 56
Collinsville smelter
    largest in world in 1916, 71
commercial production, U.S., 16
consumption, per capita, 18
consumption, zinc, 107
continuous sheet galvanizing, 9
Cook-Norteman, 10
copper tract, 1, 65
corrosion, 215, 224
Cowper-Coles, Sherard, 6
Crawford, 1
Critical Materials Stockpile Act of 1939, 82
crustal abundance, zinc, 94

cuprite, 1
custom mills, 123
cyclone devices, 207

Dagner, 240
Daniell cell, 207
Davy, Sir Humphrey, 2
De Laval, Gustave, 6
Defense Production Act (DPA) of 1950, 83
differential flotation, 38, 210
Dingley Law, 112
distillation, 35
distillation process, 104
DMEA, 208, 211
Dor, 211
Dor hydraulic press, 5
Dor press, 29
Dorr, 211
Ducktown district, 73
dusts
    electric arc, 49, 106
Dwight-Lloyd, 214

EAF dusts, 220
Eagle Mine, 55
East Tintic district, 74
Economic Stabilization Plan (ESP), 11, 48
electric-arc-furnace dusts, 106
Electrofax, 10
electrogalvanizing, 9
electrogalvanizing, advantages, 91
electrolysis, 8
electrolytic process, 103
electrolytic zinc
    early development, 36
electrolytic zinc process, 63, 103
electrostatic separation, 6
Electrothermic, 9
employee issues, 216
English furnace, 1

English process, 16
Esperanza Classifier, 217
European Producer Price (EPP), 117, 119
European Zinc Cartel, 116
exploration, zinc, 97

Fauvelle. *See* Beart
flashlight, 6
flotation, 102
flotation process, 7
Fowler, Dr. Samuel, 66
Franklin Mine, 65
    Franklin Furnace, 24, 31, 33, 45
Free Trade Agreement (FTA), 12, 115
French process, 3
Friedensville, 71
furnace, 181, 182, 186, 188, 231
    blast, 234
    Boetius, 189
    Borgnet, 190
    Bruckner, 191
    Cappeau. *See* Ropp Furnace
    Carinthian, 195
    Chenhall, 198
    Dor furnace. *See* Dor
    Dowson Producer, 212
    English Zinc furnace, 217
    Faber du Faur, 218
    Francisci furnace, 223
    Hegeler, 231
    horseshoe. *See* horseshoe furnace
    Indiana, 235
    Iola furnace, 237
    Kammerling, 239
    Keil, 239
    Keller, 239
    muffle, 231. *See* McDougall. *See* Haas Furnace
    reverberatory, 197, 218. *See* Davis Furnace

roasting, 229. *See* Douglas Furnace
Spencer, 239
Wetherill, 215

Galvani, 1
galvanism, 1
galvanizing, 204, 213, 224, 230
  dry, 213
  galvanizing of iron, 1
GATT, 235
General Agreement on Tariffs and Trade (GATT), 115
Gillman district, 55
Goethite process, 11
Golconda Mine, 53
grades, 227, 230, 232, 236
Greens Creek Mine, 53

Harmonized Trade Schedule (HTS), 115
health, 228
heavy–media, sink–float
  first unit, 73
Hegeler, 3, 4, 5, 6, 231
Hegeler roasting furnace, 30
Henckel, 1
Hoover, Herbert, 9
horizontal retort, 2, 3
Horse Head, 3, 4
hot-dip galvanizing, 2, 10
Huff electrostatic machines, 33
Huff separator, 233
Hyde. *See* flotation process

ILZSG, 234
Imperial Smelting Process (ISP), 10, 105
import quotas, 45, 46
International Lead and Zinc Study Group (ILZSG), 45
Interstate-Callahan Mine, 56
Iola furnace, 29

ISRI, 237

Jacoby, 237
jarosite process, 11
Jersey Miniere Zinc Co., 73
Jones, Samuel, 3
Joplin, 3
Joplin Ore Scale, 123

Kansas
  natural gas, 59
Karsten, 2

Lanyon, 6
Le Clair, 3
lead slag fuming, 9, 63
Lead-Zinc Small Producers Stabilization Act of 1961, 85
Leclanche, 4
Leclanche wetcell, 4
Letrange, 5, 36
Lithopone, 4, 9, 94. *See* Lithophone
LME, 8, 117, 119, 124

Macquisten Tube flotation, 7, 57
Macquisten Tube process, 34
Mattheissen, 3, 4, 5, 6
metal grades, 119
Metaline district, WA, 77
Metals Week
  price, 119
methods, mining, 98
Mexican Trade Agreement, 114
Mine Hill, 65, 67
mineral depletion allowance, 7
Mineral Leasing Act of 1920, 8
mining
  Tri-State, 25
Mining Law of 1872, 4
Morning Mine, 56, 57
Muntz, 2
MVT, 96, 97

Mississippi-Valley-type, 51, 57, 58, 59, 61, 71, 73, 78

National Defense Stockpile, 12, 48
Naval Appropriations Act of 1937, 82
new scrap, 106
New York prices, 118
North American Free Trade Agreement (NAFTA), 115

OECD agreement on scrap, 124
old scrap, 106
OME, 208

Paracelsus, 1, 15
Park City district, 74
Parkes process, 3, 4
patent paint, 4
Pehrson, 18
portable pipe shaft system, 24
Premium Price Plan, 10, 41
pricing
    delivered-basis, 118
    LME basis, 120, 124
    weighted average basis, 119
processes
    American process, 22.
    Baelen, 184
    Bartlett, 185
    batch, 186
    Belgian, 186
    cementation, 21
    chlorine dezincing, 198
    Cire-Perdue, 199
    Cook-Norteman. *See* galvanizing
    direct process, 210
    electrolytic, 232
    flotation, 179, 190, 204, 207, 221, 223, 232, 233. *See* Delprat. *See* De Bavay
    fuming, 224
    Harris process, 230
    Hyde flotation, 34
    indirect, 235. *See* French process
    iron removal, 237
    Jarosite, 11
    Jarosite process, 238
    jet process, 238
    Kivcet, 240
    Macquisten Tube, 34
    metal-based, 22
    plasma-based, 235
    Rigg, 184
    roasting, 231
    Sanders flotation, 34
    sink-float, 208, 233 '
    Wetherill process, 22
production, world trends, 17

Red Dog, 53
redistillation, 37
regulations, 200
    Economic Stabilization Program, 215
roasting furnaces, 31
Ropp furnace, 31
rotary kiln, 40
Ruberg, 1, 2

safety, 197
Samuel Wetherill, 72
Seal grades, 93
selective flotation
    early impact, 38
Sendzimir, 9
shower roasters
    suspension roasters. *See* roasters
Silver King Mine, 53
sinter, 226
sintering, 8
    calcine, 39
    early development, 39
slag fuming, 40, 42, 50
smelter schedule, 122

smelters
  changes, 34
  closures, 85
  greenfields, 86
smelting
  historical references, 21, 14, 15, 16, 17
  natural gas, 16
smelting technologies, 103
Sorel, 1
spelter, 1, 2, 3, 4, 111, 230, 240
  Bergenpoint, 187
  Bertha, 187
  first production in U.S., 22
  origin, 15
spelter tariffs
  industry protection, 114
spelter, specifications
  metal grades, 90
splash condenser, 10
St. Lawrence district, 68
St. Louis price, 118
Sterling Hill, 22, 24, 65
Stirling, Lord, 1
stockpile
  Korean War, 83
Strategic and Critical Stockpile Revision Act of 1979, 84
strategic stockpile, 9
sulfuric acid, 5
Superfund, 86
supplemental stockpile, 45, 83

Tainton process, 9
tariff, 111, 113
  agreement with Mexico, 114
Tariff Act of 1909, 113
Tariff Act of 1913, 113
Tariff Commission, 45
tariffs, 8, 9, 10, 12
  import quotas, 235
  TSUS, 1989, 230

The New Jersey Zinc Co, 3, 66
The Strategic and Critical Stock Piling Act, 10
The Zinc Convention, 116
Trade Schedule of the United States (TSUS), 115
traveling-grate furnace, 8
treatment charges (TC), 122

U.S. Bureau of Mines, 7, 12
U.S. Geological Survey, 5, 12
Upper Mississippi Valley mining district, 78
uses, 21, 109
  brass, 42
  during war, 37, 82
  early 1900's, 32
  reducing agents, 43
  standard weights and measures, 21

vertical retort, 9, 10, 11, 105
Viburnum Trend, 61, 63
volatility of zinc, 21
voltaic pile, 2

Waelz, 9
Waelz kiln, 40
Waelz process, 7
Weaton-Najarian, 9
Weights and Measures Opening, 65
Wetherill, 3, 12
  magnetic separator, 24, 31, 33
Wetherill process, 3, 22, 31, 40
Wetherill separator, 6
Wifley table, 6, 31, 33
women in zinc mining, 32

Zamak alloys, 9, 92
zinc blende, 1, 6
zinc chloride, 32
Zinc Club, 117
zinc consumption, per capita, 18

Zinc Convention, 7
zinc deposits
    types of, 95
zinc dust, 9, 92, 111, 118
zinc minerals, 94

zinc oxide, 125
zinc penny, 12
zinc powder, 93
zinc-air batteries, 8